Smart Power Integration

Smart Power Integration

Mohamed Abouelatta
Ahmed Shaker
Christian Gontrand

iSTE

WILEY

First published 2022 in Great Britain and the United States by ISTE Ltd and John Wiley & Sons, Inc.

ISTE Ltd
27-37 St George's Road
London SW19 4EU
UK

www.iste.co.uk

John Wiley & Sons, Inc.
111 River Street
Hoboken, NJ 07030
USA

www.wiley.com

Library of Congress Control Number: 2022939122

British Library Cataloguing-in-Publication Data
A CIP record for this book is available from the British Library
ISBN 978-1-78630-837-5

Contents

Preface

Smart integration is a process in which an existing system infrastructure is upgraded through the integration of multiple technologies, for example, automated sensors, advanced automated controls and forecasting systems. A smart grid allows for interaction between the consumers and enables optimal use of energy and communication systems based on price preferences and system technical stresses, without forgetting the environmental aspect.

The continuous reduction in dimensions and the need for increasingly high power density have highlighted the need for ever more efficient structures. Smart power technology has been developed to meet this demand. This technology makes specific use of (L)DMOS devices, offering new solutions because of its unique high voltage and high current characteristics. The operation of these devices is accompanied by a number of phenomena. Good modeling makes it possible to account for these phenomena and predict the physical behavior of the transistor prior to production. To this, we add an axis that has become unavoidable: the entanglement between devices, circuits, connections and substrates.

(Micro)grid designs have evolved significantly in recent years with the incorporation of information and communication technology (ICT) solutions, such as artificial intelligence (AI) and machine learning (ML). A smart microgrid, equipped with sensors and automation controls, can efficiently perform load profiling and forecasting, generation management, load prioritization, etc. A go-to example is the vehicles that are quickly becoming a center of communication, navigation and connectivity. Automotive solutions will integrate with smart city infrastructures, personal devices and in-vehicle services to become part of a connected whole.

This book introduces different domains and tools and allows the reader to develop their understanding of smart power systems through real studies. Knowledge of high school mathematics is sufficient to progress through these studies.

Mohamed Abdelhamid Abouelatta
Ahmed Shaker Ahmed Zaki Ghazala
Christian Gontrand
January 2021

1

Overview of Smart Power Integration

1.1. Introduction

Since 1965, integrated circuit (IC) technology has followed Moore's law which states that the number of integrated devices doubles every 18 months. This growth is partly due to an increase in the size of ICs that can be produced. However, the dominant effect is due to the reduction in feature size of component devices that are integrated. The reduction of feature size tends to bring advantages of increased speed and the possibility to operate at lower voltages, allowing reduced power consumption. These advantages make technology shrinkage very attractive for technical performance reasons, as well as cost.

However, there are many applications where voltage cannot be reduced for external reasons. There are three areas where this is the case: power electronics, automotive applications and wide dynamic range circuits. In such applications, system integration of high voltage, analog and digital circuitry on a single IC is attractive in order to gain advantage in terms of miniaturization, reliability, efficiency and cost. However, in order to make these gains, the conflict of reducing voltage due to technology feature size has to be resolved with the requirements for operation at continued relatively high voltage.

The different operation and interface requirements of high voltage, analog and digital require a technology development optimized for these system requirements. Different technologies have been developed to address these applications, such as smart power and various bipolar-CMOS-DMOS (BCD) processes.

Smart power integrated circuits (PICs) that monolithically integrate low-loss power devices and control circuitry have attracted much attention across a wide range of applications. These ICs improve system reliability, reduce volume and weight and increase overall efficiency. Considerable effort has been put into the development of smart power devices for automotive electronics, peripheral computer appliances and portable equipment, such as cell phones, video cameras, and so on.

Commonly used smart power devices are the lateral double diffused MOS Field Effect Transistor (LDMOSFETs) and lateral insulated gate bipolar transistors (LIGBTs) implemented in bulk silicon or silicon on insulator (SOI). The main challenges in the development of these devices are obtaining the best trade-off between specific ON-resistance $R_{ON,SP}$ (R_{ON} × area) and breakdown voltage (BV), and shrinking feature size without degrading device characteristics.

1.2. Smart PIC applications

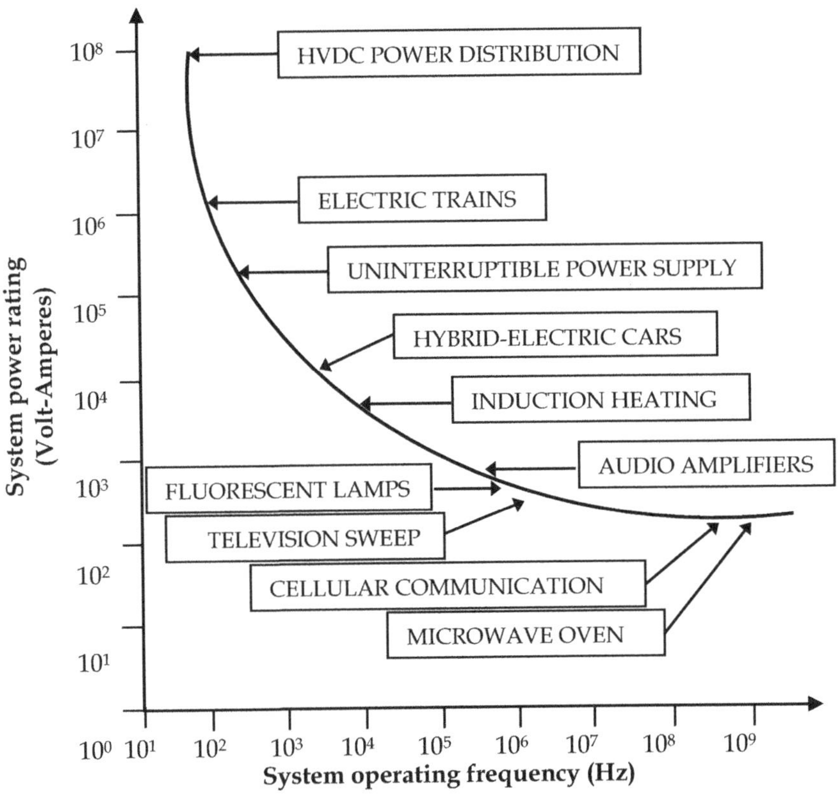

Figure 1.1. *Applications of power devices*

Smart PIC technology is expected to have an impact in all areas in which discrete power semiconductor devices are currently being used. It is anticipated that this technology will open up new applications based upon the added features of smart controls. In Figure 1.1, applications of power devices are shown as a function of operating frequency. Another classification approach of these applications involves current and voltage handling requirements, as shown in Figure 1.2. Some of these applications are listed in the following subsections.

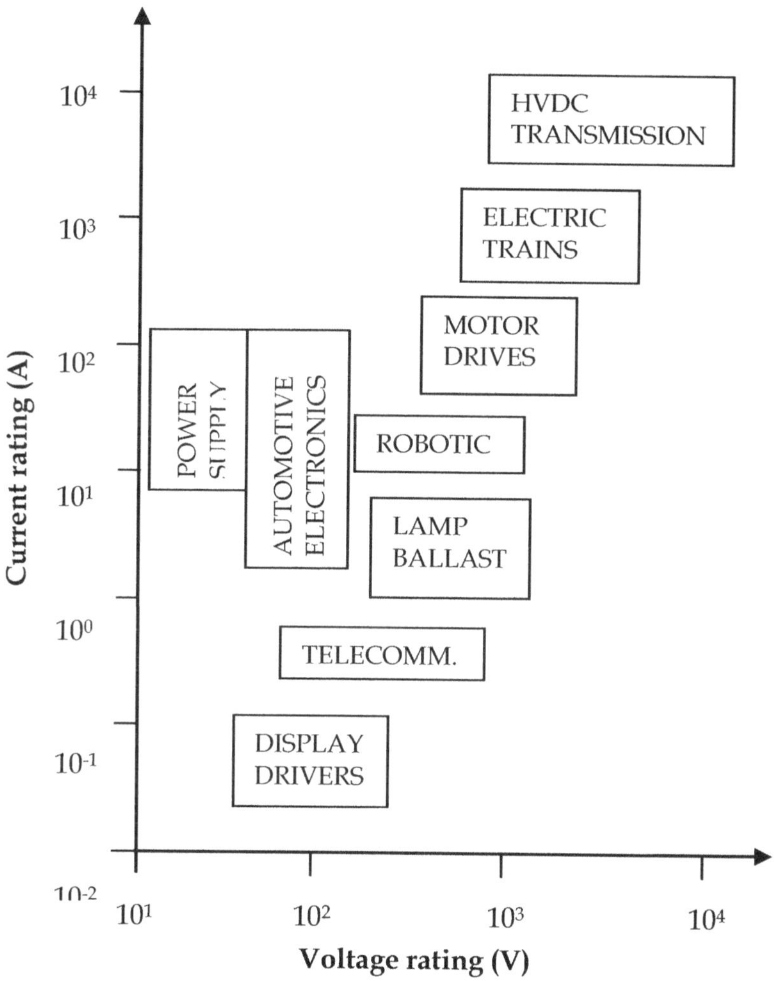

Figure 1.2. *System ratings of power devices*

1.2.1. *Flat panel displays*

The popularity of portable electronic products such as cell phones and notebook computers has generated significant demand for flat panel displays. These displays are usually liquid crystal displays (LCD) or electro-luminescence (EL) panels arranged in a matrix with large number of column and row drivers (e.g. 640 × 480 for VGA resolution). Although the required voltage may be high, the current level is low (usually in the mA range). Smart PICs with as many as 80 output channels have been fabricated on a monolithic chip.

1.2.2. *Computer power supplies and disk drivers*

Computer systems are developing continuously in terms of speed and processing capabilities. This is made possible by using higher density Very Large-Scale Integration (VLSI) technology. However, the increased power requirement has resulted in an increase in the physical size of the power supply. In 1976, the CPU board and power supply each represented one-third of the total physical volume of a computer system. By the 1990s, the power supply had grown to 50% of the physical volume while the CPU board had shrunk to about 20%. To reverse this trend, it is necessary to develop smart PIC technology to improve the density and hence the volume of the power supplies.

1.2.3. *Variable speed motor drives*

Variable-speed motor drives are being developed to reduce power loss in all applications. The improvement in performance requires smart power technology that can operate at relatively high frequencies with low power losses. This translates to a low ON-state voltage drop at high current levels, fast switching speed and rugged operation. For smart PIC implementation, additional consideration, such as level shifting to and from high voltages, over-temperature, over-current, over-voltage and short-circuit protection are more critical.

1.2.4. *Factory automation*

Advanced numerical control and robotic systems require efficient smart PIC technology to create a distributed power control network under the management of a central computer. The smart PICs for this application must be capable of providing AC or DC power to various loads, such as motors, solenoids, arc welders, and so on. They are also required to perform diagnostic, protection and feedback functions.

1.2.5. *Telecommunications*

One of the high-volume markets for smart power technology is in telecommunications. The technology required for these applications must be capable of integrating multiple high-voltage, high-current devices on a single chip. At present, this has been achieved using MOS devices fabricated using dielectric insulation. Improvements are required to reduce the cost of the dielectric insulation fabrication process. Ongoing development on direct wafer bonding has been promising in terms of providing a cost-effective process.

1.2.6. *Appliance controls*

The main benefit of using smart PICs in appliance control is to provide improvements in performance and efficiency. Onboard sensors can also provide more precise controls (e.g. temperature settings). Simple domestic appliances, such as toasters, washing machines and irons, are appearing with smart PICs for this reason.

1.2.7. *Consumer electronics*

Smart PICs are required for a large variety of entertainment systems such as CD players, tape recorders, VCRs, etc. For example, a monolithic motor control IC that regulates the speed of the motor, while minimizing power losses, is essential to all battery-operated consumer entertainment systems. Development of improved lateral power devices with greater power density is necessary to increase the efficiency of this technology.

1.2.8. *Lighting controls*

Traditional fluorescent lights use a mechanical ballast (transformer) for start-up. The electrical characteristics of fluorescent lights vary drastically from start-up to normal operation. In order to improve the efficiency and overall lifetime of a lighting system, more precise control is needed. The cost of electronic fluorescent light ballasts implemented using smart PICs can easily be justified by the resulting savings in energy and maintenance. In addition, incorporating smart PIC technology enables lighting to be controlled by a central computer, further enhancing energy savings in commercial buildings.

1.2.9. *Smart homes*

The concept of smart homes is attracting increased attention as a result of advances in smart power technology that are driving down the cost of the control module. A smart home system requires the development of a multiplexed network with smart power modules to control loads such as ovens, furnaces, air conditioners, lights and small appliances.

1.2.10. *Aircraft electronics (Avionics)*

The concept of fly-by-wire, where the hydraulic actuators in an aircraft are replaced by electromechanical actuators, is gaining acceptance among manufacturers. Success in development will depend on the availability of smart PIC technology to perform the control that is small in both size and weight. The power switches must be extremely reliable and capable of operating at high voltage and current levels with low ON-state voltage drop. In this regard, MOS-gated devices are essential for compact PICs.

1.2.11. *Automotive electronics*

One of the biggest anticipated markets for smart PICs is automotive electronics. Between the 1960s and 1970s, there was a slow uptake of discrete devices and analog ICs for automotive applications. In the 1980s, digital ICs and microprocessors were incorporated. In the 1990s, smart PICs were already being used to create a multiplexed control network in cars to reduce the size and weight of the wiring harness. The smart PIC modules control loads such as lights and motors, while providing protection functions. This has greatly enhanced fault management and diagnostic capability.

1.3. Historical view of the MOS power devices

The basic operation of the MOSFET involves the formation of a conductive channel at the surface of the semiconductor, below an insulator, by the application of a voltage to a gate electrode. The first MOSFET structure reported in 1960 was not designed to support high voltages or handle high-current levels and, furthermore, thin metal electrodes were used that had poor current-handling capability.

In the 1970s, it was recognized that the vertical architecture is required to handle high voltages and currents to produce a power device. The vertical structure enables the use of thick source and drain electrodes. The first power MOSFET structure was fabricated using a V-groove process, as shown in Figure 1.3.

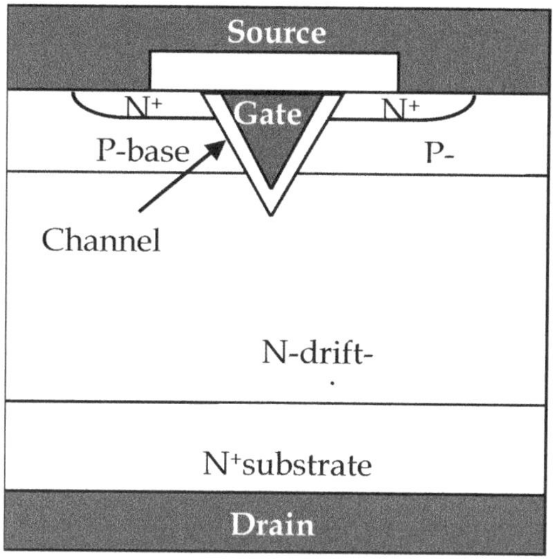

Figure 1.3. *V-groove MOSFET structure. For a color version of this figure, see www.iste.co.uk/abouelatta/smartpower.zip*

The V-MOSFET structure fell out of favor because of manufacturing difficulties. In addition, the sharp corner at the bottom of the V-groove was found to degrade the breakdown voltage.

The first commercial power MOSFET was the vertical diffused (VD) MOSFET of the mid 1970s, which is illustrated in Figure 1.4.

The non-uniform current distribution enhances its resistance (JFET region), making the internal resistance of the VD-MOSFET structure larger than the ideal specific ON-resistance of the drift region. The large internal resistance of the VD-MOSFET structure provided motivation for the development of the trench-gate power structure in the 1990s.

Moreover, the U-MOSFET is shown in Figure 1.5. There is no JFET region in the structure, enabling a reduction in internal resistance when compared to the VD-MOSFET structure.

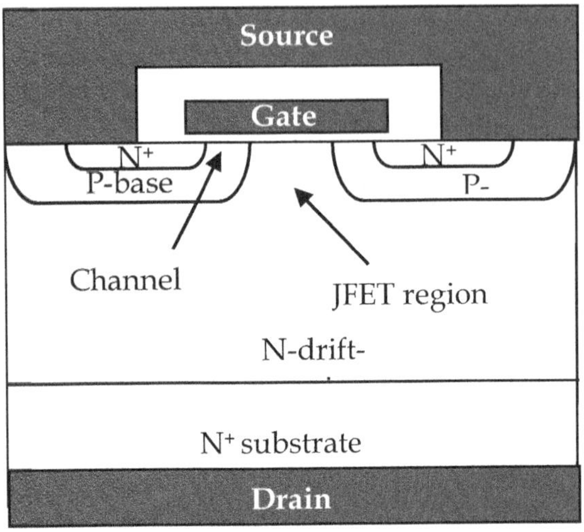

Figure 1.4. *VD-MOSFET structure. For a color version of this figure, see www.iste.co.uk/abouelatta/smartpower.zip*

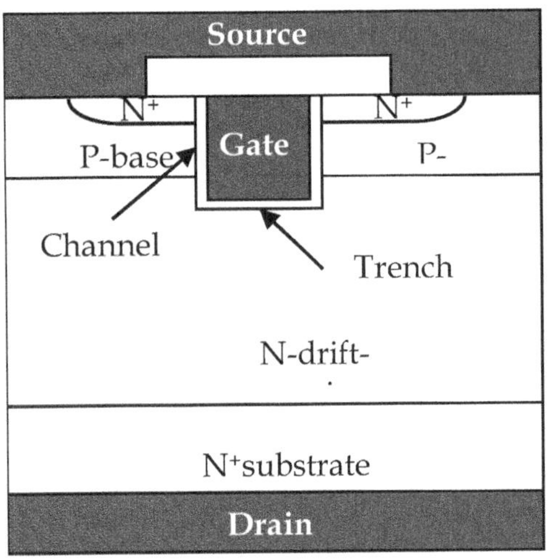

Figure 1.5. *U-MOSFET structure. For a color version of this figure, see www.iste.co.uk/abouelatta/smartpower.zip*

One significant drawback of the vertical devices is the fact that it is difficult to include multiple power devices on the same monolithic chip. The lateral structure allows all terminals to be accessed from the top surface of the chip. The current flows laterally from the drain along the surface through the MOS channel and up into the source, hence the name LDMOSFET, as shown in Figure 1.6.

LDMOSFET generally suffers from a higher specific ON-resistance due to the longer current path. Furthermore, the blocking voltage of the LDMOSFET depends critically on the curvature of the P-body to the N-drift region junction. In order to obtain high blocking voltage, it is necessary to use a low doping concentration in the N-drift region. However, this directly contradicts the low specific ON-resistance requirement. Therefore, it must be optimized to achieve high breakdown voltage with low $R_{ON,SP}$.

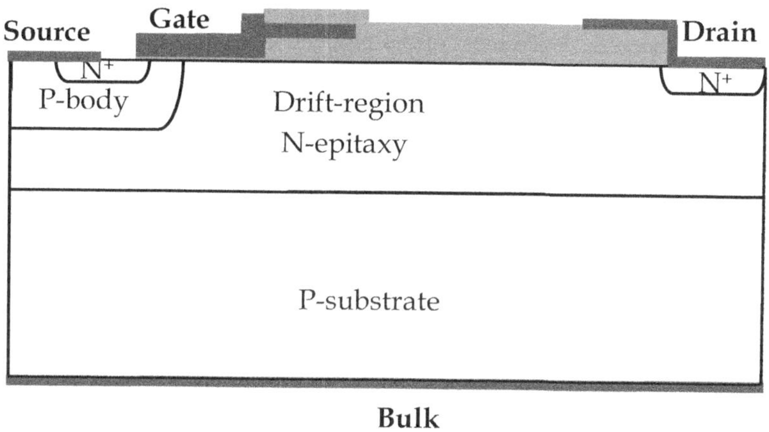

Figure 1.6. *LDMOSFET structure. For a color version of this figure, see www.iste.co.uk/abouelatta/smartpower.zip*

1.4. Smart PIC fabrication processes

Smart PIC fabrication processes are classified into two categories: dedicated processes and compatible processes.

1.4.1. *Dedicated processes*

A dedicated smart PIC technology refers to a fabrication process with the optimization of the power devices as the highest priority. The performance of the low-voltage CMOS devices is usually compromised. S.G.S. Thomson is a long time advocate of such a trend with their family of BCD processes.

1.4.2. *Compatible processes*

The compatible approach is to integrate the power devices into an existing process. While this would invariably lead to a greater degree of compromise in the performance of the power devices, it is a significantly more cost-effective approach. The goal would be to minimize the number of additional steps that have to be introduced to the original process. This will ensure low production costs since most existing processes are already fine-tuned and are running at high volume.

The blocking voltage and specific ON-resistance ratings can be optimized by selecting the appropriate doping profiles. The compatible approach is currently being adopted by many manufacturers for their existing CMOS processes, especially for out-dated processes. Since most of the production volume is being migrated to more advanced processes (e.g. BiCMOS processes), in order to maintain the existing process lines, new applications have to be found. Smart PIC technology is the ideal way to inject new life into these soon to be obsolete processes.

1.5. Insulation techniques

In PIC technology, power transistors are controlled by low-voltage BiCMOS circuitry. Insulation between these two types of components – i.e. high power devices and control/logic circuitry – is necessary in order to avoid crosstalk and ensure systems operate as they should.

Under certain conditions, a negative bias can be applied on the drain contact with respect to the substrate of laterally diffused MOS (LDMOS), as a result of the switching of an inductive load. If the technology employs junction insulation, this may lead to an unwanted injection of minority charge carriers into the substrate. The resulting currents may cause the entire PIC device to malfunction. Minority carriers injected into the substrate can cause signal disturbances in the low power circuitry or latch-up in the CMOS structures.

Several insulation techniques have been employed in order to protect the low-voltage CMOS circuitry from substrate currents that originate from power transistors.

1.5.1. *Self-insulation*

Self-insulation is characterized by a very simple process flow; it is based on the CMOS process but includes a few additional steps. However, as all devices in the same chip have to share the positive supply voltage, additional parasites are introduced. Siemens smart SIPMOS technology is an example of this.

1.5.2. *Dielectric insulation*

SOI insulation, where active devices are completely surrounded by a non-conducting dielectric layer, is shown in Figure 1.7. This technology provides excellent insulation; however, since the dielectric is a poor thermal conductor, this may lead to serious thermal problems for the power devices. Additionally, the cost of producing SOI wafers is significantly higher than it is for conventional silicon wafers.

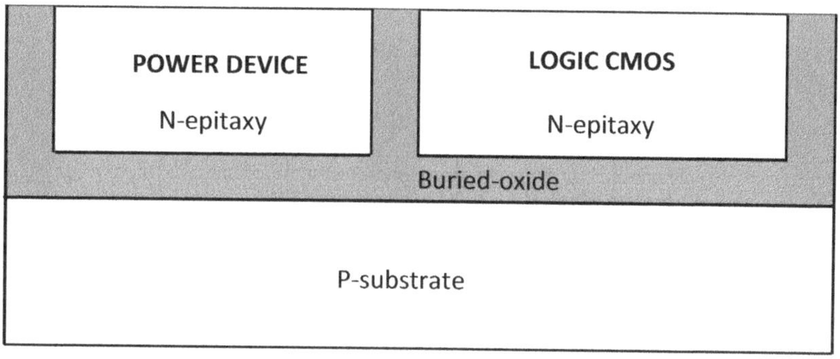

Figure 1.7. *Dielectric insulation*

1.5.3. *Junction insulation*

Junction insulation (JI) does not suffer from such serious self-heating problems, but the electrical insulation is much poorer. Standard junction insulation structures consist of a reversed biased P-N junction, as shown in Figure 1.8.

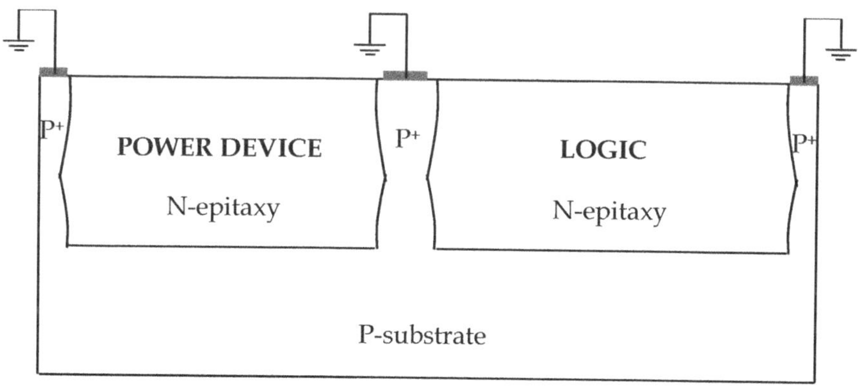

Figure 1.8. *Standard junction insulation*

1.5.4. *Advanced junction insulation techniques*

To enhance the performance of junction insulation structures, advanced techniques are used.

1.5.4.1. *Passive JI*

In order to make the insulation more effective, an extra N-region is added and enclosed by two deep P-diffusions. The N-region is permanently tied to the highest potential of the circuit, as shown in Figure 1.9. Its purpose is to remove the minority carriers as close to their point of injection as possible.

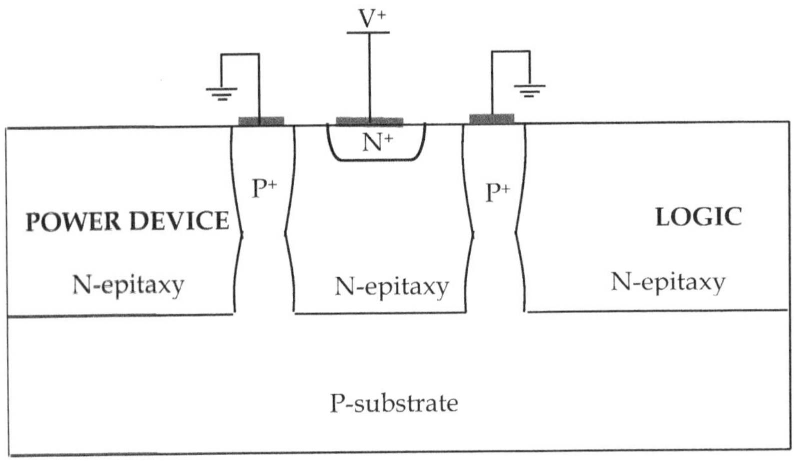

Figure 1.9. *Passive junction insulation*

1.5.4.2. *Active JI*

A significant step forward has been made by introducing the multi-ring active analogic protection (MAAP) JI, as shown in Figure 1.10. This self-triggered JI technique is able to reduce the current that reaches the low power circuitry by up to several orders of magnitude compared with the passive JI.

In the MAAP device, ND and P2 are shorted and left floating, as shown in Figure 1.10. This time, the on-chip metallization is implemented on the chip instead of on external wiring to reduce the interconnect resistance R_{via}.

Once injected from the drain side, electrons flow through the substrate and some are collected by ND, reducing the ND’s potential. This negative potential is now transferred through the metallization to P2, which adds to its built-in potential. P2 is now on lower potential than P1 and the resulting electric field will reject the substrate minority carriers injected from the drain of the power device.

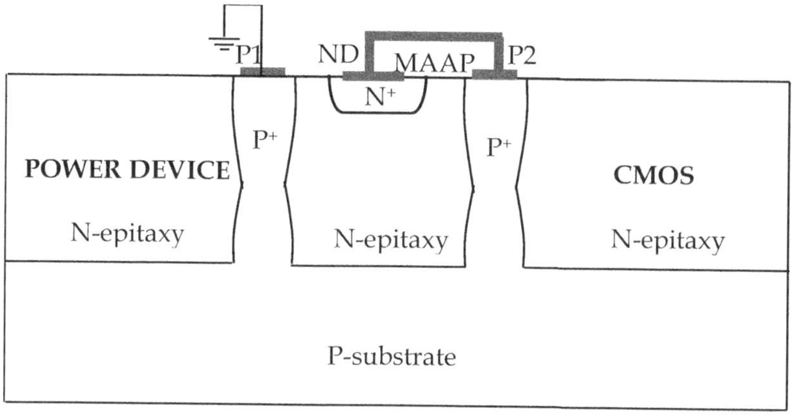

Figure 1.10. *Active junction insulation*

1.5.4.3. *Active pull-down protection*

Active pull-down protection combines junction insulation with a bipolar transistor that biases a P-substrate contact according to the injected current level. This insulation structure is highly effective, but its complexity and number of variables may lead to a significant increase in the number of iterations at the design stage.

1.5.4.4. *Multiple structures*

In an attempt to further improve the blocking capabilities of junction insulation, the use of multiple or combined structures has been investigated, keeping the surface area of the insulation device as small as it is for the single structures. This means that the width of the P and N diffusion zones is decreased in order to retain the same total area.

1.6. Motivation of the book

Smart PICs that monolithically integrate low-loss power devices and control circuitry have attracted much attention across a wide range of applications. These ICs improve system reliability, reduce volume and weight and increase efficiency. Considerable effort has been put into the development of smart power devices for automotive electronics, peripheral computer appliances and portable equipment, etc.

Commonly used smart power devices are the lateral double diffused MOS transistors (LDMOSFETs). The main challenges in the development of these devices are obtaining the best trade-off between specific ON-resistance $R_{ON,SP}$ and breakdown voltage BV, and shrinking the feature size without degrading device characteristics. A lot of work has been done on both characterization and modeling of LDMOSFETs in the literature starting from the late 1970s. Still, there is some confusion concerning the origin of special phenomena such as, quasi-saturation and explanation of the capacitive behavior of the HV device as a function of the biasing conditions. Unfortunately, this matter was not treated in a thorough manner, as most of the time only a few characteristics that do not cover the entire voltage domain were presented.

Device simulation with technology computer-aided design (TCAD) tools has been proven to play an important role for design engineers and researchers in analyzing, characterizing and developing new devices. It saves time and lowers the cost of design when compared to the experimental approach. In addition, it allows physical effects to be seen clearly in the semiconductor devices of new device concepts. Therefore, a clear understanding of TCAD tools in the analysis of power semiconductors is essential in order to obtain accurate and reliable simulation results.

This book deals with novel power semiconductor concepts for smart power applications. New device structures are suggested and studied to improve the device characteristics related to traditional power devices. Two- and three-dimensional device simulations are performed to study new device concepts. These new structures are implemented in 0.35 μm BiCMOS technology.

An important motivation point is the planar integration of HV LDMOSFETs with low-voltage CMOS devices. A new electrical insulation scheme between the power devices, as well as between the power devices and low-voltage CMOS devices is suggested, using a deep trench to reduce the insulation distance among the high-voltage devices considerably, and hence to reduce the total chip area drastically compared to technologies with a standard junction insulation scheme.

In recent years, an emerging technology was the three-dimensional (3D) integration. This technology uses through-silicon vias (TSVs) and re-distribution layers (RDLs) to interconnect multiple active circuit layers. 3D integration offers significant improvements over two-dimensional (2D) ICs on performance, functionality and integration density, and promises a solution to the so-called "wiring crisis" problem. Furthermore, 3D integration also provides new architectures for sophisticated ICs and facilitates the integration of heterogeneous materials and devices. The important motivation question is: can this 3D technology be used for the implementation of smart PICs?

In this work, ISE and SENTAURUS-TCAD tools, general purpose device simulators, are used for the simulation of established and newly proposed power semiconductor devices and integration schemes.

From a modeling point of view, two model types are presented in the literature. Several attempts were made to build compact models for the HV devices. Some authors focus on the construction of the intrinsic MOS model and treat the drift part as a trivial extension. Others focus on the modeling of the drift part only. To fulfill this target, very simple expressions are needed for both the drift and intrinsic MOS parts. Most of the time, this results in poor accuracy when compared with real device characteristics. In the second category are the macro-models that are built by merging several spice elementary devices in a subcircuit and using the result as a black box for the transistor model. The proposed LDMOS devices are used to construct an interface circuit to convert the low-level logic voltage to high-level voltage.

The upheavals of the electronics of power will come from the capacity of the component makers to propose silicon devices (in the future, other materials that can withstand the high temperature, such as silicon carbide and gallium arsenide, will be added), integrating a system and a function. The integration of power was born from the desire to increase the reliability of the systems, and simplify their assembly and maintenance, while also reducing the cost. The progress of the electronics in the coming years will be focused primarily on the projections of the integration of power. According to the power to be commutated, today, the integration of power can be classified in two dies: monolithic integration and hybrid integration.

In practice, we want the readers to be aware of the various fields and bricks of the integration of power, which is essentially multi-field and whose main concept is the approach system.

2

Modular or Hybrid Integration

2.1. Introduction

In this chapter, we develop projections of the power modules (switch of high power) containing insulated gate bipolar transistors (IGBTs) of intelligent power modules (IPMs) and application-specific intelligent power modules (ASIPMs), as well as fully customized hybrid solutions.

Over the last few years, the range of power modules has expanded. Of note are the advanced IGBT modules that compete with the gate turn-off thyristors (GTOs), invented by General Electric. Unlike conventional thyristors, GTOs are fully controllable switches that can be turned on and off by their third lead: the gate electrode. The development of non-punch through (NPT) technology, IGBTs and the massive parallelization of chips (up to more than 20) makes it possible to lay out modules in the range of 5 KV and 5 kA. These ranges of power make it possible to consider applications for electric traction.

Unfortunately, traditional solutions (with bonding) do not support thermal fatigue induced in such applications. Research is directed toward cases pressed for the best held with thermal fatigue, a better dissipation of the heat and an increased facility of assembly. Work on the cases has led to economic solutions that have a profit in facility of assembly and reliability. In parallel, manufacturers are endeavoring to improve hybrid modules by reducing internal inductances, thermal resistance, volume and weight. This will require materials with better properties (dielectric insulation, thermal conductivity, etc.). Some examples include replacing the traditional aluminum base plate with a metal matrix composite (MMC) such as Al/SiC, replacing the alumina Al_2O_3 substrate with aluminum nitride (AlN) or diamond and integrating the radiator (which becomes the base plate) or microphone-coolers. A recently suggested solution is based on the suppression of the dielectric substrate and the placement of a massive dielectric exchanger of AlN, on which the copper collector is thermo-compressed. The disadvantage of this encapsulation lies in the high cost of machining the AlN.

IPM modules contain the power stage, the drivers and protect the devices. They can be compared with monolithic smart power solutions, but those relate to a much higher power range (I < 30 A, U < 1200 V). The IPM provides complete solutions for applications such as the inverter. IPM modules are, however, relatively expensive or inflexible devices and depend on the manufacturer (not second source manufacturers). The hybrid circuits give a profit in clutter, maintenance, confidentiality, etc. One considers several technologies according to the range of power: PCB (Printed Circuit Board – printed circuit board with components assembled on the surface CMS) (750 W), SMI (Metal Isolated Substrate) (4 kW), DCB (Direct Contact Bonding, largely used in IGBT modules) (> 10 kW). The research tasks in this field join those of the modules.

	MOSFET transistor	**Bipolar transistor**
Conduction types	– Majority carriers	– Bipolar (two types of carriers in play)
Advantages	– Speed – Simple control circuit (and for low consumption) – Robustness	– Low-voltage drop for ON-state
Drawbacks	– High-voltage drop in static (low current density for high voltage MOSFETs)	– Control circuit rather complex (and with significant consumption) – Slow in destocking (although fast in switching)
Quality available on the market in discrete device (single chip)	– From 1 to 70 A – From 30 to 1000 V – High current with high voltage	– From 1 to 100 A – From 20 to 1500 V

Table 2.1. *Advantages and disadvantages of the MOSFET and bipolar transistor*

2.2. IGBT technology evolution

2.2.1. *IGBT presentation*

The IGBT was developed as a result to associate, on the same silicon crystal, the speed of the unipolar Metal Oxide Semiconductor Field Effect Transistor

(MOSFET) with the low voltage drop in the ON-state of the bipolar transistor (see Table 2.1). There are two types of IGBT: N-channel or P-channel. The first can be described as an N-channel power MOSFET structure, to which an additional P^+-type layer constituting the collector (or anode) of the IGBT (see Figure 2.1) is added on the rear side. IGBTs therefore belong to the PNPN family of four-layer components in the same way as thyristors and GTOs. The P-channel IGBT is much less common because this type of device has fewer interesting electrical characteristics and a smaller safety area than its N-channel counterpart. Here, only the N-channel IGBTs are discussed.

2.2.1.1. *The IGBT structure*

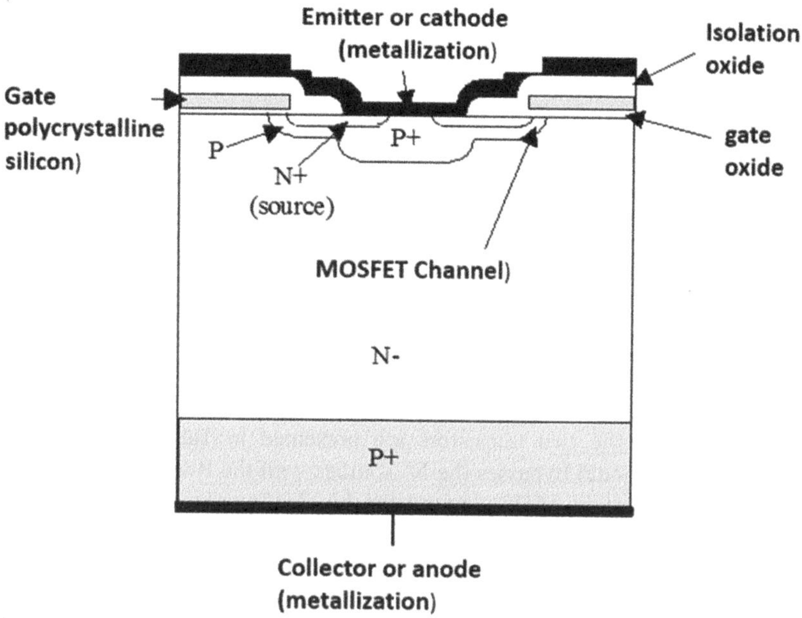

Figure 2.1. *Generic structure of the IGBT*

2.2.1.2. *Equivalent schematic*

It is customary to understand how a component works by associating it with an equivalent schematic. The first equivalent diagram (see Figure 2.2) is simplified because it does not highlight the Junction Field Effect Transistor (JFET) that appears between the cells under the gate oxide or parasitic capacitance.

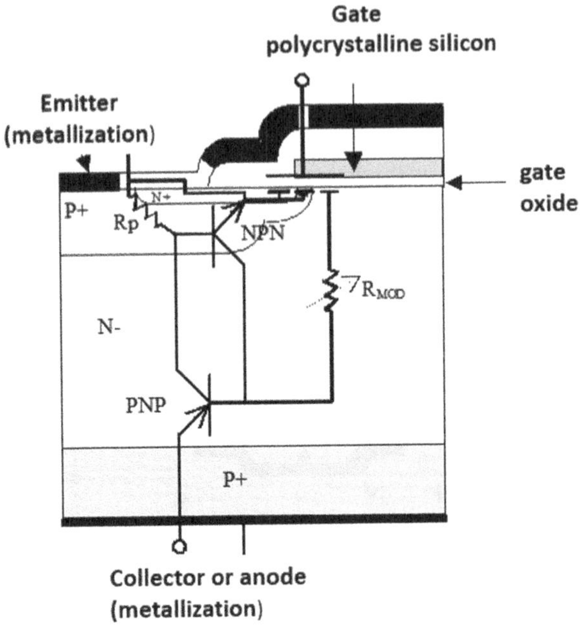

Figure 2.2. *IGBT equivalent schematic*

The stack of four layers $P^+N^-PN^+$ shows two bipolar transistors "head to tail". The components of the two transistors are presented in Table 2.2. The emitter metallization (or cathode) bypasses the N^+ sources with the P-well and theoretically eliminates the conduction of the upper NPN bipolar transistor. The resistance R_p represents the resistance of the box P under the source N^+. R_{MOD} is the resistance of the N-lightly doped layer to hold the blocking voltage.

In the case of the MOSFET, conduction is provided by the majority carriers only. The value of the resistance of the central zone N^- is the highest as the MOSFET component is high voltage (low level of doping and high thickness of the N^- layer). The IGBT has the advantage of involving both types of carriers in the conduction (electrons and holes). The resistance of the central area R_{MOD} is then modulated, and the voltage drop in the passing state is lowered compared to that of the MOSFET for the same voltage quality.

By removing the non-essential elements of the diagram in Figure 2.2, such as the resistors R_p, R_{MOD} and the parasitic transistor NPN, we obtain the equivalent circuit as shown in Figure 2.3. The N-channel IGBT then appears as a PNP bipolar transistor driven by an N-channel MOSFET in a darlington-type fixture. This equivalent model seems very simplified, but it makes it possible to understand the static and dynamic operation of the IGBT, not forgetting that the MOSFET and bipolar junction transistors are located on the same crystal and that their behaviors are therefore very much related.

	Principal bipolar transistor	**Parasitic bipolar transistor**
Type	PNP	NPN
Emitter	P^+ layer (Collector or anode of the IGBT)	Source N^+
Base	Central layer N^-	Well P^+/P
Collector	Well P^+/P	Central layer N^-

Table 2.2. *Internal bipolar transistor of an IGBT structure*

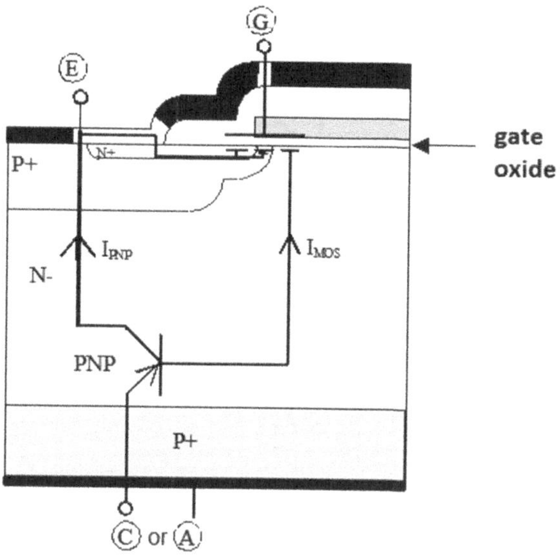

Figure 2.3. *Simplified equivalent schematic of the IGBT*

The IGBT goes into conduction when its gate voltage becomes greater than that of the threshold and is polarized positively between the collector (anode) and emitter. The static current is broken down into a current of holes injected by the IGBT collector (P^+ layer on the back) and a current of electrons fed through the MOSFET channel (see Figure 2.4). The N^- zone, the base of the internal bipolar transistor PNP, is then in strong injection. The ratio of these two currents to the emitter electrode (I_p/I_n or I_{PNP}/I_{MOS}) constitutes the current gain β_{PNP} of the internal bipolar transistor PNP. The current in the IGBT can decompose as:

$$I_C = I_{PNP} + I_{MOS} = (1+\beta_{PNP}).I_{MOS} = \frac{1+\beta_{PNP}}{\beta_{PNP}}.I_{PNP} \qquad [2.1]$$

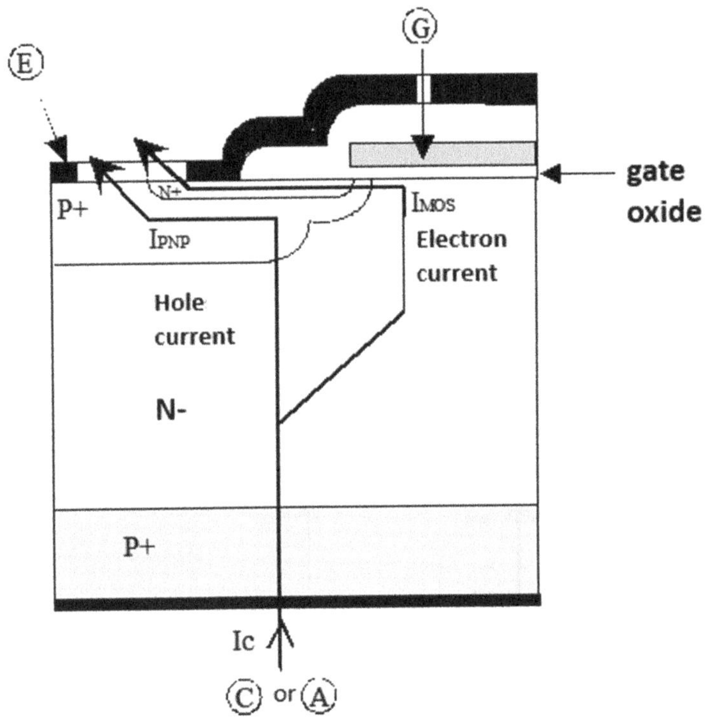

Figure 2.4. *Decomposition of hole current and electron current*

The β_{PNP} gain (after resolving the diffusion equation) of the PNP bipolar transistor depends mainly on two terms: the injection efficiency of the transmitter γ and the transport coefficient α_T in the base. By neglecting the strong injection in the N^- base, these two coefficients can be expressed by

$$\gamma = \frac{1}{1 + \frac{D_{nE}.N_{DB}.L_{pB}}{D_{pB}.N_{AE}.L_{nE}}.th\left(\frac{W_B}{L_{pB}}\right)} \qquad [2.2]$$

$$\alpha_T = \frac{1}{ch\left(\frac{W_B}{L_{pB}}\right)} \quad [2.3]$$

Here, D_{nE} and D_{pB} are, respectively, the diffusion coefficients of the electrons in the emitter and holes in the base, N_{DB} and N_{AE} are the doping levels of the base N^- and the emitter P^+, W_B is the N^- base width, and L_{pB} and L_{nE} are, respectively, the diffusion lengths of the holes in the N^- base and the electrons in the P^+ emitter. By playing on the value of the gain β_{PNP}, manufacturers propose families of IGBTs with electrical behavior that closely resembles the bipolar transistors if the gain is high, or the MOSFETs if the gain is weak. Generally, the gain β_{NPP} has a value less than unity.

2.2.1.3. *Trigger of the parasitic thyristor*

By reconsidering the complete equivalent schematic in Figure 2.2, the IGBT may look like a thyristor but never works like it. Current control and command are achieved by the gate voltage (MOSFET portion) that regulates the flow of electrons. The parasitic bipolar transistor NPN must never be activated as this may trigger the ignition of the parasitic thyristor structure, and the current control in the IGBT by the gate voltage will be lost. The equivalent diagram reveals a resistance R_p that represents the resistance of the P layer under the source N^+. This resistor, placed between the base and the emitter of the NPN bipolar junction transistor, plays a key role since the triggering of the parasitic thyristor structure will occur if the N^+ emitter injects electrons into the base P of the NPN transistor; this will occur when the voltage drop across R_p exceeds 0.6 V. The turning current enabling the NPN bipolar transistor can be expressed by

$$I_{initiation} = \frac{(1+\beta_{PNP})}{\beta_{PNP}} \cdot \frac{0.6}{R_p} \quad [2.4]$$

with

$$R_p = \frac{L}{S} \cdot \frac{1}{q.\mu_p.N_A} \quad [2.5]$$

where β_{PNP} is the current gain of the internal bipolar transistor PNP, q is the charge of the electron, μ_p is the mobility of the holes in the P well under the source N^+, N_A is the doping level of the P well, L is the length of the source N^+, and S is the surface area under the source N^+.

The reversal phenomenon occurs when the two bipolar transistors work simultaneously. The following relationship is then verified:

$$\alpha_{PNP} + \alpha_{NPN} = 1 \qquad [2.6]$$

where α_{NPN} and α_{PNP} represent, respectively, the common base current gains of the NPN and PNP internal bipolar transistors.

Figures 2.5 and 2.6 represent the simulated current lines before and after the appearance of the locking of the parasitic thyristor. Figure 2.5 shows the current lines in a conventional operation: the electrons pass through the channel under the gate oxide, the holes pass through the junction between the P/P$^+$ well and the N$^-$ layer, and then flow under the N$^+$ source. When the thyristor structure is triggered, the NPN transistor comes into play and its emitter (source N$^+$) injects electrons into the well (see Figure 2.6). The channel then no longer plays its role of regulating the flow of electrons. The four-layer structure, composed of the two bipolar transistors, is triggered and the reversal appears (see Figure 2.7); the current density increases until the destruction of the device by generalized reversal to the entire silicon pad or thermal runaway.

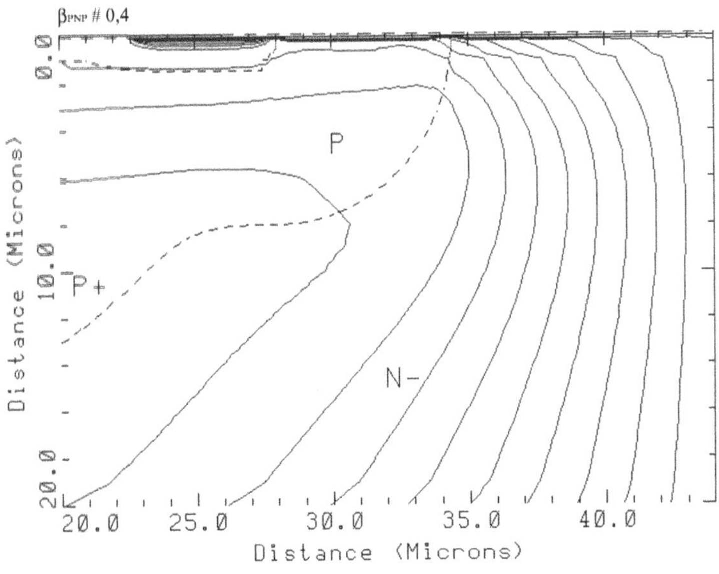

Figure 2.5. *Current lines in normal mode in a half-cell of IGBT (before reversal), $J_c \cong 100$ A/cm², MEDICI simulation (10% of the total current in each tube). For a color version of this figure, see www.iste.co.uk/abouelatta/smartpower.zip*

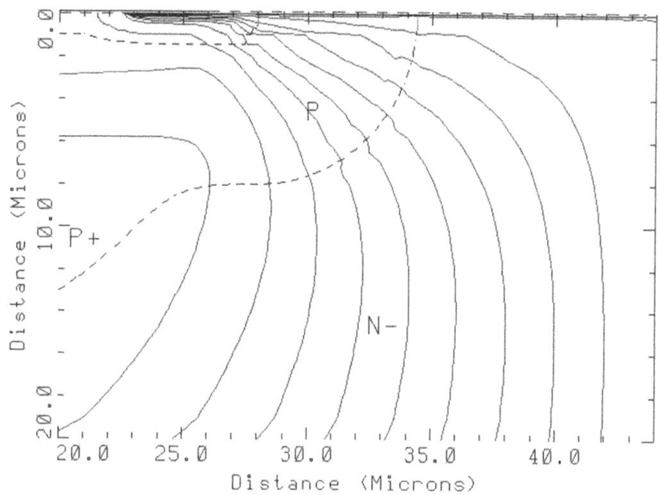

Figure 2.6. *Current lines after reversal in half a cell of an IGBT, $J_C \cong 200$ A/cm², Sentaurus simulation (10% of the total current in each tube). For a color version of this figure, see www.iste.co.uk/abouelatta/smartpower.zip*

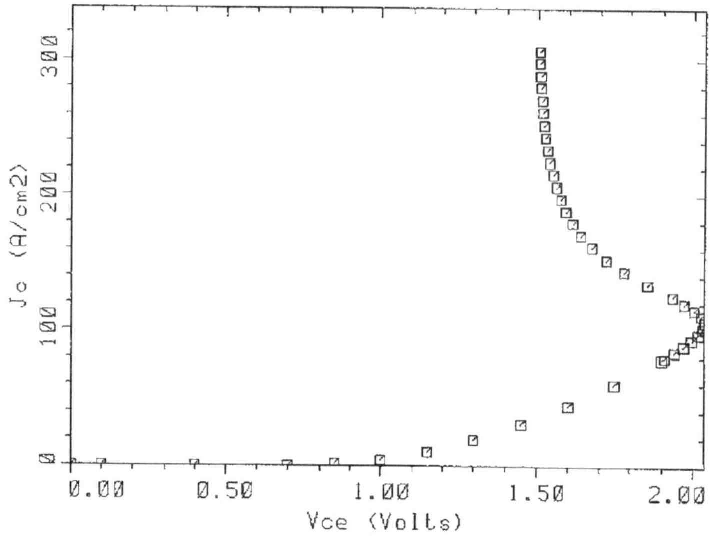

Figure 2.7. *Static curve $J_c = f(V_{ce})$ with flipped appearance (Sentaurus simulation)*

Solutions to prevent triggering of the parasitic thyristor are well known by the manufacturers; it is necessary to either decrease the current of the holes by lowering the gain β_{PNP} of the PNP internal bipolar transistor or increase the conductance of the box P under the sources N^+. If we want to keep the same modulation of the N-layer, then we must keep the same gain β_{PNP}. The increase in the conductance of the P well is the main issue to prevent flipping. For this, it is necessary to have an N^+ source length that is as short as possible associated with a doping level of the high P well. Manufacturers use the principle of a double P/P^+ diffusion to increase the level of doping between and under sources, while trying to preserve an adequate level of doping under the gate oxide so as to not modify the threshold voltage. On this particular point, a "counter-doping" step to restore the correct doping level in the channel region under the oxide can be implemented. The principle is to carry out an additional donor doping (N) at the channel to annihilate the presence of acceptor-type impurity.

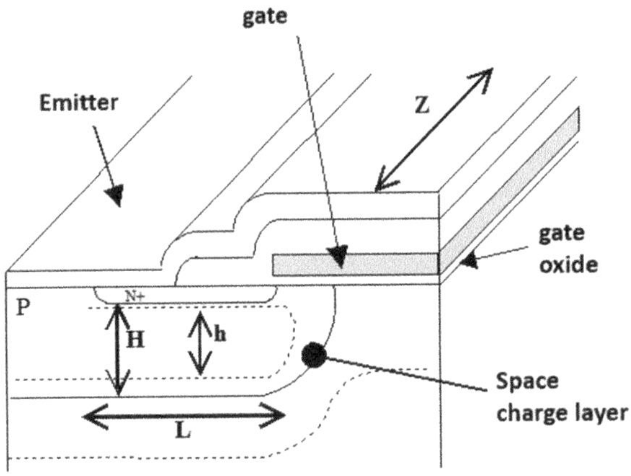

Figure 2.8. *Appearance of the space charge layer*

The flipping can be static or dynamic. In an inductive load switching phase or when a short-circuit-type fault occurs, the current level for which dynamic switching appears is less than the static due to the spread of a space charge zone, which confines the hole carriers under the N^+ sources in a narrower current tube (see Figure 2.8), and the creation of an additional capacitive current flowing under the N^+ sources. The conductance of the P zone under the source N^+ decreases because the section concerned by the transport of the holes is truncated during the spreading of the space charge zone under the source. There is an additional donor doping (N) at the level of the channel to annihilate the presence of acceptor-type impurity atoms (P). Unfortunately, this technique degrades the mobility of electrons in the channel.

We will see in section 2.2.1.4 that the shape of the cells also has an influence on the triggering of the parasitic thyristor.

2.2.1.4. *Cell geometry*

Like the power MOSFET transistor, the IGBT is a multicellular component for which it is possible to imagine different technological solutions on the shape of the cells. The influence of the geometry of the cells has been studied in terms of sensitivity to the flipping. Work has shown a comparison of square, circular, continuous and interrupted cells (see Figure 2.9). It appears that the band structure is the one that prevents the appearance of the parasitic thyristor latch for a given P/P^+ box geometry.

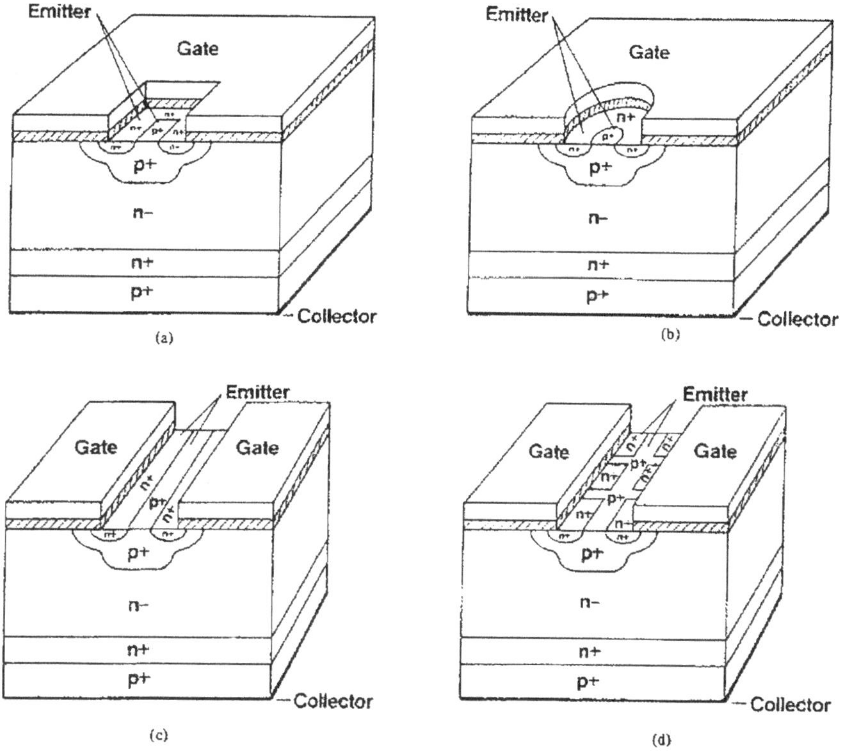

Figure 2.9. *Different cell forms: a) square, b) circular, c) in continuous tapes and d) in interrupted tapes*

Today, manufacturers choose their cell form using patents and technological knowledge. They know how to optimize the size of the cells, the distance between the cells, the geometry of the P/P$^+$ box and the gain, β_{PNP} of the PNP internal transistor to the IGBT, so that the reversal no longer appears under the normal conditions of component use. Thus, the current level for which the switching would appear is outside the safety area provided by the manufacturer.

2.2.1.5. *Compromise speed: low voltage drop in the ON-state*

The IGBT is positioned between the power MOSFET and the bipolar transistor. For high voltage ratings, the IGBT allows much higher current densities than the MOSFET due to its bipolar conduction that modulates the resistance of the N$^-$ layer. On the other hand, the IGBT is slower than the MOSFET because of the presence of the charge stored in the N$^-$ base of the PNP internal transistor. For the nominal current density, the central zone N$^-$ is in strong injection ($n \approx p >> n_i$). Since the transistor is in a floating base, this charge cannot be evacuated by an external contact. The decay of the current during the ON-state phase of the device then comprises two phases (see Figure 2.10). We first observe a sudden drop in the current corresponding to the disappearance of the current MOS (current of electrons), then a current tail. This current drag corresponds to the evacuation of the charge stored in the N$^-$ base. As a first approximation, we can consider that the evacuation of the stored charge follows an exponential type law:

$$Q(t)=Q(0).\exp\left(-\frac{t}{K.\tau_a}\right) \qquad [2.7]$$

Here, $Q(0)$ represents the static stored charge, τ_a is the ambipolar lifetime of the carriers in the N$^-$ base, and K is a coefficient that takes into account the evacuation of electrons to the transmitter of the PNP internal transistor (anode of the IGBT).

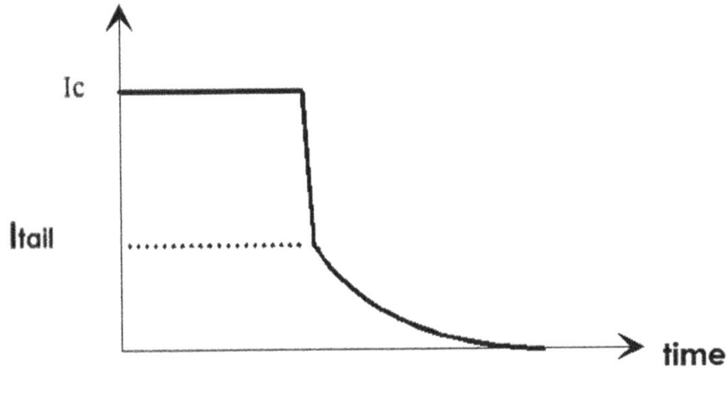

Figure 2.10. *Tail of current during the opening phase*

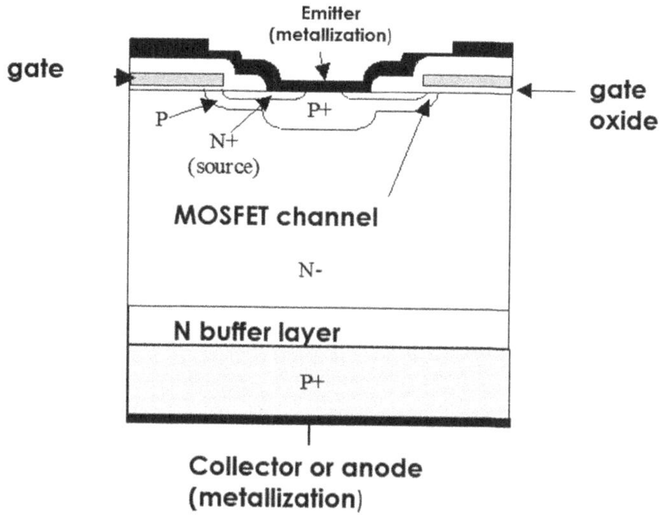

Figure 2.11. *IGBT with N buffer layer*

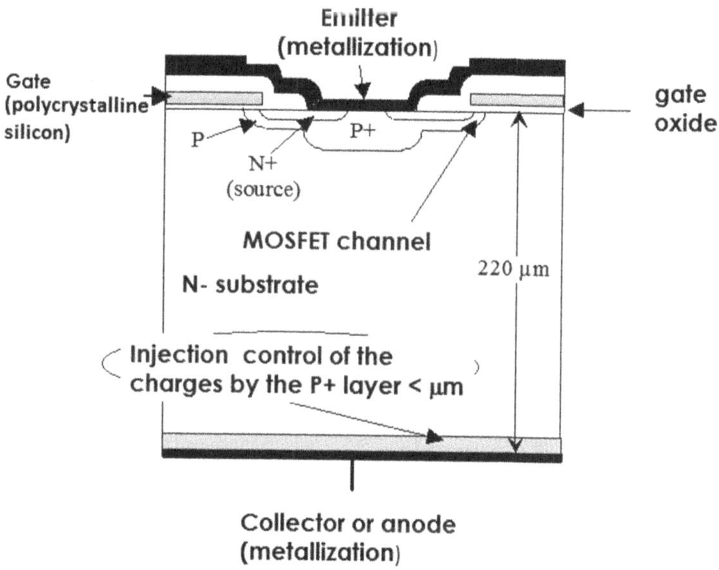

Figure 2.12. *IGBT with load injection control, referred to as a "homogeneous structure"*

We note that it would be interesting to reduce the gain β_{PNP} in order to reduce the current tail, turn-ON time and switching losses. However, we must not forget that what we gain on one side is lost on the other; indeed, lowering the gain β_{PNP} will obtain a faster IGBT, but will degrade the voltage drop in the ON-state because the resistance of the central area will be less modulated. Thus, many manufacturers offer several series of IGBT, each with different compromises between speed and static voltage drop.

2.2.2. *Epitaxial structure with buffer layer and reduction of carrier lifetime*

2.2.2.1. *Origin*

The epitaxial IGBT is designed from the N^- channel power MOSFET structure whose N^+ substrate has been replaced by a P^+ substrate approximately 300 µm thick, strongly doped with acceptor atoms (very low resistivity) constituting the manifold (see Figure 2.13). The area with high resistivity to hold the blocking voltage is achieved by the epitaxial growth method. Then the technological steps to achieve the surface part (the gate oxide, the channel by a double diffusion, polycrystalline silicon gate, metallization, etc.) are similar to those necessary for the design of the power MOSFET. This initial structure is improved by the addition of an N-type epitaxial buffer layer made before the epitaxy of the N-type resistive layer.

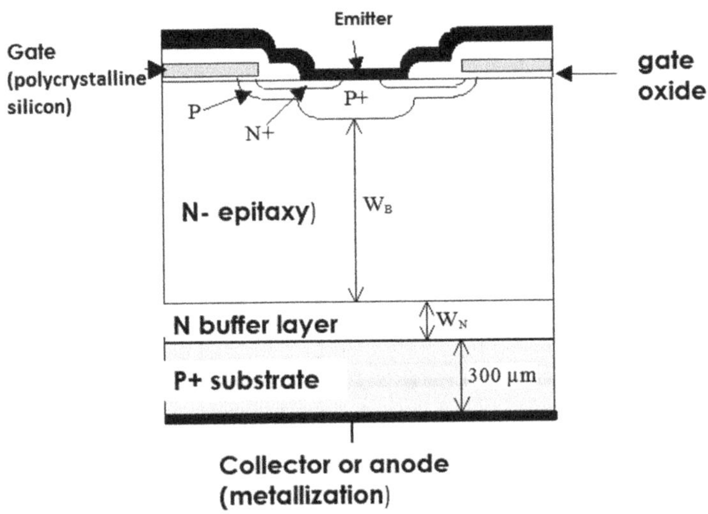

Figure 2.13. *IGBT with epitaxial structure with buffer layer and reduction in carrier lifetime*

In addition, manufacturers using this technology are now performing an additional technological step to reduce the lifetime, in order to improve the electrical characteristics. Finally, this first IGBT family can be characterized as IGBT with an epitaxial buffer layer structure and reduction in carrier lifetime.

2.2.2.2. *Buffer layer Influence*

Its presence introduces a modification in the optimization of the thickness of the epitaxial layer N^- necessary to ensure the drift resistance. If we consider the initial structure without buffer layer (see Figure 2.14), the width of the N^- zone must be sufficient so that, under full blocking voltage, the space load zone developed from the junction P well P^+–epitaxy N^- does not reach the P^+ collector layer. In this case, the following relationship should be verified:

$$V_{\max} = \frac{q}{2.\varepsilon}.N_D.W_B^2 \qquad [2.8]$$

where N_D is the doping level of the base and W_B is the width of the base (see Figure 2.13).

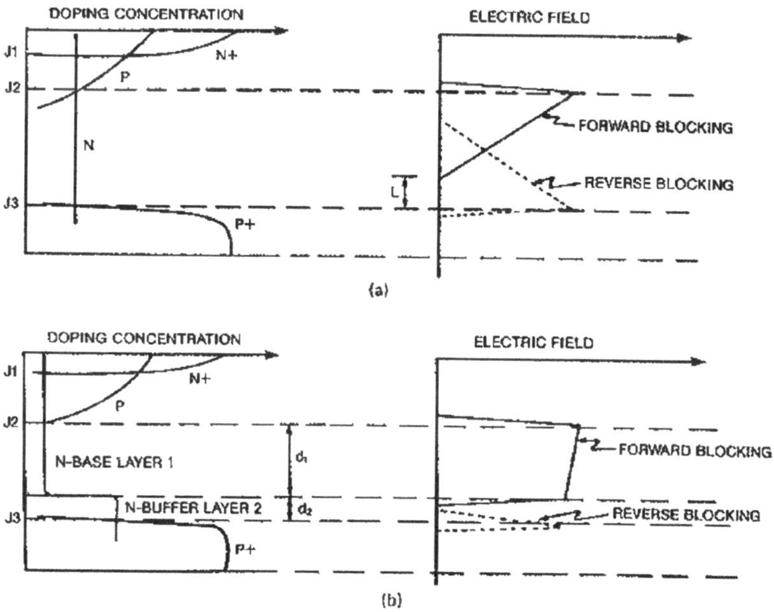

a) without buffer layer b) with buffer layer

Figure 2.14. *Comparison of the spread of the space load zone between an IGBT without buffer layer and an IGBT with buffer layer*

The presence of the buffer N layer makes it possible to screen the electric field (see Figure 2.14). The thickness of the epitaxial N^- zone can be reduced. The intrinsic resistance of this N^- zone is thus lowered. The buffered IGBT is optimized to hold a positive blocking voltage. It has a much lower negative bias voltage. The buffered layer IGBT is also known as a "punch-through" or "piercing" structure because the space charge area can spread over the entire N^- epitaxy, the piercing problem being annihilated.

The introduction of the buffer N layer makes it possible to reduce the value of the current gain β_{PNP} of the PNP transistor by lowering the injection efficiency γ. The charge stored in the epitaxial zone N^- is then less important and the resistance R_{MOD} (see Figure 2.2) is less modulated. Consequently, even if the thickness of the epitaxy is smaller, the static performances are degraded. Figure 2.15 shows that the IGBT with the thickest buffer layer (20 µm) has the highest voltage drop in the ON-state for a given current density.

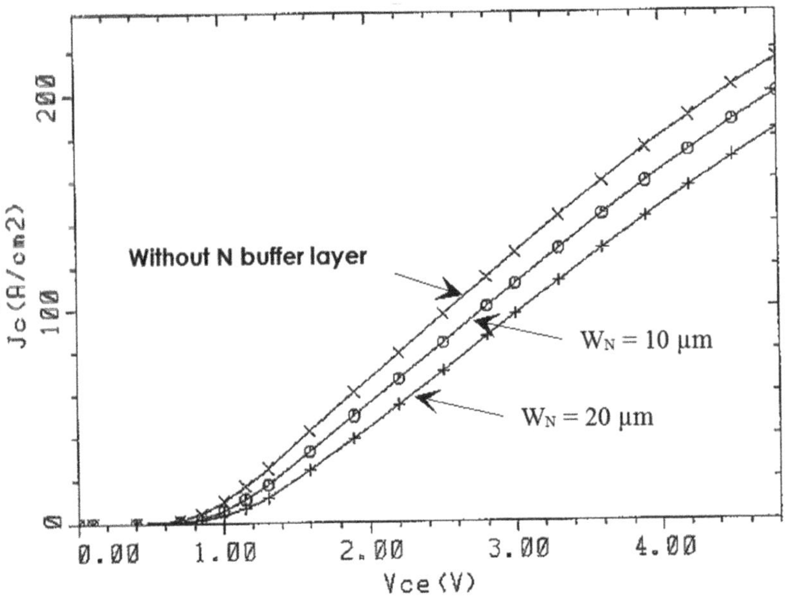

Figure 2.15. *Static curve $J_c = f(V_{ce})$ for a buffer layer with a thickness W_N = 0, 10, 20 µm and a doping level $N_D = 3 \times 10^{16}$ cm^{-3}. Central layer: $\tau_n = \tau_p$ = 1 µs, $N_D = 2 \times 10^{14}$ cm^{-3}, W_{N^-} = 80 µm, V_g = 15 V (Sentaurus simulation)*

The use of the buffer layer is now widespread among all manufacturers using epitaxial technology. This buffer layer has a variable geometry (doping, thickness) according to the compromise speed/desired static voltage drop. We give an example in Figure 2.16, which represents a resistive micro-sounding reading. This curve, on an IRGPC50S device, allows us to measure the following values: W_N = 11 μm, $N_D = 4 \times 10^{17}$ cm^{-3}.

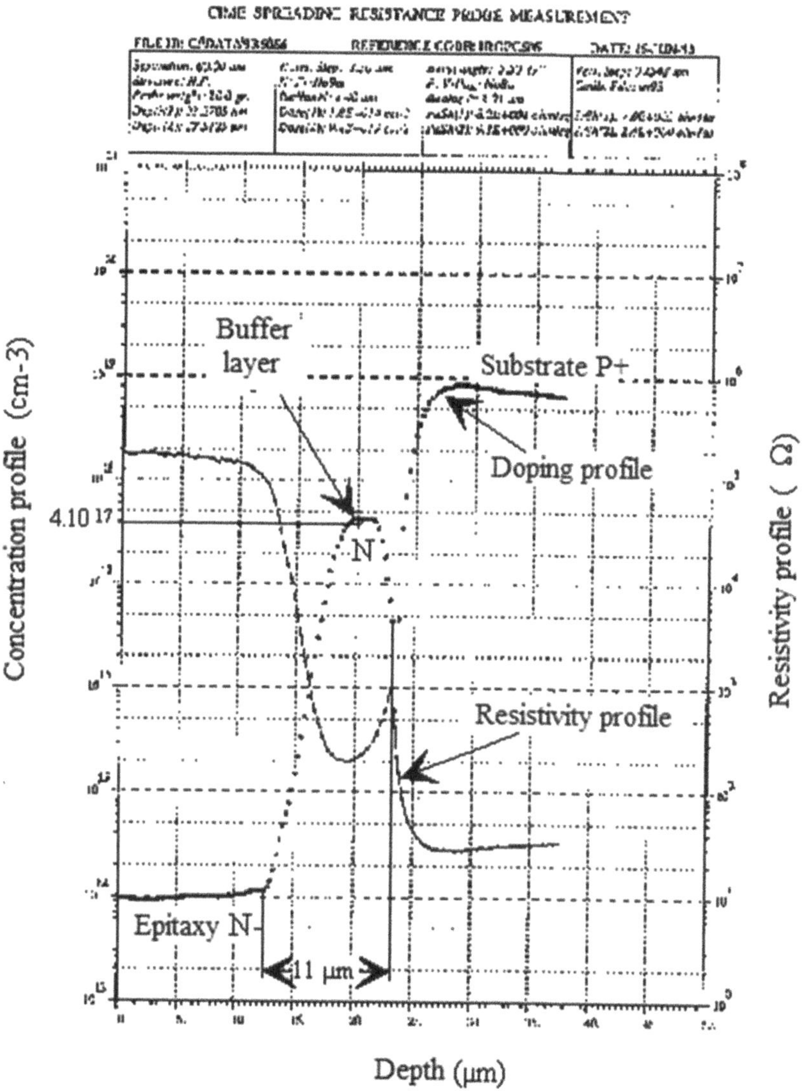

Figure 2.16. *Resistive micro-sounding reading on an IRGPC50S: resistivity and volume concentration profiles*

2.2.2.3. *Carrier lifetime influence*

In order to adjust the β_{PNP} gain, the manufacturers not only add the buffer layer, but also include an additional technological step to reduce the carrier lifetime in the N^- zone, in order to accelerate carrier recombination. The reduction in the lifetime of the carriers makes it possible to lower the transport coefficient in the base α_T, and thus reduce the stored charge (see Table 2.3). The turn-ON time of the IGBT will be reduced; on the one hand, the initial stored charge $Q(0)$ will then be smaller and, on the other hand, the carrier lifetime will be shorter. The β_{PNP} gain decreases as carrier lifetime decreases.

In epitaxial structures, the reduction of carrier lifetime is never really localized and applies to both the central zone and the buffer layer, as well as the emitter of the bipolar junction transistor PNP. Thus, the decrease in the transport coefficient α_T is accompanied by a reduction in the injection coefficient γ. Reducing the lifetime has the same effect as increasing the thickness of the buffer layer or raising the doping level of this buffer layer (see Figure 2.17).

Technological modifications	Inserting an N buffer layer	Reduction in carrier lifetime
Impact on model performance	Reduction of the injection coefficient γ	Reduction of the transport coefficient in the base α_T and the injection coefficient γ

Table 2.3. *Technological changes and consequences on epitaxial IGBT*

2.2.2.4. *Influence of the increased conduction area*

In epitaxial buffer-layer technology and carrier lifetime reduction, some manufacturers are adding an increased conduction zone (patent). The principle is to overload N between the P/P^+ box under the gate (see Figure 2.18) in order to increase the conductance of the N^- zone and reduce the pinch effect by JFET. The use of the increased conduction zone makes it possible to increase the density of cells without degrading the voltage drop in the ON-state. Figure 2.19 shows an optical microscope photo of cells of an IRGPC50S IGBT. The size of the hexagonal cells is 16 μm × 20 μm. A polishing followed by a chemical revelation of the junctions made it possible to estimate the distance between the caissons P/P^+ as being approximately equal to 4 μm.

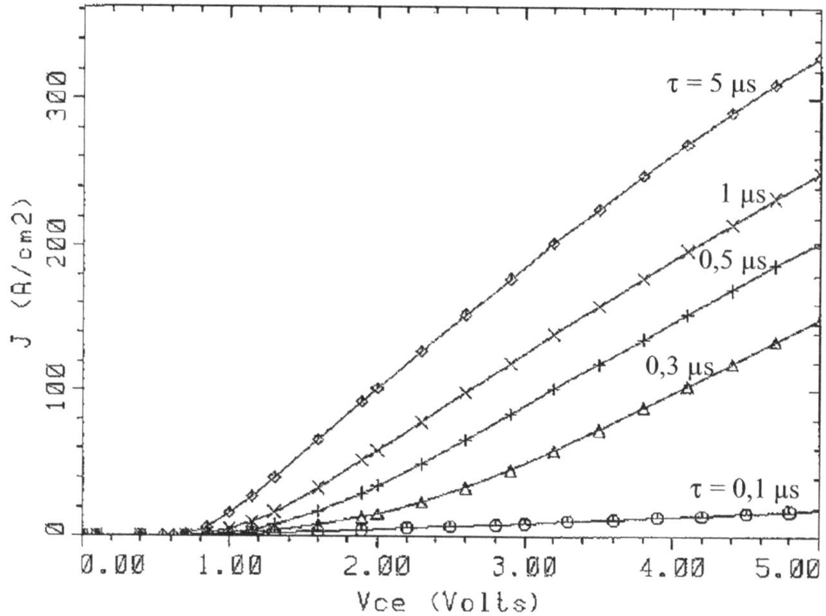

Figure 2.17. *Static curve $J_c = f(V_{ce})$ for the following lifetime in the central area $\tau_n = \tau_p$ = 0, 1; 0, 3; 0, 5; 1; 5 μs. Buffer layer: $N_D = 3 \times 10^{16}$ cm^{-3}, W_N = 20 μm. Layer zone: $N_D = 2 \times 10^{14}$ cm^{-3}, W_{N^-} = 80 μm. V_g = 15 V (Sentaurus simulation)*

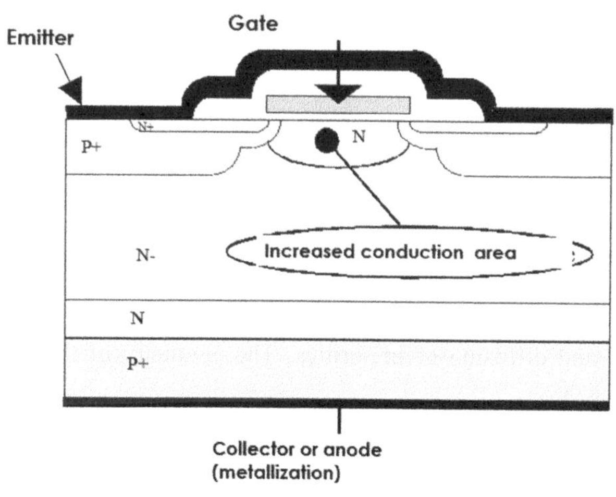

Figure 2.18. *Presence of increased conduction zone in epitaxial technology*

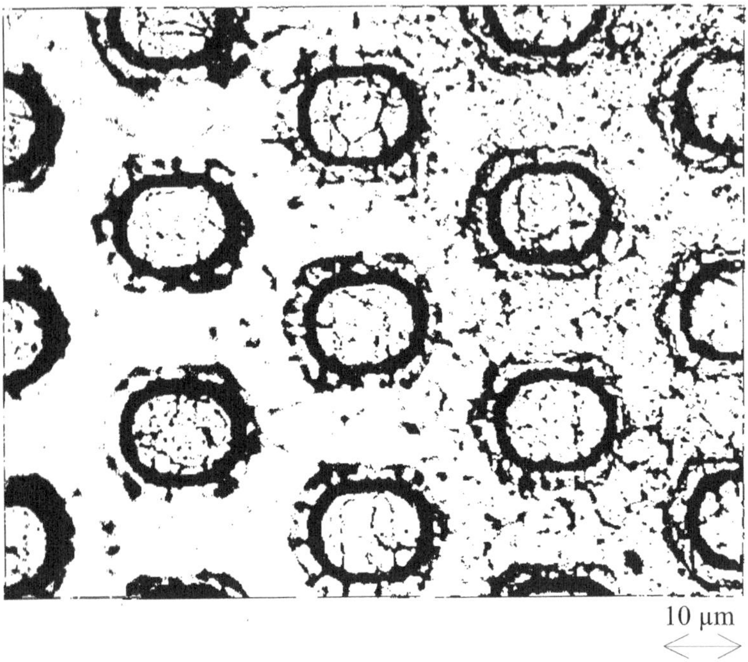

Figure 2.19. *Surface view of the cells of an IGBT IRGPC50S, photo taken under an optical microscope, G * 1250*

2.2.3. *Homogeneous structure with control of load injection*

This approach uses the technique of controlling the injection of charges because of a thin emitter (weak γ). This implies that the base substrate is N^- type, which constitutes the central layer of high resistivity. To benefit from sufficient rigidity, the minimum thickness of this substrate must be of the order of 220 μm. This thickness makes it possible to have a device with a withstand voltage greater than 2000 V. Thus, IGBT homogeneous structures are disadvantaged in terms of voltage drop in the range of small withstand voltages (<1,200 V). The P^+ type collector on the rear face of the chip (the emitter of the PNP internal transistor) is produced by the implantation and diffusion of impurities. The geometry of this P^+ layer has a considerable influence.

IGBTs with control of the injection of charges have a thick N^- base (of the order of 220 μm minimum). To have a voltage drop in the low passing state, the lifetime of the carriers is kept naturally high. However, the ON-state phase requires a low stored load. In this technology, the injection efficiency γ is lowered by playing on the geometry of the P^+ layer (the emitter of the internal bipolar PNP transistor). The stored load is expressed as follows:

$$Q = \frac{I_c W_B^2}{2D} \cdot \left(\gamma - \frac{1}{1+b} \right) \qquad [2.9]$$

Here, I_c represents the current in the IGBT, W_B is the thickness of the base, D is the ambipolar diffusion coefficient of the carriers in the base and b is the ratio of the mobility of the electrons and the mobility of the holes (μ_n/μ_p). The hypotheses of obtaining the preceding equation are the following: no recombination in the base (infinite carrier lifetime) and base in strong injection. The previous equation shows that the stored load is related to injection efficiency γ. The charge injection control will tend toward $1/(1 + b)$. If we consider this condition verified, the injection efficiency γ can be written as:

$$\gamma \approx \frac{1}{1+b} + \frac{b}{1+b} \sqrt{\frac{Q_E}{(1+b)} \frac{1}{I_c} \frac{4D_P}{W_B^2}} \qquad [2.10]$$

with

$$Q_E = q.N_A.S.W_E = q.\text{dose} \qquad [2.11]$$

where D_P represents the diffusion coefficient of the holes in the base, q is the charge of the electron, N_A is the doping level of the P^+ emitter of the PNP transistor (anode of the IGBT), S is the surface area and W_E is the thickness of the P^+ emitter of the PNP transistor. The injection efficiency is entirely controlled by the implanted dose to produce the P^+ layer.

The pace of the current during the opening phase of an IGBT, with homogeneous technology with charge injection control (see Figure 2.20), is characterized by a rapid decrease in the current favored by a low gain β_{PNP}. On the other hand, there is still a tail of current that now extends for several microseconds because carrier lifetime is high.

To conclude our work on IGBT with homogeneous technology and control of the injection of loads, we will specify that it belongs to the family of IGBT called "Non Punch Through (NPT)" or "non-piercing" because the space charge zone must never spread over the entire N^- zone to avoid its penetration into the bipolar transistor.

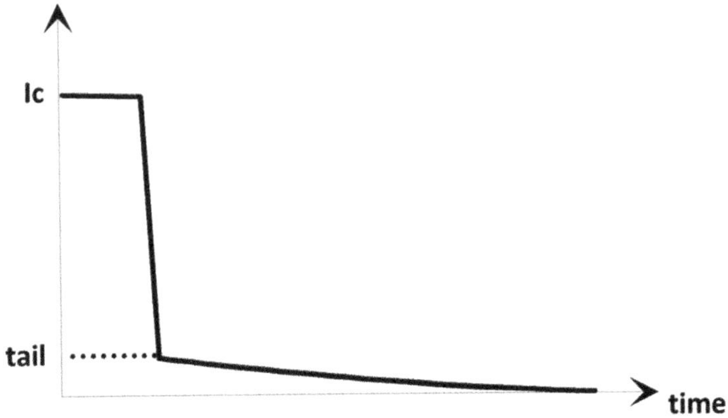

Figure 2.20. *Current tail during the opening phase of an IGBT with homogeneous technology with load injection control*

2.2.4. *Silicon direct bonding-IGBT*

In order to optimize the geometry and doping of the buffer layer, silicon direct bonding (SDB) technology has been developed. We are looking for a thin and very doped buffer layer, and this is not compatible with the multiple technological steps, and in particular, epitaxy.

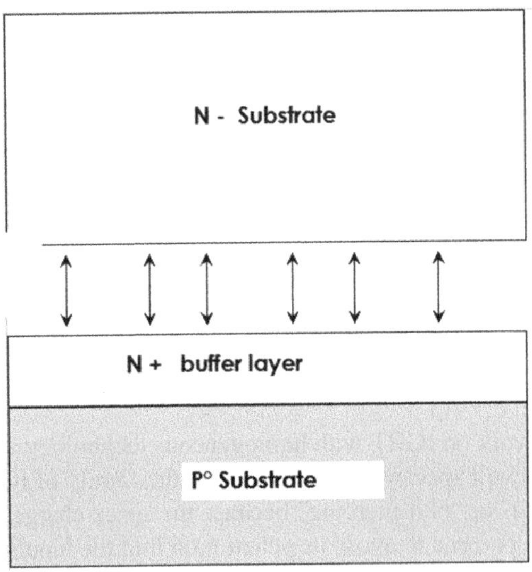

Figure 2.21. *SDB-IGBT technology: Two substrates assembling (principle)*

SDB-IGBT technology uses two wafers: the P^+ substrate and the N^- wafer, which are welded to one another by contact, silicon on silicon (see Figure 2.21). Before the "bonding", the buffer layer is processed on the P^+ substrate. In the end, the section of the SDB-IGBT is similar to that of the PT-IGBT (see Figure 2.11), with an optimal buffer layer geometry. This technology makes it possible to reduce the implanted dose, or even eliminate the step of controlling carrier lifetime.

2.2.5. *IGBT trench*

In order to increase the integration of the cells of an IGBT (to decrease the static voltage drop for a given current density), trench-IGBT technology was developed (see Figure 2.22). The channel is now vertical. The width of the trench is a few micrometers. This technology can be combined with the SDB process.

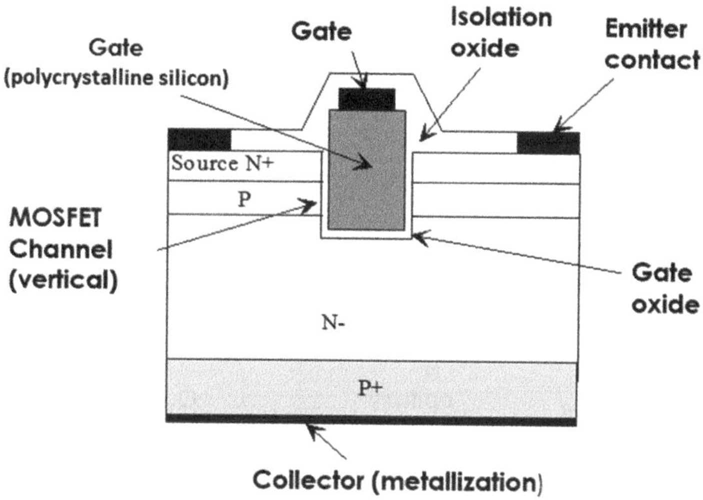

Figure 2.22. *Trench technology (principle)*

2.2.6. *Lateral IGBT*

Figures 2.23 and 2.24 present the views of a lateral IGBT and its equivalent circuit.

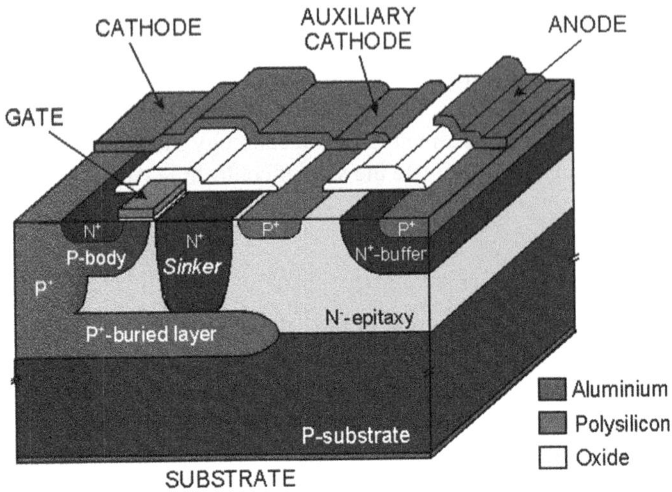

Figure 2.23. *Lateral IGBT. For a color version of this figure, see www.iste.co.uk/abouelatta/smartpower.zip*

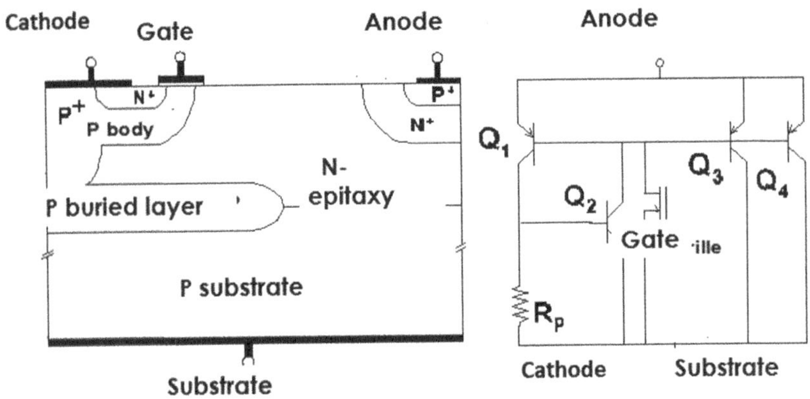

Figure 2.24. *Cut and equivalent electrical schematic of the lateral IGBT*

2.3. Assembly technology

Figure 2.25 shows the "DCB" (Direct Copper Bonding) hybrid assembly technology used in most IGBT-based power modules.

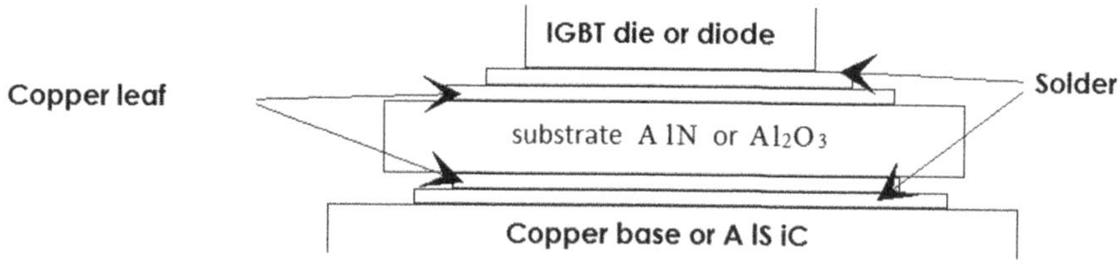

Figure 2.25. *DCB power module assembly technology*

2.4. Thermal aspect

The increasingly advanced integration of components dealing with power inevitably leads to an increase in the heat balance. One of the current barriers for the industry is heat dissipation, and more precisely, thinking of new forms of packaging. What is more, the problems of electrical insulation impose dielectric substrates with as little conduction of heat as possible. Since the thermal resistances induced by current encapsulations (conductivity of materials and contact resistances at interfaces) are too high, it is necessary to study new materials and assembly technologies (see Figures 2.26, 2.27 and 2.28).

Let us take an example of a power device of the IGBT (see Figure 2.29 and 2.30), and the current commercial hybrid modules (several chips in parallel), which can exceed powers of the order of 3 kV × 1 kA. A power component is a heat source of the order of 100 W (which is a considerable amount).

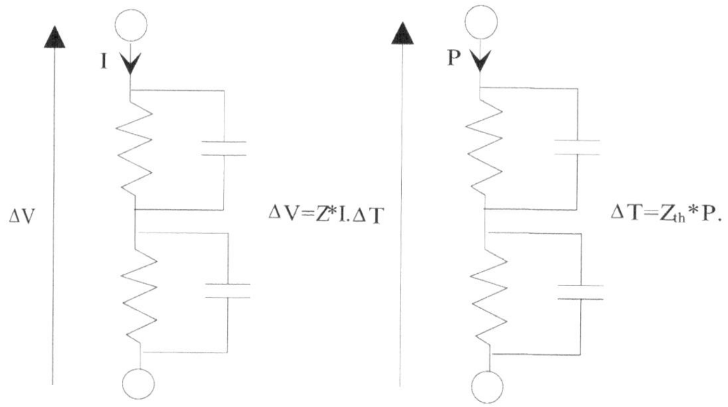

Figure 2.26. *Analogy between electrical/thermal resistance*

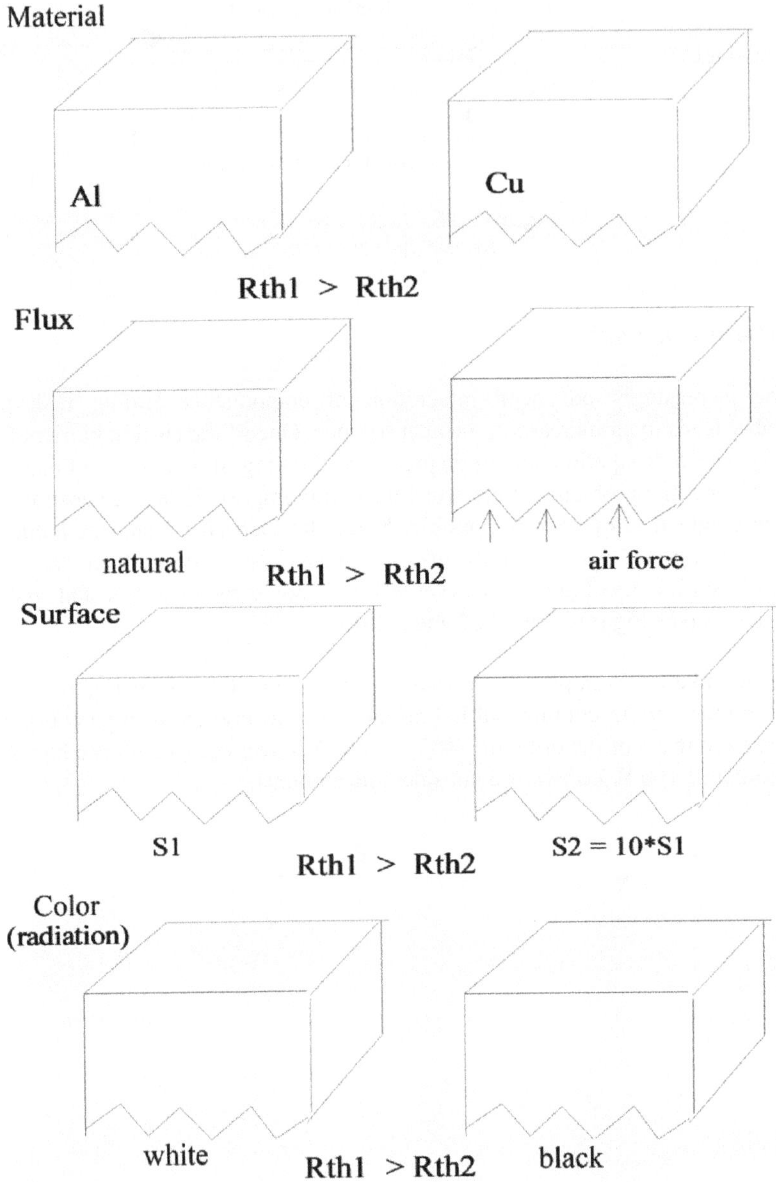

Figure 2.27. *Thermal resistance comparisons*

2.4.1. *Thermal impedance*

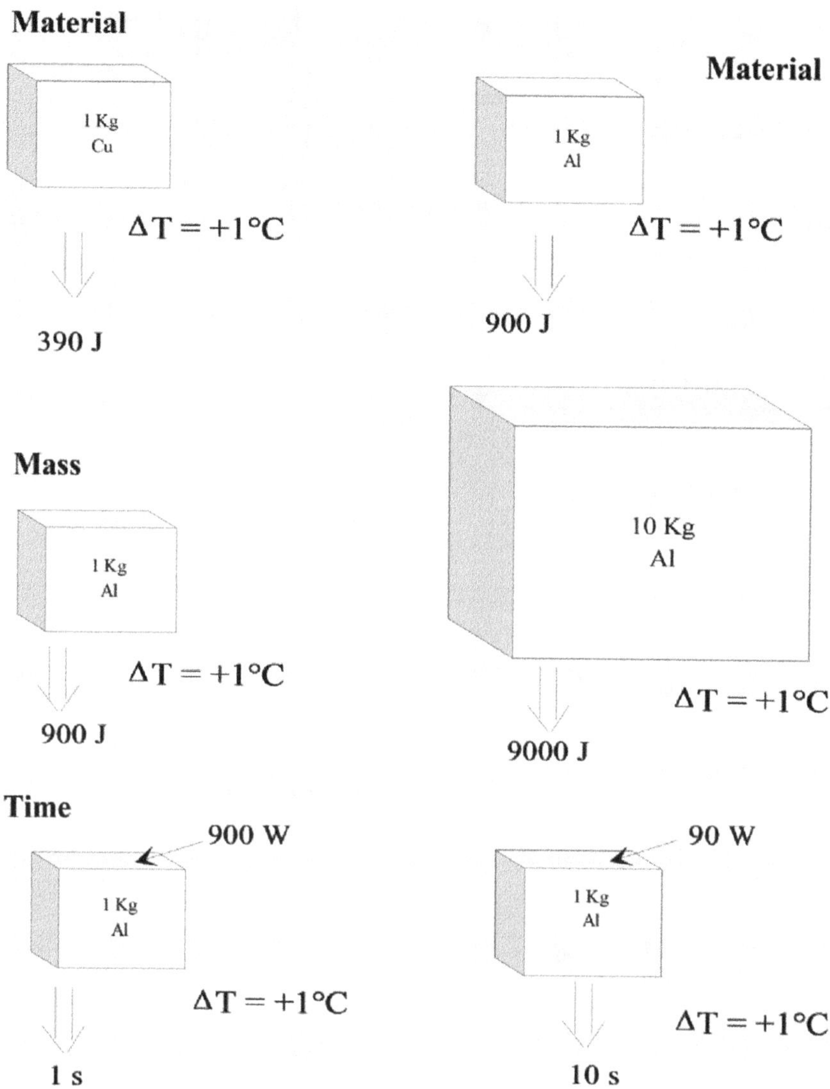

Figure 2.28. *Power module example*

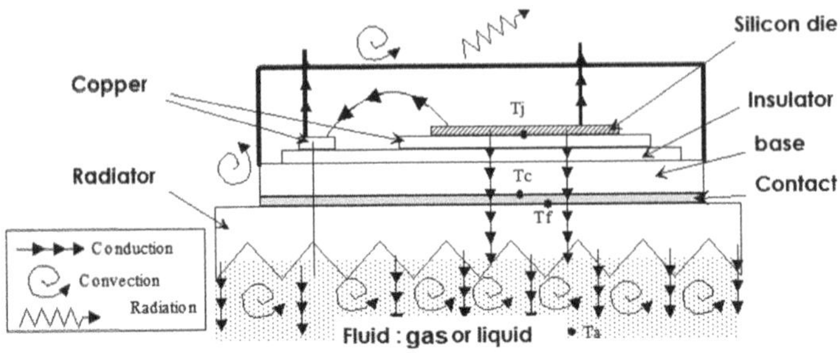

Figure 2.29. *Simplified cutting of a power module with possible heat paths*

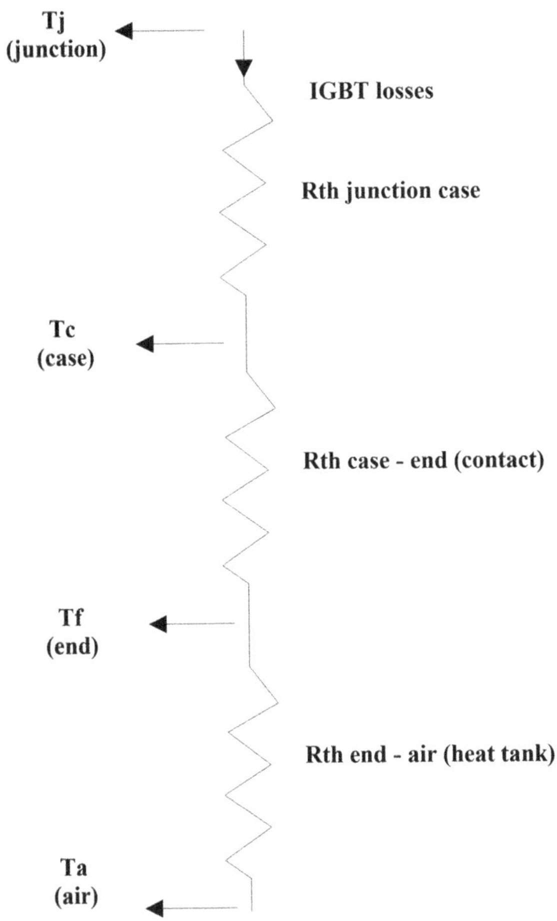

Figure 2.30. *One-dimensional simplified thermal model*

The thermal resistances must be as small as possible and it will be necessary to check the flatness and roughness of the basic module. To improve the thermal contact between the module and the heat sink, it will be useful to add silicon grease. The most preferable case being a contact-type press-pack, that is, a stack under pressure. This is only possible for certain power components of thyristor, GTO or diode type. Press packs for IGBT appear to be very expensive due to the complexity of grid bonding. Example of a digital application: $R_{thj-c} = 0.022°C/W$, $R_{thc-f} = 0.018°C/W$, $T_j = 120°C$ (it is recalled that the limit typically given on the datasheets is 125°C, the physical limit being 150°C):

$$\Delta T = T_j - T_a = P \cdot \left(R_{th\,j-c} + R_{thc-f} + R_{th\,f-a} \right) \qquad [2.12]$$

What should be the thermal resistance of the heat sink when the losses at the component are of the order of 500 W? By developing the equation above, it is calculated as

$$R_{th\,f-a} = 0.12°C/W .$$

2.5. Applications fields

2.5.1. *IGBT power modules for electric traction applications*

2.5.1.1. *The reliability aspect*

Power modules dedicated to electric traction must have a low internal inductance (less over voltages during switching), withstand power and temperature cycling. The two main aging failure mechanisms during thermal cycling are as follows: fatigue at the level of the welds between the ceramic and the base of the module and the bonding detachment.

2.5.1.1.1. Weld fatigue

The problem arises from a difference between the coefficients of thermal expansion (CTE) and the materials that is too significant (a “mismatch”). The difference between the copper base and the Al_2O_3 (alumina) ceramic in CTE is about 10 ppm/K (ppm: % per million). For example, a 55 mm long base with $\Delta T = 100°C$ undergoes a length variation of 55 µm ($55 \cdot 10^{-3} \frac{10}{10^6} 100 = 55\ \mu m$).

	Interface substrate – silicon		Interface substrate – embase		
Base and substrate materials	Mismatch CDT – silicon substrate (ppm/k)	Difference in longitudinal expansion (µm)	Mismatch CDT base substrate (ppm/k)	Substrate waist (mm × mm)	Difference in longitudinal dilatation
Cu and AlN	1	1	13	28 × 28	40
AlSiC and AlN	1	1	3.5	55 × 55	20
Cu and AlN	1	1	13	55 × 55	71
Cu and Al_2O_3	4	5	10	55 × 55	55

Table 2.4. *Comparison of the effect of thermal expansion in various modules*

The consequences include mechanical stress, which can lead to cracks, and poor thermal contact. It should be noted that this generates a runaway; the temperature increase introduces cracks that lower the thermal dissipation, which in turn tends to increase the temperature until the appearance of a defect.

If alumina (Al_2O_3) is replaced by AlN (more expensive), the result is even worse because the difference in CTE is about 13 ppm/K. In the previous example (55 mm at 100°C), the variation in length is 71 µm. Alumina with a better thermal characteristic can only be used for small substrate sizes.

One solution would be to use an AlN substrate with an AlSiC base (CDT difference of the order of 3.5 ppm/K, again with the same example, the variation in length would be 20 µm). There may also be thermal fatigue between the silicon chip and substrate, but this interface is generally less critical because the size of this seal is smaller (the surface of a power chip is less than 2 cm^2).

The difference of CTE between Al_2O_3 (alumina) and silicon is of the order of 4 ppm/K, and the difference of CTE between AlN (nitride) and silicon is of the order of 1 ppm/K (see Table 2.4).

2.5.1.1.2. Bonding detachment

The origin of the problem is the same: it comes from a difference in the coefficient of thermal expansion between the silicon and the bonding of aluminum (18 ppm/K). The solution would be to use an alloy weld between aluminum and molybdenum, as shown in Figure 2.31.

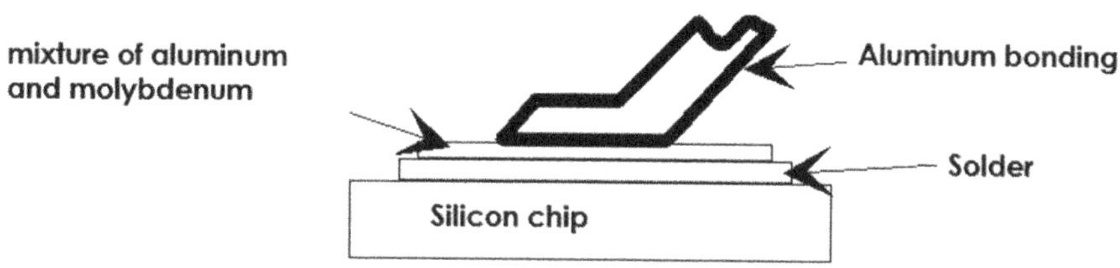

Figure 2.31. *Molybdenum and aluminum layer to improve contact between silicon chip and aluminum bonding*

2.5.1.1.3. Parallelization of power chips

The parallelization of IGBT-type power chips is facilitated when they have a positive temperature coefficient. This results in self-balancing of temperatures and currents in the various devices in parallel.

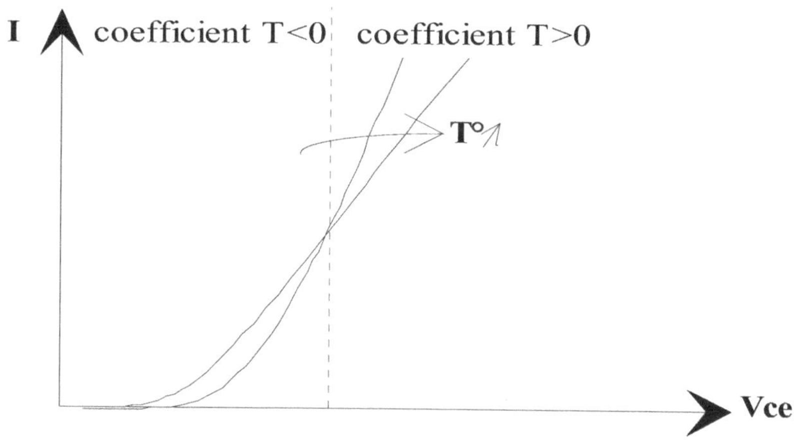

Figure 2.32. *IGBT static characteristics and temperature coefficient*

If we assume an assembly of two IGBTs in parallel, they have the same collector-emitter voltage (V_{ce}); if the temperature coefficient is positive, the component that heats up will automatically be relieved by current (see Figure 2.32) and its temperature will go down.

2.5.2. *IPM for low- and medium-power applications*

The IPM family offers hybrid assemblies of power chips, their drivers and their protection systems (voltage, current, temperature), to offer a system solution integrated in a single housing. In this chapter, we present an IPM of the Fuji family RH (see Figures 2.33 and 2.34).

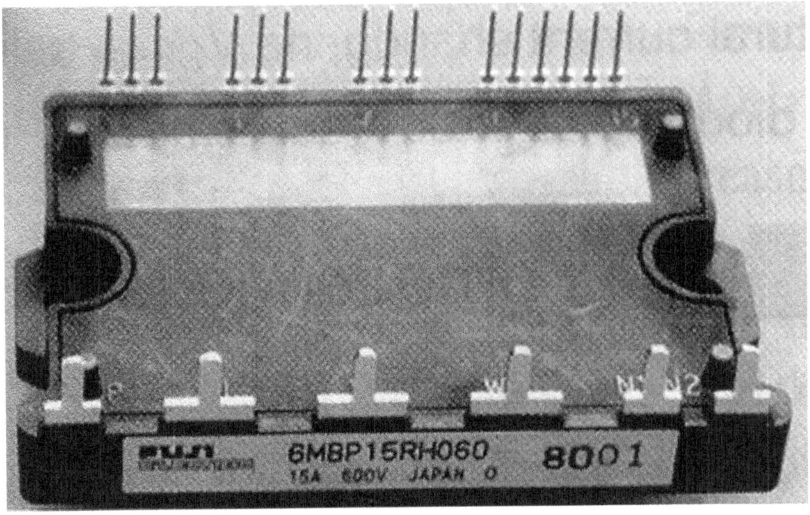

Figure 2.33. *Mini IPM of Fuji*

Figure 2.34. *Internal design of the IPM*

The Fuji 6MBP15RH-060 reference IPM contains three inverter arms (see Figure 2.35) and uses an alumina substrate with DCB technology (see Figure 2.25). This IPM contains an integrated circuit for controlling the bottom three IGBTs and an integrated IGBT control circuit for the top. The protections included in this IPM are applied to:

– the temperature on each IGBT;

– a supply voltage of the integrated circuits that is too low;

– the overcurrent at the level of each IGBT;

– the overcurrent on the continuous bus.

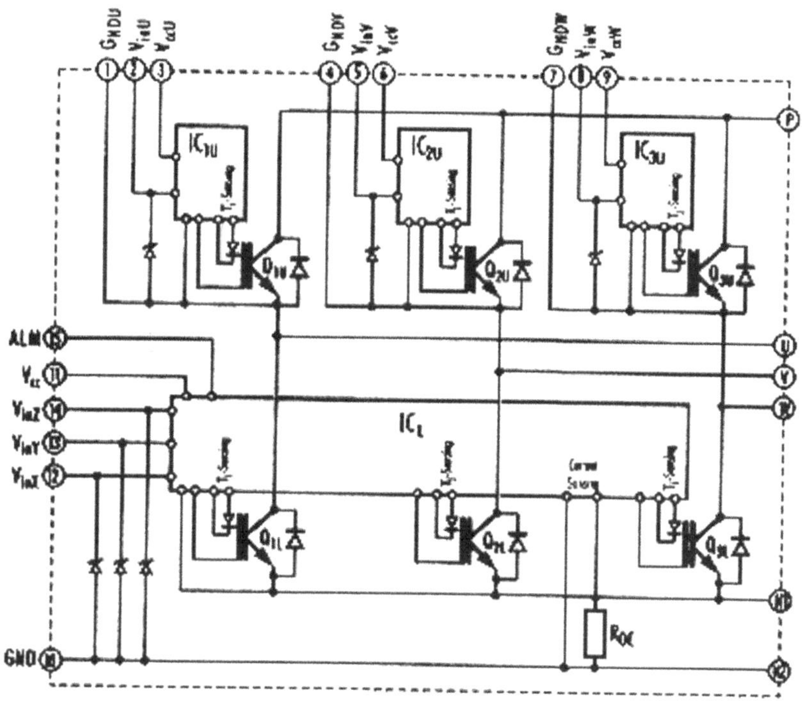

Figure 2.35. *Block schematic of the IPM*

Each IGBT has an internal temperature gauge, which uses a monolithic sensor carried out on a polysilicon diode on the surface of the IGBT. The drivers are designed to be connected to the exits of optocouplers. Each integrated circuit is protected from an overpressure (or a negative supply voltage) because of a Zener diode. Resistances to turn the IGBT ON and OFF correspond to internal resistances of the MOSFET of the integrated circuit driver. In the event of a defect, the drivers ensure a “soft” turn OFF to avoid overpressures (see Figure 2.36).

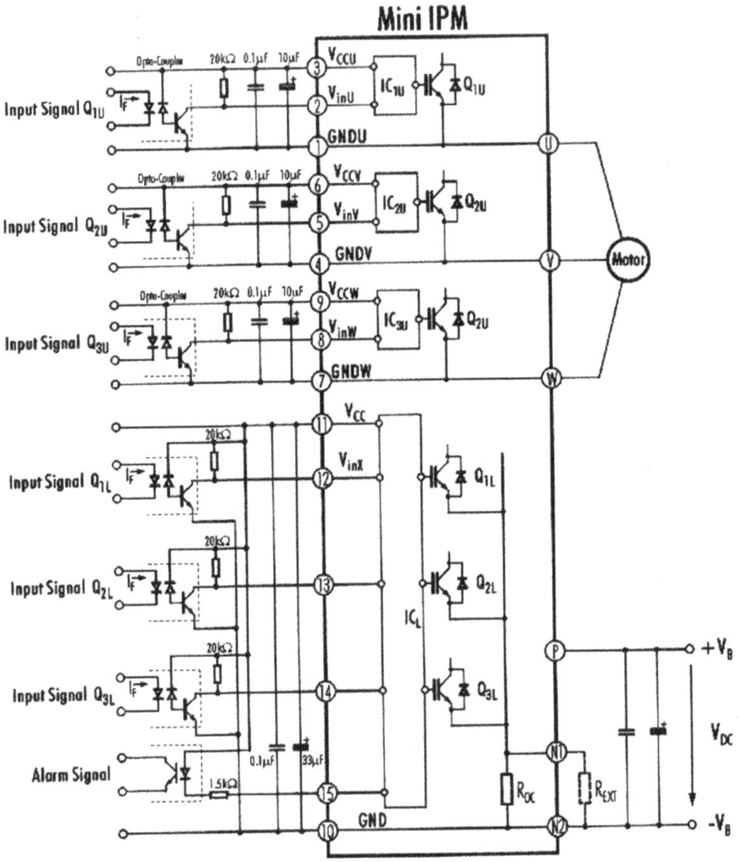

Figure 2.36. *The IPM in an application*

3

Monolithic Integration

3.1. Functional integration and smart power

Progress in monolithic integration has advanced in recent years and two routes can be distinguished: the integration of "smart power-type" systems and functional integration.

Monolithic smart power integration is the integrating of the analog electronic parts and digital and power stages on the same silicon crystal, obtaining electrical characteristics comparable to those using discrete products. Beyond this definition, smart power integration enables the design of self-protected systems in terms of temperature, current and voltage. The same silicon crystal supports the power stage and controls electronics, protection, diagnosis, communication, etc. This requires insulation of the low-signal logic and analog part from the power stage (insensitivity to strong di/dt and dv/dt). Smart power integration is limited to a switched power of the order of kVA.

In the current range of a few amperes, and for a voltage of a few hundred volts, metal oxide semiconductor (MOS), bipolar or even BiCMOS technologies with junction insulation could be used. For higher switched powers, it is currently impossible to integrate the thyristor or the vertical IGBT; much work has been focused on the development of lateral power components. In particular, these include the L-IGBT (lateral insulated gate bipolar transistor), which has a capacity that is three to five times greater than a MOSFET of equivalent size. In parallel, there are new types of insulation being studied: SOI (silicon on insulator) or "bonded Si/SiO_2/Si wafers". The first involves isolating silicon by a deposition of an oxide; this is practiced today on all surfaces of the chip and, in the future, it will be necessary to be able to create local SOI structures. The second makes it possible "to stick" various types of substrates (silicon or oxide). Smart power turns to increasingly complex circuits that use VLSI and ULSI technologies. At the present time, the rules of production are not scaled down from 1.5 to 2 µm (U > 600 V) and 0.6 µm (U < 100 V) and this because of the "vertical complexity" ("vertical" in the sense of perpendicular to the surface) of the components.

Manufacturing and development costs for smart power components are high because all of the components are used with the most complex technology. The manufacturers direct themselves toward a flexible approach by proposing programmable smart power components ("Super Smart Power"). While this includes battery chargers for mobile telephones, the main application of smart power is the car (lighting and other advanced systems for managing the vehicle equipment: ABS, headlights. etc.), but also includes fluorescent lamps, robotics, medical equipment, etc.

Functional integration seeks to develop new functions that are not possible in smart power integration technology. It is based on interaction between semiconductor layers (or between components) and not on their interconnection. Thus, the thyristor and the TRIAC could be regarded as the first examples of functional integration. More recently, the bipolar association of technology and MOS revealed the IGBT, which is also a new example of functional integration. Functional integration provides new opportunities by associating traditional components, such as diodes, P or N MOS transistors, Zener diodes, etc., with vertical components. An example is application-specific discrete (ASD) technology, developed by SGS-Thomson. We are therefore able to acknowledge the benefits of limiting device current, component protection and, more generally, of the circuits connecting themselves directly to the alternative network.

3.2. Transition from low-voltage technology (CMOS) to high voltage

Functional integration is useful in the development of switch functions (in the sense that they do not exist in discrete components). The largest innovation is based on the development of a monolithic dual thyristor, this being conceivable by benefitting from an optimized association of semiconductor areas to the quite specific characteristics: the concept of functional integration, even.

3.2.1. *Introduction*

3.2.1.1. *MOS transistor presentation*

This is a question of demonstrating how to adapt a 5V CMOS technology to a high voltage. The body of the system will continue to use the devices of the supported technology. Thus, the many circuit libraries already developed could be used.

3.2.1.1.1. Principle and operation

The MOSFET is of double interest as it takes up less silicon surface and consumes less electric power when it is used in CMOS configuration.

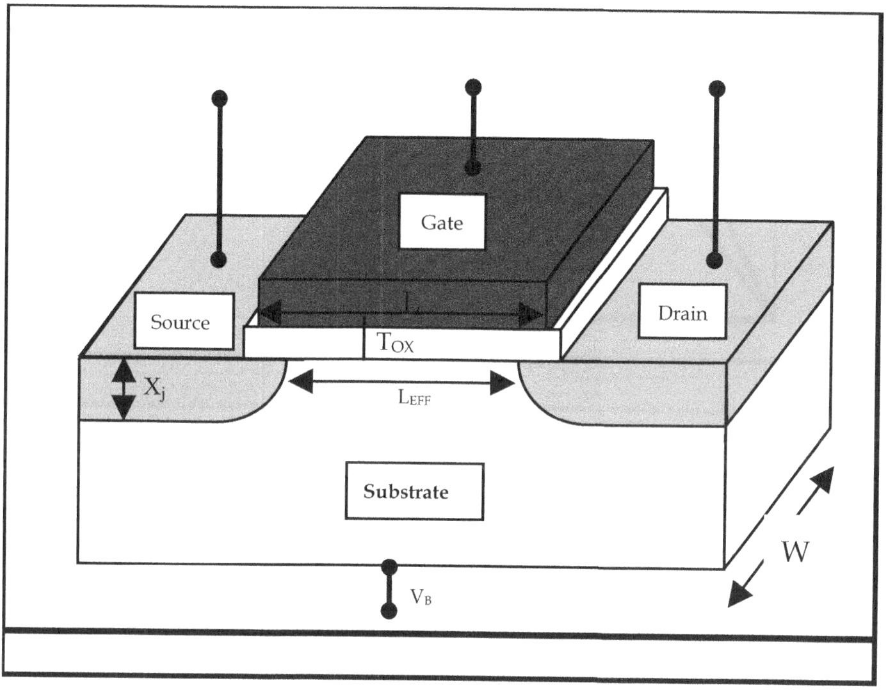

Figure 3.1. *MOS transistor structure*

To describe the operation of this device, we will take the case of an NMOS transistor. The substrate is then of P type (boron doped), and the sources and drains of the N type. The application of a positive voltage on the grid initially forms an area below the grid that is depleted of majority carriers, with the holes being pushed back by electrostatic effect in the bulk. This involves the electrons in the bulk moving toward the interface. When the density of electrons exceeds the density of holes present in silicon on the surface, the weak inversion mode starts. Beyond a certain voltage, known as the threshold voltage (V_T), a weak variation of the potential involves a high density of negative charges present in the inversion layer. The mode of strong inversion is reached. This layer (channel) links the N source and drain regions with a conducting channel. It is localized on a few tens of angstrom thickness. It constitutes the channel of the transistor and the transistor is then in ON state.

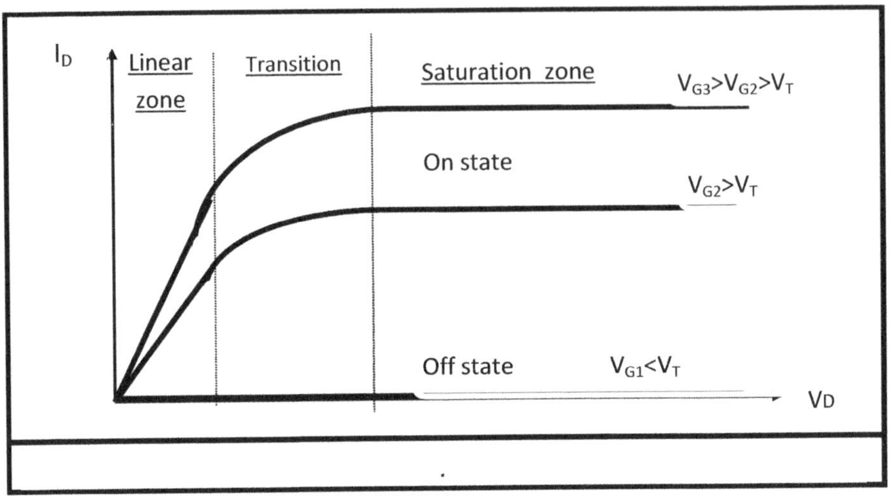

Figure 3.2. *Current drain voltage characteristics*

The resistance of the channel is a function of the single gate voltage for small drain voltages only. As V_D increases, the potential difference between gate and channel decreases from the source to the drain, resulting in a decrease in the charge of the channel, and thus an increase in resistance. The variation of the current I_D as a function of the drain voltage V_D becomes nonlinear; it is the transition mode (Figure 3.3(a)).

When V_D reaches the so-called saturation voltage, denoted by V_{Dsat} and defined as the value of V_D, when the potential difference between the gate and channel becomes equal to the threshold voltage V_T, at which the inverting charge disappears at the drain. We can say that there is a pinching of the channel (Figure 3.3(b)). The potential present at the end of the channel on the drain side then remains constant, and the intensity of the drain current I_{Dsat} no longer increases. We have reached the saturation regime. For higher V_D values, the pinch point moves up to the source (Figure 3.3(c)).

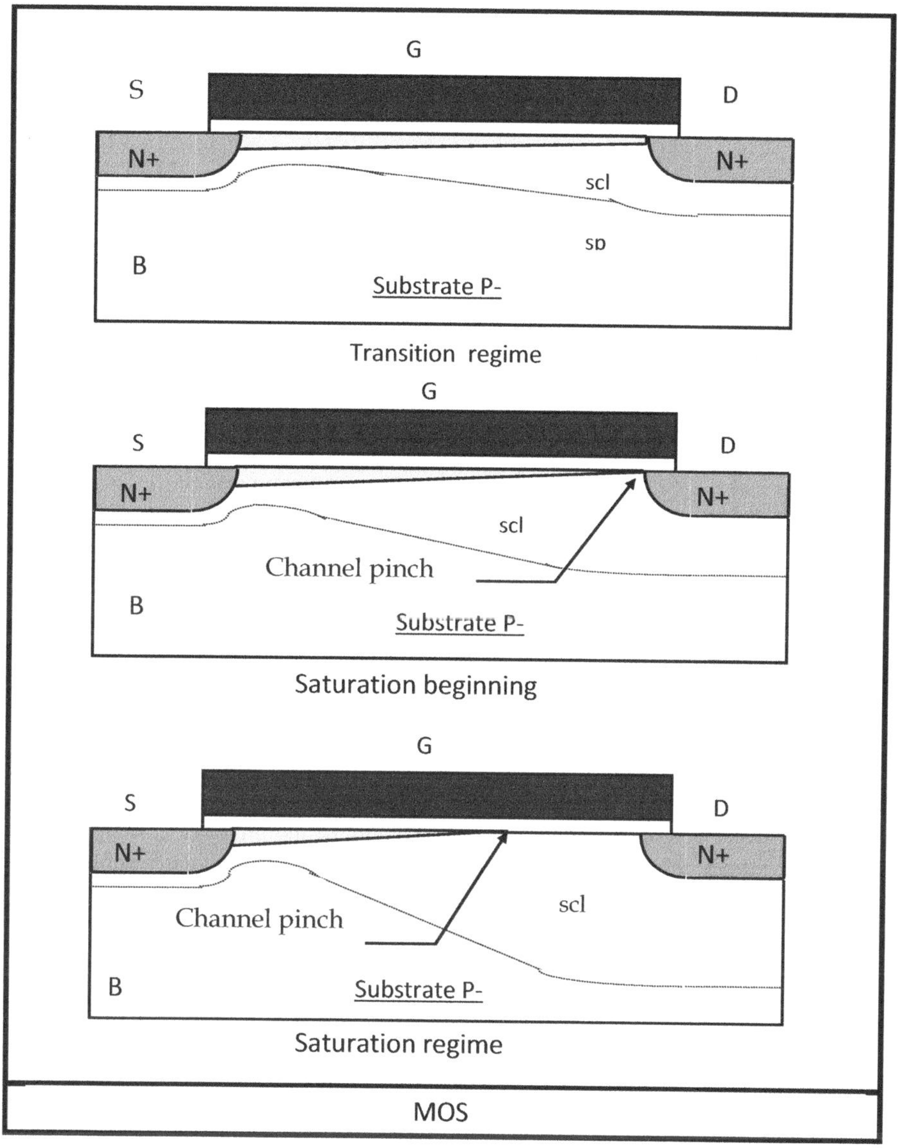

Figure 3.3. *MOS transistor conduction*

3.2.1.1.2. Threshold voltage

Figure 3.4 shows the diagram of the energy bands of a MOS capacitor.

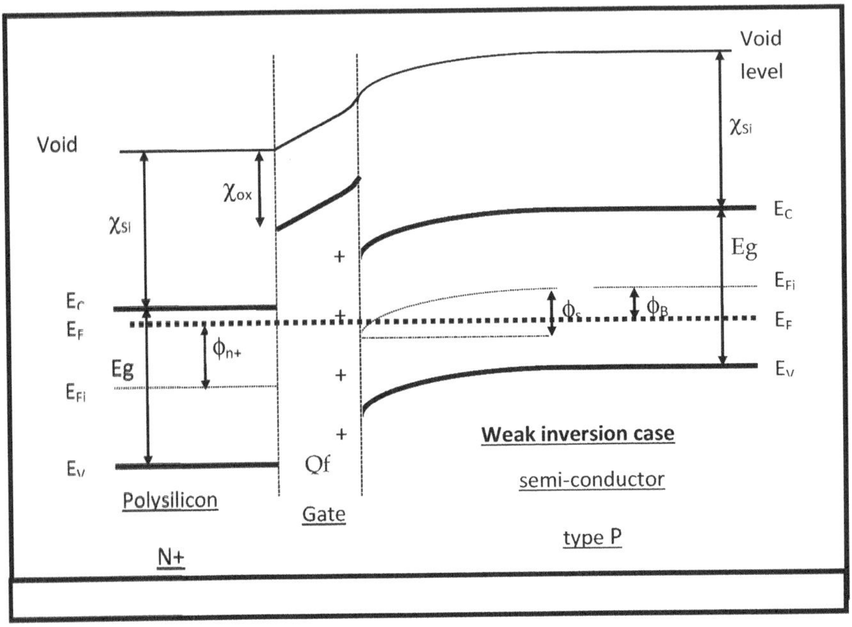

Figure 3.4. *Energy band diagram of an MOS structure*

The threshold voltage V_T is defined as the gate voltage required to achieve strong inversion. Strong inversion occurs when the density of the minority carriers present at the oxide/silicon interface is equal to the density of the majority carriers present in the substrate. This corresponds to the following condition:

$$\phi_S = 2.\phi_B \quad [3.1]$$

Hence

$$V_T = \frac{\sqrt{2\varepsilon_S q N_D 2\phi_B}}{C_{OX}} + 2\phi_B + V_{FB} \quad [3.2]$$

3.2.1.1.3. Drain current

To be able to calculate the drain current analytically, it is useful to approximate the current of the gradual channel (the variation of the longitudinal electric field is negligible with respect to that of the transverse electric field). The density of free charges present in the inversion layer, at distance X from the source, is equal to the difference between the density of charges related to the capacitive effect, less the density of fixed charges present in the silicon space charge zone.

$$Q_n(x) = Q_{OX}(x) - Q_{Si}(x) \qquad [3.3]$$

The surface potential ($\varphi_s(x)$) to the inversion can be approximated by $2\phi_B+V(x)$, with V(x) representing the potential difference between point x and the ground.

The conductivity of the transistor is then obtained by integration over the entire length of the channel.

$$g = \frac{W}{L}\int_0^L Q_n(x).\mu_n(x)dx \qquad [3.4]$$

For very weak V_D, in the case of the linear mode, the density of electrons in the inversion layer can be only regarded as a function of gate voltage. By taking V(x) = 0, we obtain:

$$Q_n(x) = -(V_G - V_T).C_{OX} \qquad [3.5]$$

Hence

$$I_D = \frac{W}{L}(V_G - V_T)C_{OX}\mu_n V_D \qquad [3.6]$$

The saturation regime is reached when the channel begins to pinch on the drain side. The saturation voltage is thus determined by the fact that the quantity of mobile charges in the drain-side inversion layer is zero, hence the following condition is obtained:

$$V_{Dsat} = V_G - 2\phi_B + K^2(1 - \sqrt{1 + \frac{2V_G}{K^2}}) \qquad [3.7]$$

The saturation current can be obtained from the above equations.

3.2.1.1.4. Sub-threshold characteristics

The gate voltage is lower than the threshold voltage, and the surface of silicon is in weak inversion. The drain current is mainly related to the diffusion of the minority carriers present in the silicon under the gate.

$$I_D = -qA.D_n \frac{n(0) - n(L)}{L} \quad [3.8]$$

where A represents the area of a vertical section to the flow of current in silicon, D_n is the diffusion coefficient of the electrons in silicon, and $n(0)$ and $n(L)$ are the densities of the minority carriers on the source and drain sides, respectively. As the carrier density in a semiconductor is related to the Fermi level position compared to the conduction and valence bands, the preceding equation can be written as:

$$I_D = -qA.D_n \frac{n_{p0} e^{\beta \phi_S}}{L} (1 - e^{-\beta V_D}) \quad [3.9]$$

with:

$$\beta = \frac{kT}{q}$$

As φ_s varies linearly with V_G, the drain current under the threshold varies exponentially with V_G. For drain voltage higher than 3kT/q, this current no longer depends on V_D.

3.2.1.1.5. Speed

In a digital circuit, the basic element is the CMOS. This structure is composed of an NMOS transistor and a PMOS transistor, with their respective drains and gates. The NMOS P substrate is biased at V_{SS}(0V), whereas the PMOS N substrate is at V_{DD} (5 V or 3.3 V depending on the technology) (see Figure 3.5). When the input is biased to V_{SS}, the NMOS is blocked. The potential at the output of the structure is determined by the conductive PMOS. A current that is passing with a small potential difference is between its drain and source. The output potential is therefore close to V_{DD}. Conversely, if the input is biased to V_{DD}, the PMOS is blocked while the NMOS is ON. The potential difference between the drain and the source of the NMOS is then low, resulting in an output potential that is close to V_{SS}. This structure then behaves as an inverter. An input to V_{SS} gives an output to V_{DD}, and, conversely, an input to V_{DD} gives an output to V_{SS}.

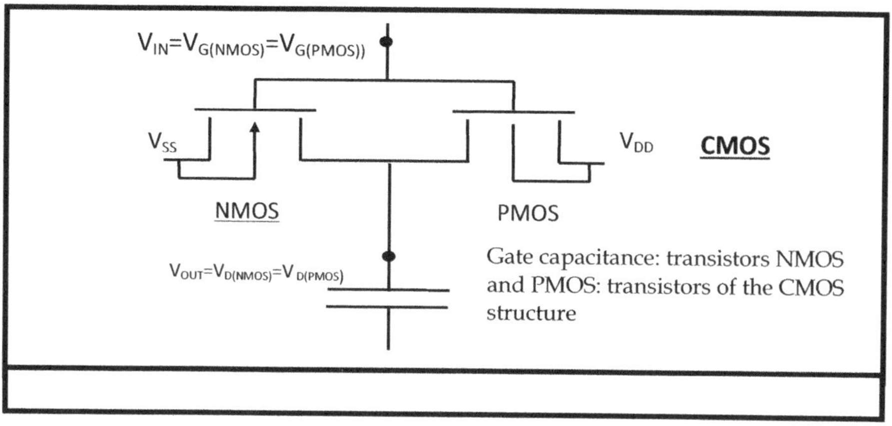

Figure 3.5. *Principle of use of CMOS structures in digital circuits*

A logic circuit consists of a number of series-connected CMOS structures. Logic information is propagated from a CMOS element toward the following elements. When the first CMOS structure changes logical state, it will become busy via its NMOS or its PMOS. This current will charge or discharge the capacitances from the gate of MOS structures of the following CMOS cell; therefore, the electric states of these new PMOS and NMOS transistors will be changed. The switching time can be expressed at first approximation according to the following equation:

$$t_{commutation} = \frac{C_{OX}.V_T}{I_{Dsat}} \qquad [3.10]$$

Hence

$$t_{commutation} = \frac{L}{W}\frac{V_T}{\mu_n.m.(V_G - V_T)^2} \qquad [3.11]$$

In order to decrease the switching time, it is necessary to have a transistor with a low MOS capacitance, a weak threshold voltage and a high saturation current. However, a low gate capacitance involves a weak current and a strong threshold voltage.

On the other hand, the capacitances of the source and drain junctions influence the speed of the circuit. These capacitances are determined by the extension of the space charge zone (SCZ) present at the electric terminals. The weaker the dopings on its terminals, the more important the extension of the SCZ is, and thus the junction capacitance becomes lower, but in order to obtain good ohmic contact between the metallization and the silicon, the doping of the contacted zones must be sufficiently high. On the other side of the junction, the N and P substrates of the P and N MOS transistors, respectively, become increasingly doped, in order to avoid the short channel effects when reducing the channel length. Dopings of the junctions N^+/substrate P and P^+/substrate N thus tend to increase, which increases the capacitance of these junctions.

3.2.1.1.6. Consumption

The second important criterion is the electric consumption of the circuit. CMOS technology makes it possible to obtain circuits with low electric power consumption because of its very large transconductance and the very large resistance at the blocked state, as well as the very large conductance in the ON state. It is necessary to distinguish two consumption modes: one in a static state and the other taking place only during the commutation of CMOS structure (see Figure 3.6).

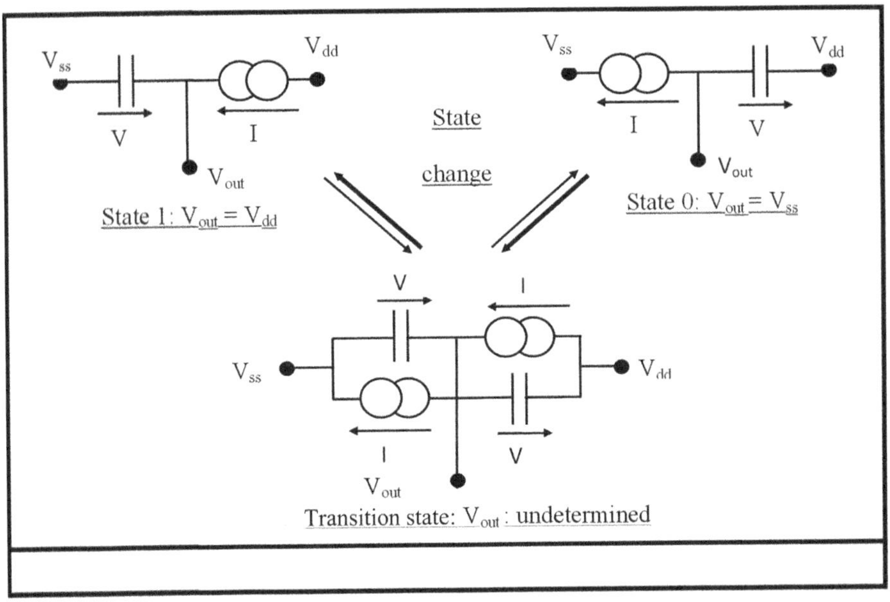

Figure 3.6. *CMOS structure commutation*

3.2.1.1.7. Reliability

The ageing of MOS structures is related to the presence of hot carriers. When the electric field is about 10^5 V/cm, the free carriers acquire an energy that is slightly higher than that of the phonons (carrying heats). When their energy reaches a certain threshold, typically 1.5 times the energy of forbidden band E_G, they have sufficient energy to create electron-hole pairs at the time of collisions with the lattice atoms. This phenomenon, known as impact ionization, becomes the dominating mechanism of loss of energy for the carriers.

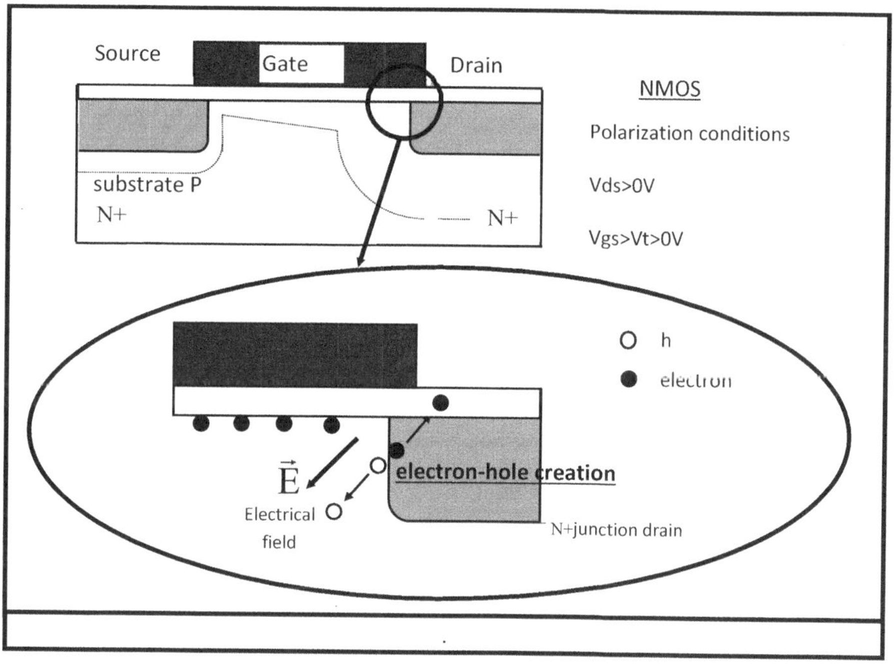

Figure 3.7. *Aging of NMOS transistors by hot carriers*

These energy carriers can then cross the oxide to be the origin of a gate current. Some can remain trapped there, locally degrading the quality of oxide via charges and states of interface. These defects created will disturb the electric characteristics of the transistor, modifying the value of the threshold voltage and degrading the drain current. The influence of the technological developments of CMOS processes on the three preceding criteria is presented in Table 3.1. A positive effect on the criterion is represented by +, a negative effect by – and no effect by 0.

	Gate shrinking			Substrate doping	Lowering voltage
	Gate oxide	W	L	Enhancement	
Speed	0	–	+	–	–
Consumption	+	–	+	–	+
Reliabilty	–	0	–	–	+

Table 3.1. *The influence of the technological developments of CMOS processes*

3.2.2. *A typical CMOS technology*

3.2.2.1. *Principal technological steps*

The key technological steps are shown in detail as depicted in Table 3.2. In addition, a cross-sectional view for the CMOS is displayed in Figure 3.8.

Device insulation

Volume insulation

- N well implantation
- P well implantation
- Well diffusion

Surface insulation

- nitride deposit for active area definition
- field implant
- field oxide growth

MOS transistor realization

Definition of the MOS capacitance

- gate oxide growth
- polysilicon deposit
- Adjustment implant of the threshold voltage (V_T ADJUST.)
- polysilicon etching for the definition of the transistor gate

Sources and drains definition

- LDD implant (Lightly Doped Drain)
- Spacers forming
- Implants N^+ and P^+

Interconnects realization

- metal first level
- metal second level
- passivation

Table 3.2. *Key technological steps in the CMOS process*

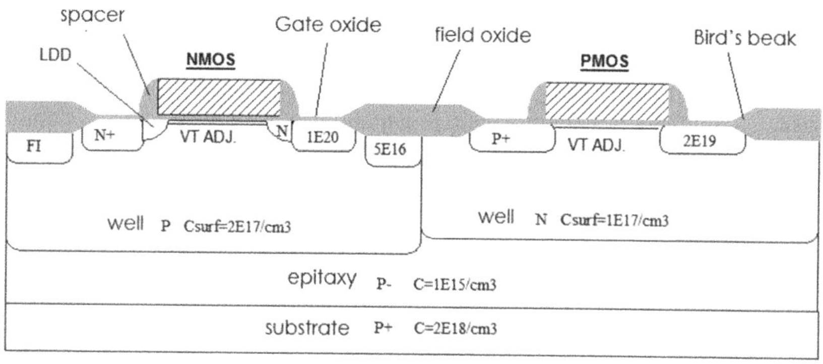

Figure 3.8. *Cross-sectional view of CMOS*

In the context of this part, only the application of the high voltage to the drain of the transistor is of interest. Given the breakdown voltages of the elementary structures, the transistors resulting from the support technology cannot support much more than 12 V applied on their drains (see Figure 3.9), the breakdown voltages of the gate oxide and the junction N-box/P-box being, respectively, 15 V and 50 V.

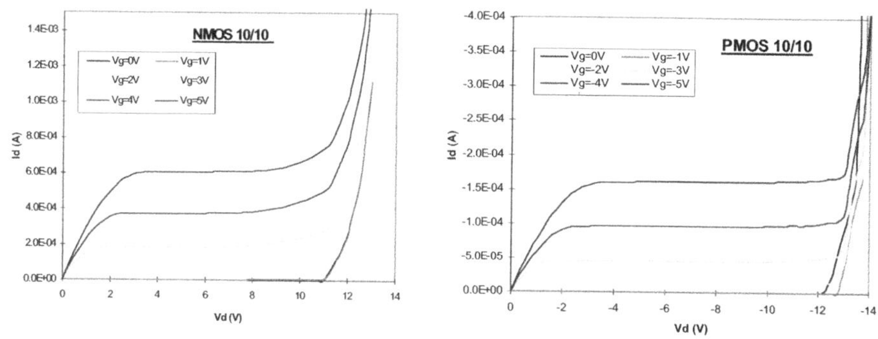

a) b)

Figure 3.9. *a) 5V NMOS characteristics. b) 5V PMOS characteristics. For a color version of this figure, see www.iste.co.uk/abouelatta/smartpower.zip*

3.2.3. *Breakdown voltage of a microelectronics structure*

In the field of microelectronics, obtaining currents of some intensity is the goal in order to have fast transistors. In the power range, not only the current but also the voltage is taken into account. With regards the breakdown voltage, because the intensity of the current is only related to the surface of the component, it is a function on the length and width of the channel of the transistor. Thus, the voltage is the basic variable in any power device. In simulations, we initially solve "Poisson", which gives the potential (and thus the field) quickly. It is thus necessary to be familiar with the equipotential lines. A local tightening of those involves the presence of an electric field more significant than elsewhere.

3.2.3.1. *Junction breakdown limit*

Figures 3.10 and 3.11 represent, respectively, the current–voltage characteristics of the N^+/P well and P^+/N^- box junctions in inverse polarity. In the case of the P^+/N^- box junction, the breakdown occurs more abruptly. This difference is explained by the fact that the rate of impact ionization of the electron is much higher than that of the hole.

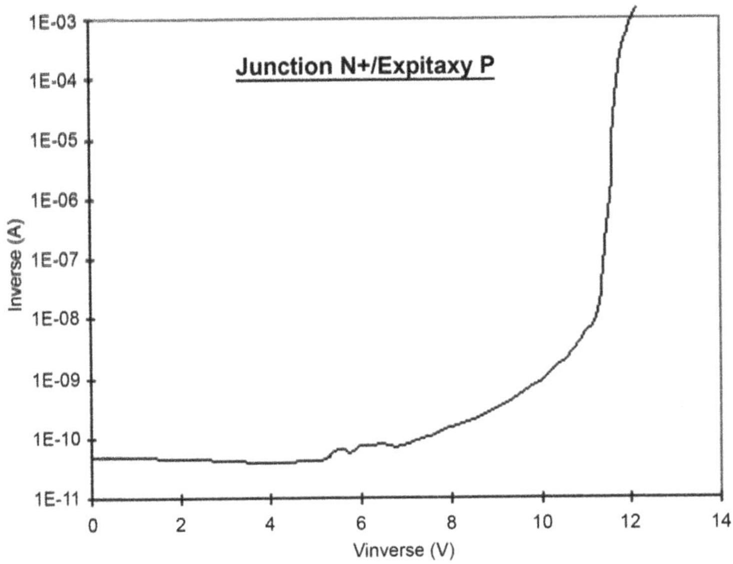

Figure 3.10. *N^+ junction characteristics*

When the electric field in the semiconductor exceeds a certain value, the carriers acquire energy such that they can create electron–hole pairs by impact ionization. The generation rate G of electron–hole pairs related to this effect depends on α_n and α_p: where α_n (α_p) is the ionization rate of the electron (of the hole), that is, the number of electron–hole pairs generated by an electron (a hole) per unit distance traveled by it; n (p) is the density of free electrons (holes), and V_n (V_p) is the velocity of the electrons (holes).

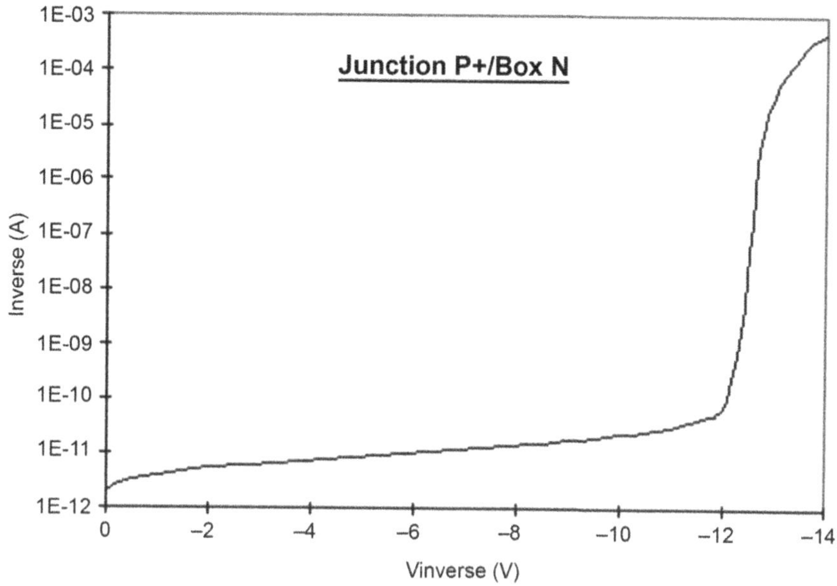

Figure 3.11. *P^+ junction characteristics*

The ionization rates α_n and α_p strongly depend on the electric field in which the carriers evolve.

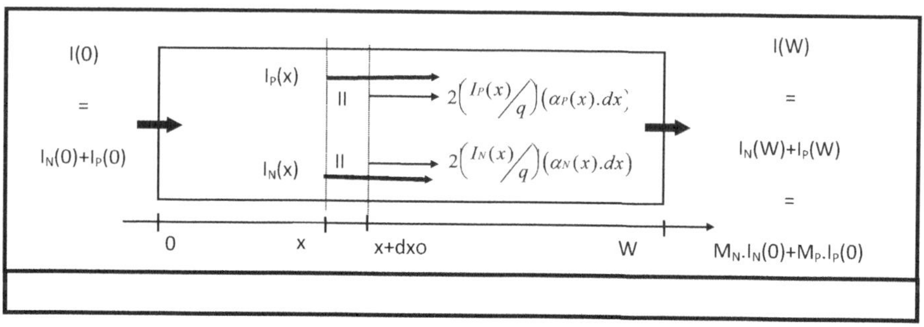

Figure 3.12. *Current generated by impact ionization along a silicon bar*

The current generated by impact ionization along a silicon specimen in one dimension is shown in Figure 3.12. In the semiconductor, between x and x + dx, electron–hole pairs are created by the hole current I_P (x) but also by the electron current I_N (x). The variation of the hole current is then expressed as follows:

$$d(\frac{I_P}{q}) = (\frac{I_P}{q})(\alpha_P.dx) + (\frac{I_N}{q})(\alpha_N.dx) \qquad [3.12]$$

The integration along the semiconductor material, assuming $I_P(W) = M_P.I_P(0)$, gives:

$$1 - \frac{1}{M_P} = \int_0^W \alpha_P.\exp\left[-\int_0^x (\alpha_N - \alpha_P)\,dx\right] \qquad [3.13]$$

Breakdown takes place when the current increases toward infinity, which corresponds to M_P tending toward the infinite value. According to the preceding equation, it is necessary that the following condition is checked:

$$\int_0^W \alpha_P.\exp\left[-\int_0^x (\alpha_N - \alpha_P)\,dx\right] = 1 \qquad [3.14]$$

This integral is called the ionization integral of the holes.

The concept of the ionization integral is most often used by the two-dimensional numerical electric simulators like the junction breakdown criterion. However, it is very tricky to evaluate the breakdown voltage. The ionization integral is instead a microscopic criterion integrated on the whole of the structure. The critical electric field is calculated starting from the ionization integral. It is the maximum value of the electric field (E_C) obtained at the time of the breakdown of the junction with the surfaces of the isopotentials closer in the peripheral junctions. The electric fields being then more important, the critical electric field is reached on the level of the side junctions first. The breakdown voltage of the spherical junction is thus less than that of the cylindrical junction, which is less than that of the planar junction.

The presence of a weaker breakdown voltage in the periphery will impose methods of protection, aiming at obtaining a breakdown that is close to that of the planar junction (see Figure 3.13).

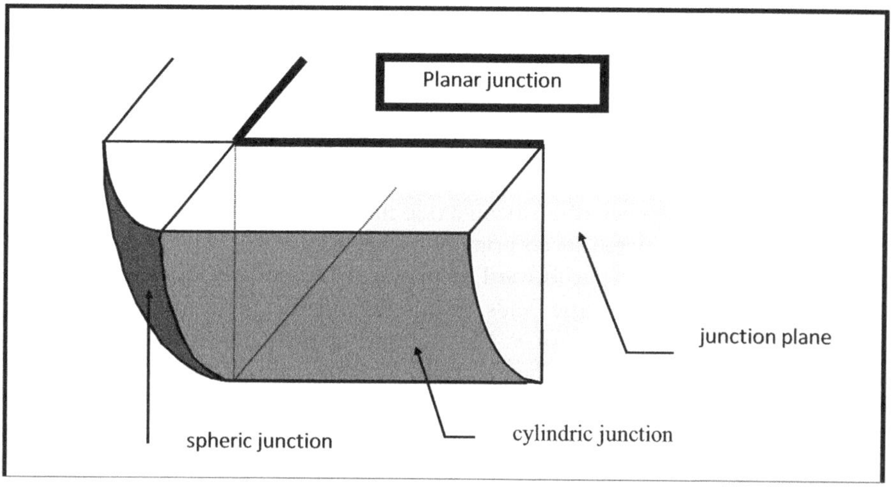

Figure 3.13. *The three components of a planar junction*

3.2.3.2. *Insulator breakdown limit*

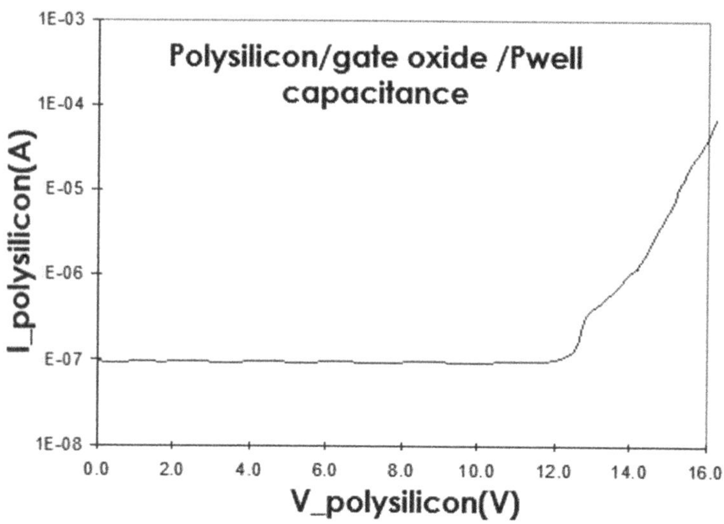

Figure 3.14. *MOS capacitance: I-V characteristics*

Figure 3.14 shows the current–voltage characteristic of a polysilicon/gate oxide/P-box MOS capacitor derived from our technology process. It is necessary to bias the gate to more than 15 V to observe an irreversible breakdown of the structure. The ability then behaves like a resistance.

In this example, the critical electric field (dry oxide uncontaminated by parasitic charges) is 10 MV/cm.

Because of the presence of a silicon/oxide interface, the phenomenon of impact ionization is not considered to be primarily responsible for the breakdown of the oxide. A theory has been put forward involving the silicon/oxide interface. Under the application of strong electric fields, from 6 to 8 MV/cm (corresponding to a gate oxide of 150 Å at a voltage of 9–12 V) some carriers have enough energy to overcome the potential barrier. The electrons thus injected into the conduction band have a potential energy of 4–5 eV greater than the energy of the Fermi level. This energy is dissipated when these electrons are captured by deep defects present in the oxide. This local energy loss is enough to break the chemical bonds and ionize the atoms. This creates irreversible defects in the structure. These local modifications of the structure influence the carrier injection conditions in the oxide. Focusing this injection according to preferred defects will lead to the creation of a preferential path of low resistivity between the silicon and the gate electrode. The oxide will crack when this path connects the two ends of the oxide. High-voltage components must overcome this breakdown of the gate oxide. For this, an increase in the thickness of the oxide is possible, which reduces the electric field at its terminals. However, the capacitance of the transistor is then greatly reduced, and the threshold voltage of the structure is higher.

3.2.4. *Improved junctions breakdown by guard techniques*

In this part, the effectiveness and characterization of guard techniques are introduced. The maximum output of such a technique is equal to the report/ratio, expressed as a percentage, of the breakdown voltage of the structure protected on the breakdown voltage from the junction vertical planes.

3.2.4.1. *Guard rings*

The yield of guard rings can reach 80%. The principle is as follows: during the implementation of the main junction to be protected, one or more regions are also implanted all around the main junction. These will be similar in geometry and type; they will act as a voltage divider along the surface of the device (Figure 3.15). By laterally separating the isopotential lines, this technique avoids the breakdown of the cylindrical and spherical junctions to tend toward the breakdown of the vertical junction.

During the opposite biasing of the principal junction to a low voltage, the extension of the space charge is small and does not extend to the floating N ring. Consequently, this ring takes the potential of the little doped P substrate. When the biasing increases, the space charge extends to the floating ring. The conditions known as "punch-through" are reached. The potential of this ring then becomes that of the first isopotential which reaches it. The isopotentials following lines must circumvent this biased ring, and thus they tend to be spread out around it, which involves a reduction in the electric field in the periphery of the principal junction.

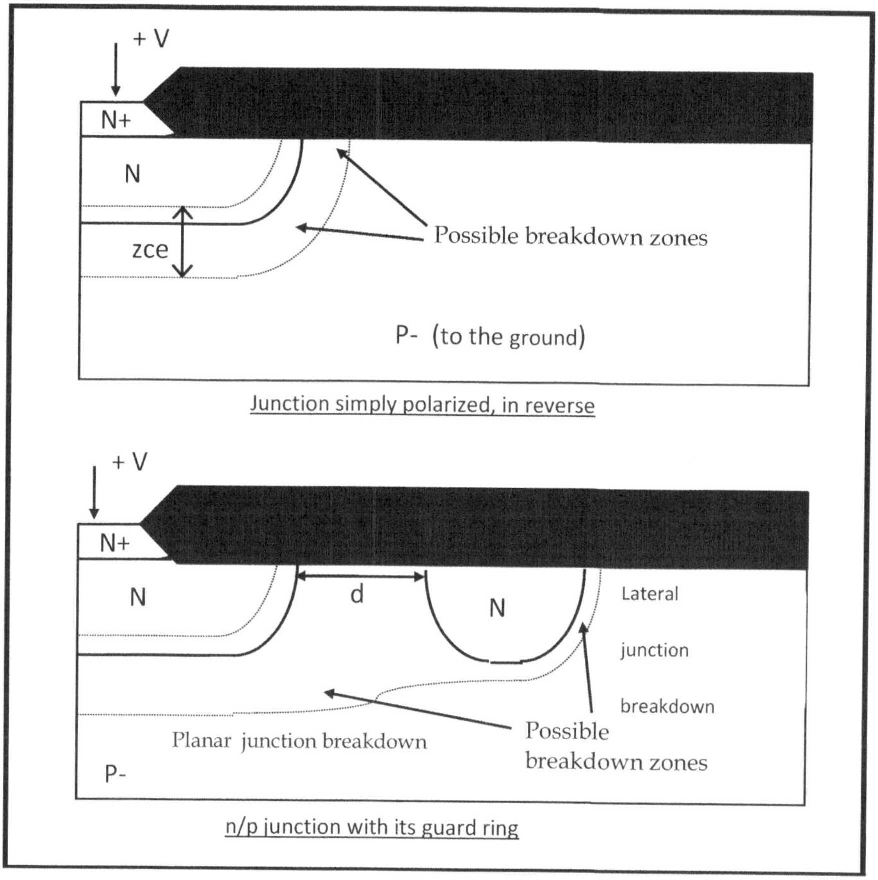

Figure 3.15. *Influence of the guard ring on the extension of the CEA*

The presence of only one guard ring is not sufficient to achieve these conditions. It is necessary to introduce other rings in order to further divide the voltage. Its disadvantage is its great sensitivity to the interface states (modification of the distribution of equipotential to the interface oxide/silicon). The conditions of punch-through are then strongly modified. To avoid this, a more recent configuration of this technique is proposed, which consists of biasing the secondary junctions. Contrary to floating rings, biased rings offer the advantage of homogenizing the distribution of the electric field on the surface of all of the rings, which decreases the risk of premature breakdown of one of the secondary junctions.

3.2.4.2. *Electrodes or field plates*

The goal of field plates is to space the equipotential lines on the level of the side junctions. This effect is obtained by the presence of a metal electrode located above the metallurgical junction (Figure 3.16).

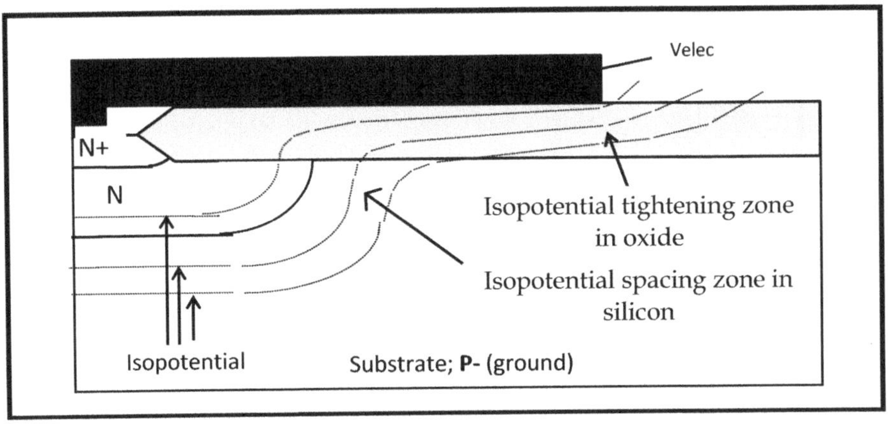

Figure 3.16. *Junction protected by a field electrode*

When a positive voltage is applied to the field electrode, the majority of the holes of the P substrate are pushed back from the surface towards the silicon bulk. This involves the appearance of a zone of local space charge on the surface, just below the electrode. Part of the negative fixed charge of the P substrate is coupled with this metal MOS capacitance electrode/oxide/silicon. On the surface, the substrate then seems to be less doped, the SCL of the biased junction in reverse extends laterally. The radius of curvature of the side junctions is thus increased. A negative voltage applied to the electrode attracts holes of the P substrate toward the surface below the field electrode. The P substrate thus appears more doped on the surface, and the extension of the space charge to this place is then reduced. The effect on the holding voltage is negative because, in this case, we obtain a curve radius narrowing of the isopotential lines.

The biasing of the electrode (which is equipotential) must be of the same sign as the polarization of the junction to be protected. The equipotential lines, to be closed again, pass by oxide that is present between the silicon and the electrode. At the electrode end, the isopotential lines can go up on the surface of the structure, which causes a tightening of some of them (especially in oxide). Important criteria are the length of the overflow of the electrode on the metallurgical junction and the oxide thickness.

3.2.4.3. *Implanted junctions or lateral junction extension*

The aim is still to increase the spacing between the equipotential lines at the level of the small radii of curvature, an intermediate box between the reverse bias and the substrate is used. This box is of the same type as the junction to protect, but its doping is much less. It is implanted at the periphery of this junction, as can be seen in Figure 3.17.

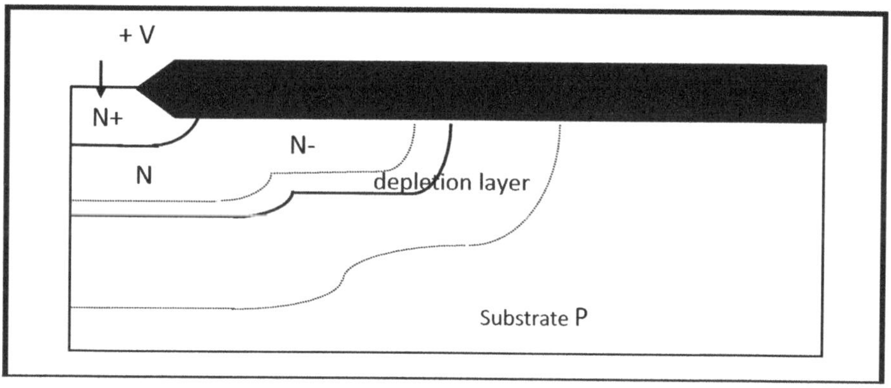

Figure 3.17. *Junction protected by a secondary implant*

The two important condition parameters for good performance of this guard technique are the amount of the secondary implant and the final geometry of the secondary box. If a peripheral box has a low doping, the box will be completely depleted quickly; the fall of potential is weak there, thus one practically turns over in the case of junction N/P^-. Regarding the geometry, without a very broad box, one can find the preceding situation. This is depleted quickly. Another important size is the depth of the secondary box, compared to the box to be protected. A secondary box that is not deep only acts on the surface of silicon. Overall, on the level of the cylindrical junctions, the isopotential lines are hardly isolated (inefficiency).

3.2.4.4. *Semiresistive layer*

A semiresistive layer is deposited between two connections, either directly on the silicon or on a previously deposited oxide, as can be seen in Figure 3.18.

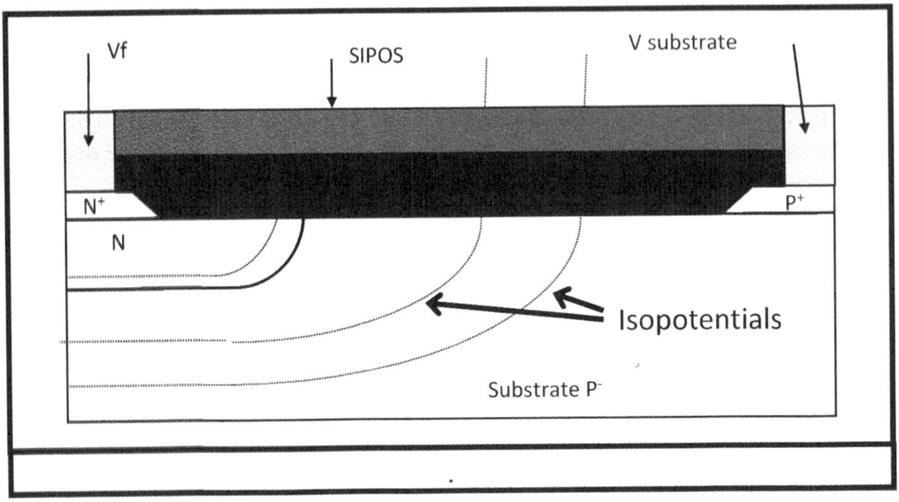

Figure 3.18. *Junction protected by a semiresistive layer*

The linear distribution of the potential on the surface of the oxide induced a quasilinear distribution of the potential to the surface of silicon, which results in lower values of the peaks of the field than the corresponding values in the structure without a semiresistive layer. The voltage behavior is improved.

3.2.4.5. *RESURF technique*

This technique owes the name RESURF (REduced SURface Field) to the fact that the equipotential lines present around the peripheral junctions are spaced in order to reduce the electric field on the surface. The breakdown then refers to the horizontal plane junction. The yield is then close to 100%, making it the most effective guard technique.

An N-shaped epitaxial layer produced on the P^+ substrate is the basis of this effect. A P well is implanted all around the junction for protection. We are in the presence (Figure 3.19) of a plane junction, an epitaxial N^- P^+ substrate and a lateral cylindrical epitaxial N^-/P-box.

The side junction has a breakdown voltage that is lower than the vertical plane junction (see Figure 3.19). In the case of a thin and slightly doped epitaxy layer the space charge regions of the two preceding junctions interpenetrate. The space charge region of the vertical junction quickly reaches the interface epitaxy/oxide. The epitaxy layer is thus completely depleted. The side field is considerably reduced, so much so that the plane junction can be broken down before the side junction.

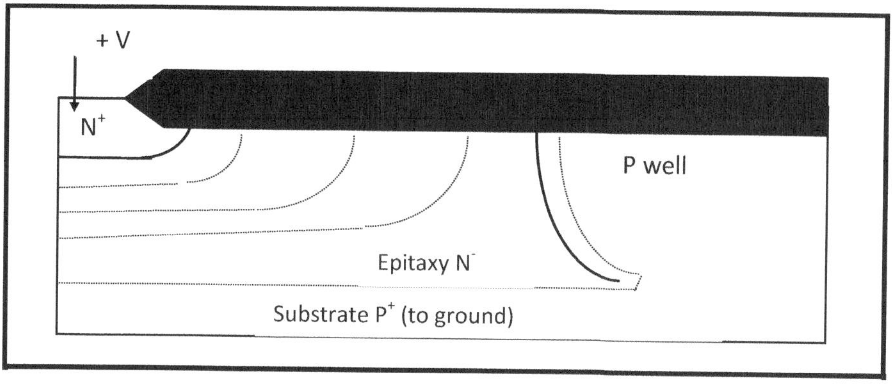

Figure 3.19. *RESURF guardtechnique*

Three critical dimensions affect the performance of the RESURF effect: the thickness of the epitaxy layer must be quite low so that the space charge region of the vertical junction reaches the interface epitaxy/oxide before the breakdown of the side junction of the epitaxy layer. The lower the thickness of the epitaxy layer, the higher the doping must be so that the RESURF effect is optimal. In the opposing case, the pinching of a slightly N-doped epitaxy under the N^+ contact can cause a limitation of the behavior in the voltage of the vertical structure N^+/epitaxy N^-/P^+ substrate. The distance between the P well and the N^+ region must be rather significant, so that the space charge region of the side junction contains a quantity of fixed sufficient charge.

3.2.5. *Improvement using electrical insulation techniques*

One of the major problems encountered in the design of a power integrated circuit is the static and dynamic insulation between the power part and the digital and analog voltage.

3.2.5.1. *Junction insulation*

The method involves using the very high resistivity of a reverse-biased junction. Since this junction consists of at least one lightly doped region, the space charge area is large enough to limit any parasitic capacitance effect. It is located either on the periphery of the device to be insulated (Figure 3.20), or it is part of the device (Figure 3.21).

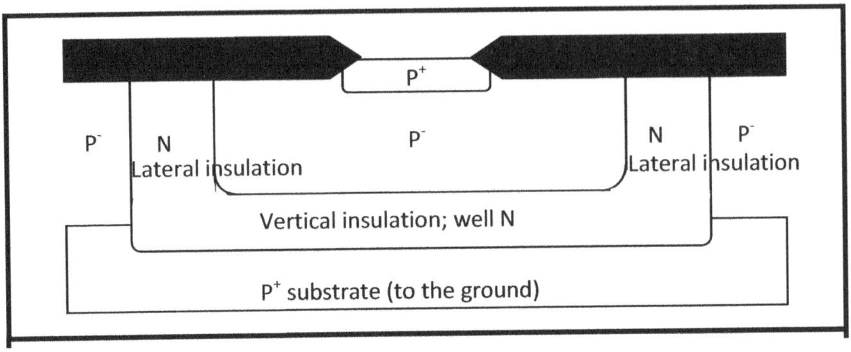

Figure 3.20. *Electrical insulation by diffused junctions*

In the case of an epitaxy P^-/P^+ substrate configuration, it is sensitive to bias with the high voltage of a double diffused P junction while keeping the P^+ substrate with the bulk. The N well formation all around the P junction allows a side insulation. One creates a horizontal structure P well with the bulk/well N floating/P well biased with the high negative voltage. Because of the distribution of the potentials throughout this structure, the potential of the floating N well is higher than that of the strongly biased well in reverse. The junction N well/P well/P^+ is then biased in reverse, with a very good side insulation. In the same way, in order to obtain a vertical insulation, a buried N well is necessary.

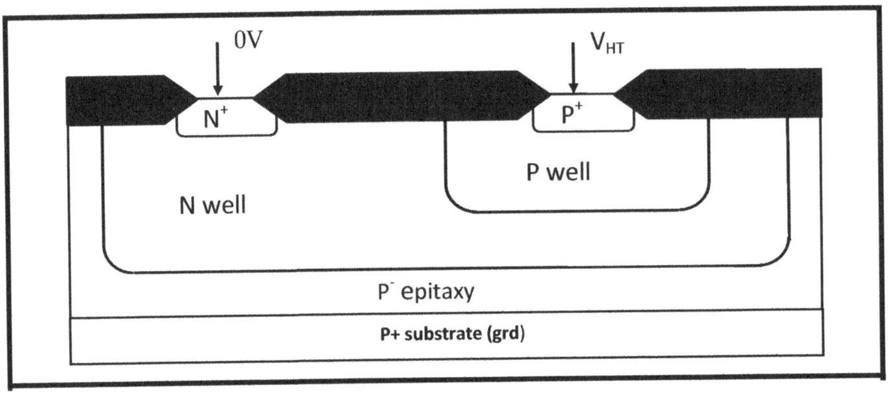

Figure 3.21. *Electrical self-insulation by active junction*

3.2.5.2. *Junction insulation*

The junction supporting a high positive voltage will be carried out by the N well established in the epitaxial P layer. The junction supporting the high negative voltage will be obtained by the formation of a low depth P well, in an N well with a much more significant depth. The negative voltage will be confined with the electric terminal P well/N layer. The N well/epitaxy junction is biased with 0 V, and the high voltage junction is electrically isolated (note: a bipolar transistor is PNP parasitic. It is necessary to limit the appearance of this parasitic transistor). The use of an epitaxy N/P$^+$ substrate configuration (the case encountered during the use of the RESURF guard technique) uses the vertical insulation of the junction epitaxy N/P$^+$ substrate. The diffused P junctions biased by high voltage all around ensure the side isolation.

3.2.5.3. *Dielectric insulation*

The second commonly used material is silicon oxide (Figure 3.22) (greatly reduced parasitic capacitance). Several techniques are used to obtain this result:

– oxide wells;

– SIMOX technology, consisting of deeply implanting oxygen atoms;

– very promising new technologies have appeared in recent years under the name silicon on insulator. The low voltage components used for logic circuits are implemented on the starting substrate. At the surface of this substrate, a thick oxide is produced and is deposited above a layer of polysilicon (the problem is its potential recrystallization). The polysilicon/oxide/substrate vertical structure allows perfect insulation between the high voltage applied on the polysilicon layer and the low voltage of the substrate. The polysilicon etching makes it possible to isolate the different power components between them.

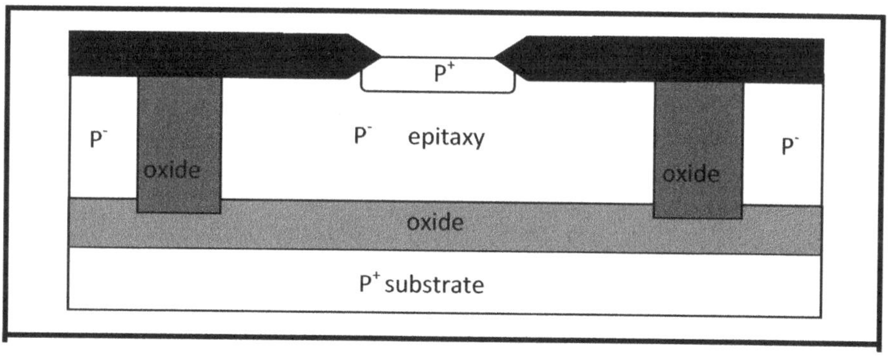

Figure 3.22. *Electrical insulation by dielectric*

3.2.6. *Review of the main MOS devices for high-voltage integrated circuits*

In order to obtain a mixed power – digital/analog technology, many new components have been developed. They have relatively significant voltage or power characteristics. They come either from power techniques or from microelectronics techniques and are either unipolar, bipolar or hybrid MOS-bipolar.

3.2.6.1. *Electrical characteristics of high-voltage MOS devices*

3.2.6.1.1. Breakdown voltage

It is necessary to take into account overshoot transient effects which may, for example, impose voltages of several hundred volts on circuits created for a few tens of volts. The problem is similar for automobiles, and even more critical for the devices intended to operate directly on the sector.

It is necessary to distinguish the voltage in the OFF state (V_{bdOFF}) from that in the ON state (V_{bdON}). In the OFF state, the HV (High Voltage) junction strongly conditions the voltage performance of the device.

The presence of the MOS structure (gate oxide of low thickness) can also limit the behavior of the structure voltage. Moreover, the polysilicon gate behaves like a field plate. In the ON state, the same limitations come into play. Moreover, the presence of free carriers considerably increases the phenomenon of impact ionization, and can also modify the distribution of the potential and thus the electric field.

3.2.6.1.2. Current intensity

This must be very weak when the transistor is in a blocked state in order to limit the dissipated power. In this case, it is determined by the leakage currents of the transistor. Although the high voltage transistors do not always have capabilities in very high power, these transistors often request that currents of a few hundred milliamperes are allowed to pass through while having a minimal voltage drop between source and drain. The consumption of the transistor in a conducting state is thus minimal. In order to relate this current intensity to the weak V_{DS}, the electric resistance parameter at the conducting state, R_{on}, is usually introduced. This is given with very weak drain voltages. Some authors even extract this parameter by taking the reverse from the tangent line slope to the curve $I_{DS} = f(V_{DS})$ with $V_{GS} = 5V$, $V_{BS} = 0V$ and $V_{DS} = 0V$. This resistance to the conducting state is usually expressed in Ohm per cm^2 of silicon area.

Currently, the most powerful components of power have a resistance to the conducting state from about 100 cm^{-2}. The resistivity of silicon strongly affects resistance in the ON state. In order to obtain a high voltage, it is important to increase this resistivity, which degrades resistance to the conducting state of the transistor. A compromise between the breakdown voltage and the resistance in the ON state is required. Once the required current intensity to the specified load and acceptable maximum consumption are known, it is possible to determine the transistor area.

	Dimension reduction			Enhancement
	Gate oxide	W	L	Substrate doping
Applied Voltage	–	0	–	–
Current intensity	+	–	+	–

Table 3.3. *The influence of the technological developments of voltage and current ratings*

The applied voltage of MOS structures is completely incompatible with the evolution of CMOS technologies. The more advanced technology is, the more difficult it will be to introduce components that can support high voltages (see Table 3.3). The introduction of stages specific to the new high voltage structures then become increasingly essential. It is always difficult to intervene in the technological process at the level of the stages, allowing the creation of the MOS capacitor. The high voltage is then only applied to the drain. It is supported by a box plug located between the drain biased at the high voltage and the substrate at ground. Microelectronic technology double diffusion is very penalizing for the intensity of current crossing the structure. The surface of the various points of contact becomes important, and focusing on the level of these points of contact, involving strong thermal disturbances.

3.2.6.2. *DMOS transistor*

The gate (MOS) to realize the buffer well that supports the high voltage is shown in Figure 3.23. The N+ source/P channel/intrinsic drain N– source side structure defines the active area of the NMOS device. The electrons first traverse a lateral path under the gate in the channel, then a vertical path in the epitaxial layer N–, before being collected by the N+ well.

When the drain is strongly biased, all the voltage is found at the edges of the box junction P/N-polarized epitaxy in reverse. Although strong electrical potentials are present under the gate at the terminals of this buffer junction, the thin thermal oxide does not snap. If the distance between the two P buffer wells is small, their space charge areas will interpenetrate. At this moment, when the space under the gate is completely depleted, the strong potentials are found deep in the silicon. In the case of very long structures, the effect of "self-shielding of the gate" makes it possible to explain the absence of oxide breakdown despite the absence of overlap of the two SCZs. Only a very small part of the isopotential lines are found in the gate oxide, the rest spreading in the silicon under the gate. The presence of the N/P-biased P-channel epitaxial junction recovers any mobile positive charge present on the N-side epitaxial side of the ECA. Thus, any presence of an inversion layer at the oxide/silicon interface, under the gate, in the N-epitaxy is impossible. The penetration of the electric field is then very important in the silicon. The voltage is mainly supported by the silicon.

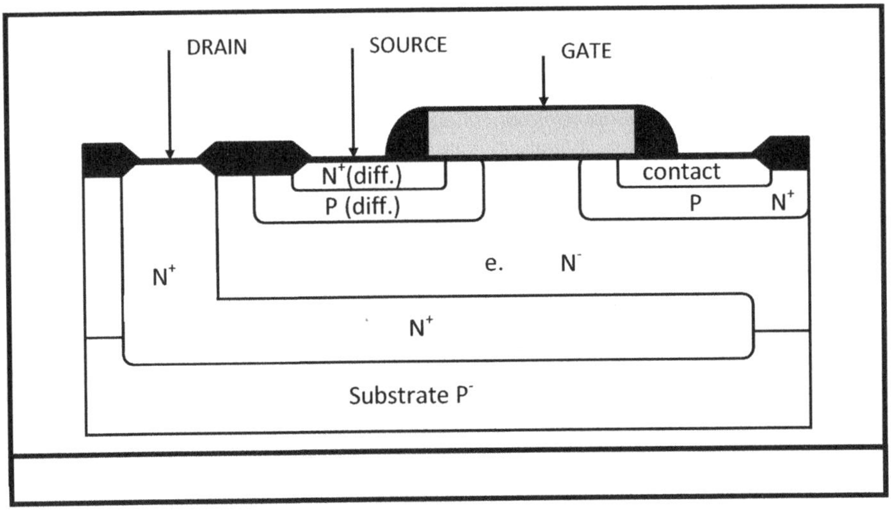

Figure 3.23. *Cross-sectional view of a DMOS structure in "up drain" configuration*

Figure 3.23 represents a section of a "vertical" DMOS transistor. The self-alignment technique of the casings – implanted with respect to the gate, which serves as a mask to their implantations – makes it possible to overcome misalignment between lithographic levels. It is possible to obtain very small channel lengths, even submicron, and to obtain significant current intensities per unit of width of the transistor. Since the direction of the electron current is vertical at the level of the intrinsic drain, it is called the VDMOS, that is, Vertical DMOS, component. This component comes from power technologies. Usually, the contact of the drain is located on the rear face, the starting substrate being N+ type instead of P–, as shown in Figure 3.22. The carriers only see a very low resistance series before reaching the drain. The realization of a buried well to collect the electrons and bring them back to the surface gave the name "up drain".

3.2.6.3. *LDMOS transistor*

The lateral DMOS (Figure 3.24) is much less power-efficient than the DMOS. A first implementation, followed by a diffusion, forms a P^- well under the gate. This defines the length of the transistor channel. Field oxide serves as a mask for the implantation of sources and drains.

The isopotential lines, to close, pass into this thick oxide, which makes it possible to overcome the early breakdown of the gate oxide which is linked to the field electrode effect. It seems that the self-protection of the gate takes place on this structure; breakdown voltages much greater than the voltage of the gate oxide have been observed. The presence of the reverse-biased high-voltage junction prevents any appearance of an inversion layer, formed of the minority carriers that are the holes. The voltage it is able to withstand is essentially limited by the avalanche breakdown of the planar junction epitaxial N-/P-box. The major defect of such a component is the biasing at high voltage of the N+ substrate via the drain. The rest of the circuit will have to be placed in a well that is perfectly isolated from the N+ substrate. This greatly affects its simplicity of integration on an ASIC technology.

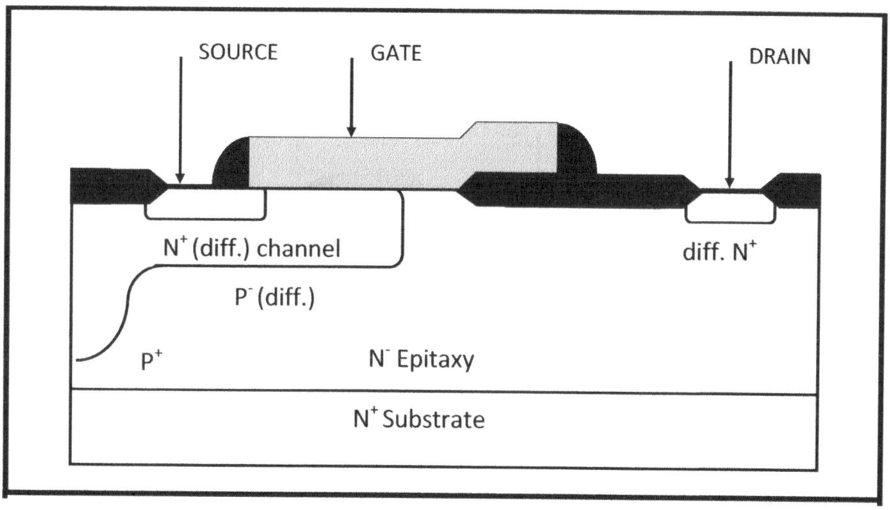

Figure 3.24. *Classical LDMOS transistor configuration*

3.2.6.4. *The LDMOS RESURF transistor*

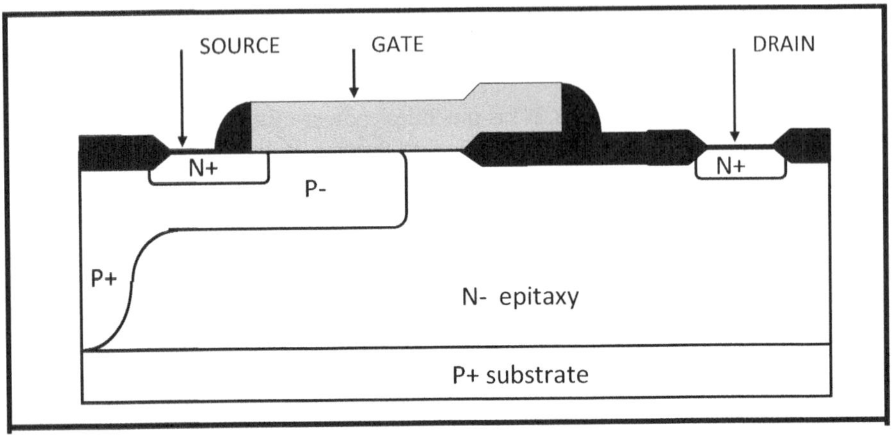

Figure 3.25. *LDMOS RESURF transistor configuration*

3.2.6.5. *Self-isolated LDMOS transistor*

In this transistor, the junction isolation is adopted. The double diffusion makes it possible to create the well, which will support the high voltage between the drain and the channel (see Figure 3.26).

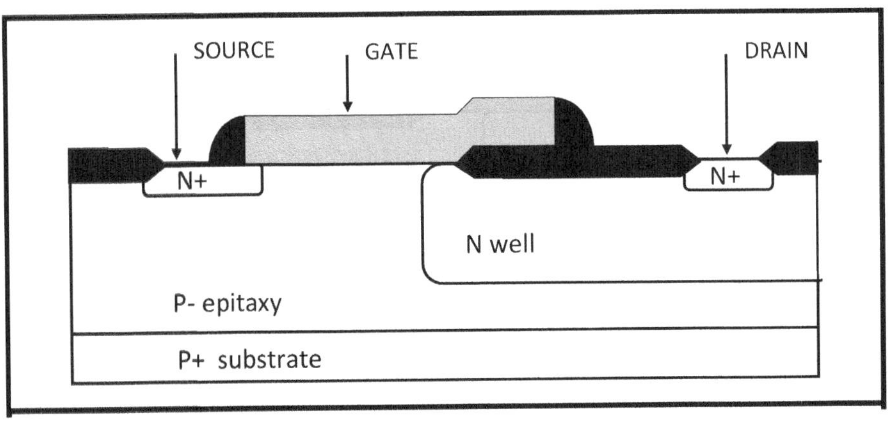

Figure 3.26. *Self-isolated LDMOS transistor configuration*

The benefit of this architecture is its ability to be integrated into a standard CMOS technological process. The device exists directly on the silicon plate P/P$^+$ epitaxial substrate. The N well of transistor LD-NMOS was that of standard

PMOS transistors. The LD-PMOS used the box NR standard as a starting substrate. The P well, established between drain and channel to form the high voltage P well/ N well junction, was confirmed.

3.2.6.6. *Low voltage/high voltage interface simulation (Figure 3.27)*

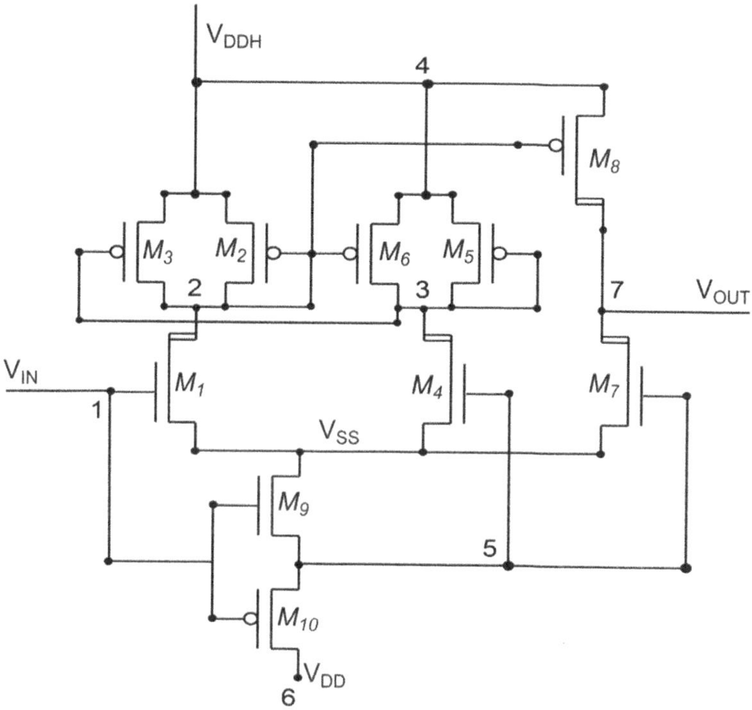

Figure 3.27. *High-voltage interface (LV/HV conversion)*

The M_7 and M_8 transistors behave like a conventional CMOS structure. The LD-NMOS M_7 transistor is biased with respect to V_{SS}, while the LD-PMOS transistor is biased with respect to V_{DDH}. Since the V_{GS} potential difference of the LD-PMOS transistor cannot be greater than 5 V – due to the low voltage of the MOS capacitor – a voltage shifter is required to control the LD-PMOS transistor. This shifter goes from an input voltage 0/5 V to provide an output voltage V_{DDH}-5V/V_{DDH}. This voltage shift is obtained with transistors M_1 to M_6. In addition, transistors M_9 and M_{10} play the classic role of a simple low-voltage CMOS inverter. The circuit then behaves as follows: when V_{IN} = 0V, the transistor M_1 is blocked, M_4 and M_7 are ON. The current flowing through the PMOS M_4 allows the gate capacities of the M_3 and M_5 transistors to discharge. As their voltage difference V_{GS} becomes increasingly negative, the transistors become increasingly ON. When V_{GS} is such that $I_{DS}(M_4)$ = $I_{DS}(M_5)$, the steady state is reached. The importance of the sizing of transistors becomes clear at this time. If M_5 does not provide enough current, it will have to be biased under extreme conditions in order to reach the stationary state. The transistor M_3 is ON, and as M_1 is blocked, the voltage difference V_{DS} (M_3) is close to 0 V. Thus, M_2, M_6 and M_8 are blocked. At the output stage, as only the M_7 transistor is ON, the load is biased to V_{SS}. When V_{IN} = 5 V, the reverse operation takes place. M_1 is ON, while M_4 and M_7 are blocked. This means that M_3 and M_5 are also blocked. In the stationary state, M_2, M_6 and M_8 become ON. The load is thus fed to V_{DDH}.

3.3. Combining analog and digital (mixed)

3.3.1. *Analog: basic functional blocks in CMOS technology and basic analog structures*

In this part, we present two basic configurations most used in analog CMOS circuitry: common source editing and cascode editing.

3.3.1.1. *Common source NMOS transistor*

The circuit shown in Figure 3.28 will be analyzed below.

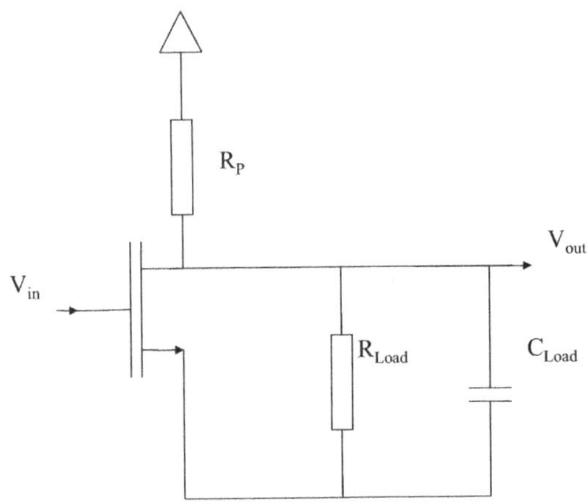

Figure 3.28. *Common source mount*

The equivalent diagram in small signals is given in Figure 3.29.

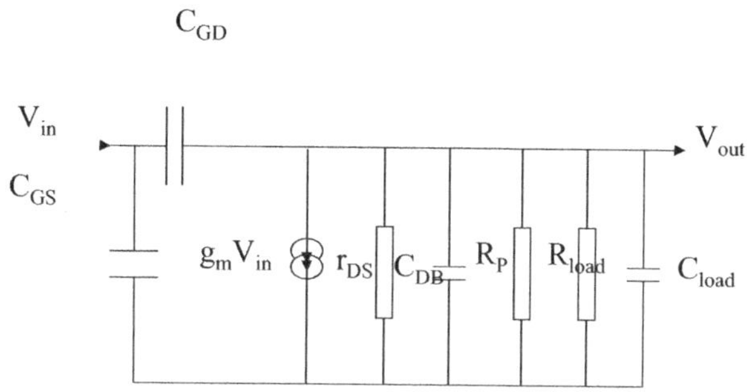

Figure 3.29. *Equivalent diagram in small signals*

We then have: $i_{cc0} = -g_m V_{in}$, short-circuit current regardless of capabilities.

The impedance seen from the output (when $V_{in} = 0$) is worth: $Z_{out} = R_{LT} = r_{DS} // R_{load} // R_P$.

Hence the gain in low frequency:

$$Av_0 = \frac{V_{out}}{V_{in}} = \frac{Z_{out}\ i_{cc0}}{V_{in}} = -g_m R_{LT}$$

or R_L, the output resistor of the transistor alone, and it is worth $R_L = R_{load} // R_P \Rightarrow Av_0 = -g_m R_L // r_{DS}$.

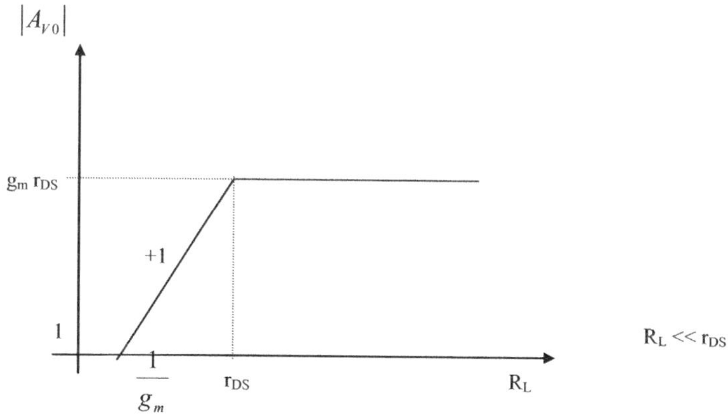

Figure 3.30. *Gain module*

The representation of the low-frequency gain module as a function of R_L on a log-log scale (Figure 3.29) shows that the maximum value achieved by this module is the product: $g_m . r_{DS}$.

3.3.1.2. *Cascode mounting*

The low-frequency voltage gain of a common source assembly is limited by the term $g_m\ r_{DS}$ (for infinite R_L). At equal transcanductance, it is the r_{DS} resistor that sets the maximum value of the output impedance and therefore of the gain at low frequencies. By nature, the maximum gain of a common source amplifier in CMOS technology is a few tens in magnitude. In order to obtain a greater amplification, we therefore seek to increase the output impedance of the assembly due to the driver transistor M; for this, we add a second one mounted in cascode (Figure 3.31).

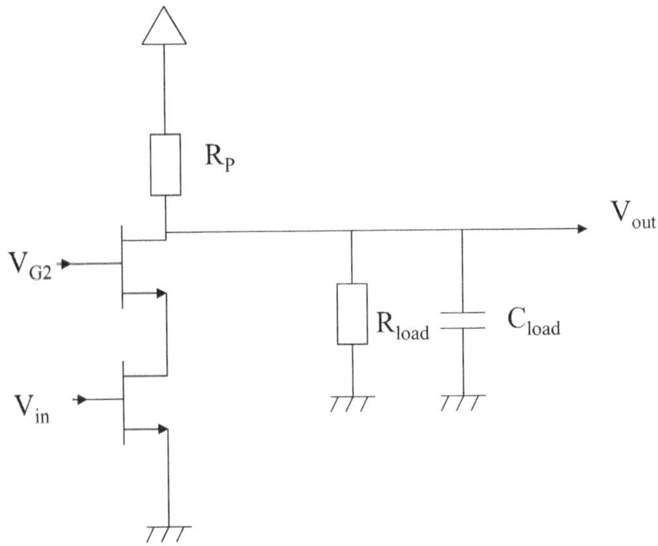

Figure 3.31. *Cascode mount*

$$V_{GS2} = V_G - V_{S2} = -V_S\,(V_G = 0)$$

Figure 3.32 shows the equivalent diagram in small alternating signals at low frequencies.

The output short-circuit current is obtained from the previous schematic (Figure 3.31) by connecting the output node to ground.

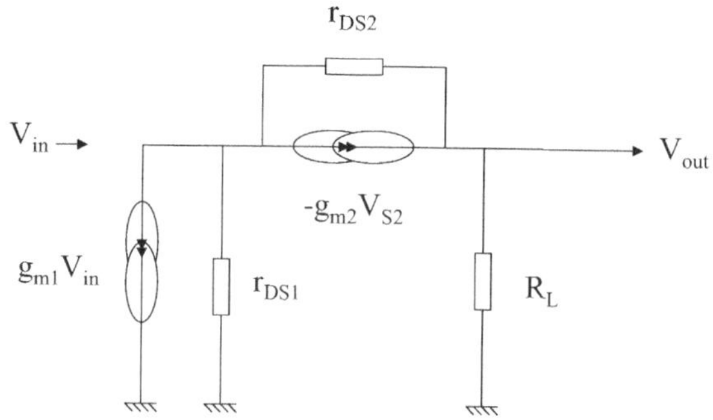

Figure 3.32. *Small signals equivalent schematic*

We find: $i_{cc0} = g_{m2}\, V_{in} + \dfrac{V_{S2}}{r_{DS2}}$ and $i_{cc0} = -g_{m2}\, V_{in} - \dfrac{V_{S2}}{r_{DS1}}$

Then $i_{cc0} = -g_{m2}\, V_{in}\, \dfrac{(1 + g_{m2}\, r_{DS2})r_{DS1}}{r_{DS2} + (1 + g_{m2} r_{DS2})r_{DS1}}$

i.e. $i_{cc0} = -g_{m2}\, V_{in}\, \dfrac{\dfrac{1}{r_{DS2}} + g_{m2}}{\dfrac{1}{r_{DS1}} + \dfrac{1}{r_{DS2}} + g_{m2}} = -g_{m2}\, V_{in}\, \dfrac{r_{DS1} // r_{DS2} // \dfrac{1}{g_{m2}}}{r_{DS2} // \dfrac{1}{g_{m2}}}$

which can be interpreted as follows: in a short-circuit and seen from the source of the transistor (cascode) M2, the current source gm_2V_{S2} behaves like a resistor of value $\dfrac{1}{g_{m2}}$.

The drive current $g_{m1}V_{in}$ is then shared between the resistors r_{DS1}, r_{DS2} and $\dfrac{1}{g_{m2}}$, The output short circuit current is the part of this drive current that passes through resistors and r_{DS2}. We actually have a simple current divider.

In the case where all the transistors are in saturation mode, we have $r_{DS} >> \dfrac{1}{g_m}$, which results in $i_{cc0} \approx -g_{m1}\, V_{in}$, the value obtained in common source assembly. Using Figure 3.33, the output impedance is obtained in small alternating signals.

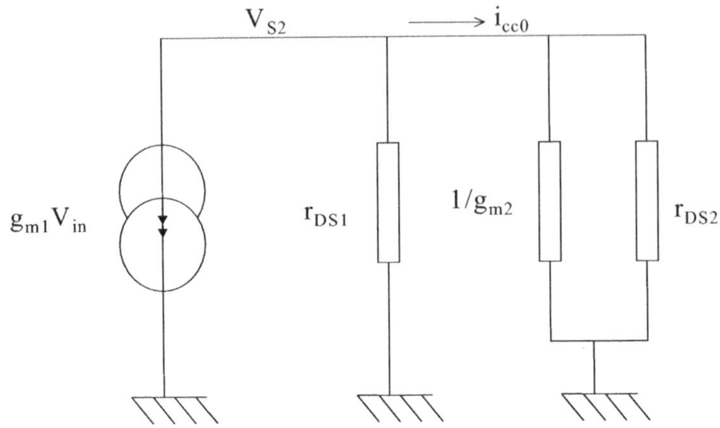

Figure 3.33. *MOS equivalent schematic*

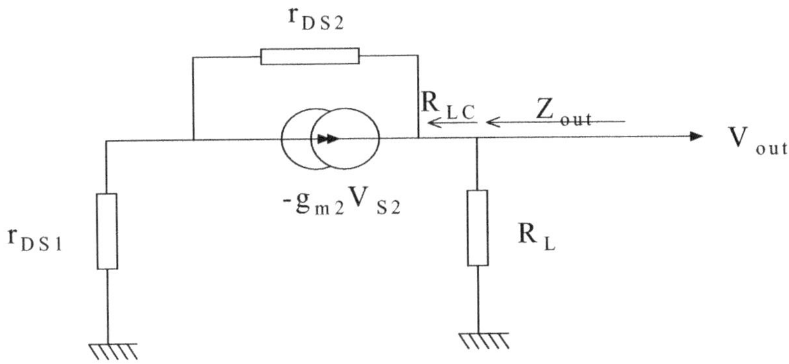

Figure 3.34. *Small signals schematic*

$Z_{out} = R_L \,//\, R_{LC}$

A quick calculation shows that $R_{LC} = r_{DS2} + (1 + g_{m2}\, r_{DS2})\, r_{DS1}\, (1 + g_{m2}\, r_{DS2})$

This is where the first cascode effect comes in: the r_{DS1} impedance seen from the output is multiplied by the factor $(1 + g_{m2}\, r_{DS2})$ in the saturation zone, this term g_{m2} r_{DS2} can be worth a few tens of ohms. Therefore, the RLC output impedance seen in the driver transistor is one or two orders of magnitude greater than the r_{DS1} output impedance of a common source circuit.

The gain in voltage at low frequency for small AC signals (Figure 3.34) is therefore:

$$A_{v0} = \frac{z_{out} i_{cc0}}{v_{in}} = -g_{m2} \frac{r_{DS1} // r_{DS2} // \dfrac{1}{g_{m2}}}{r_{DS2} // \dfrac{1}{g_{m2}}} (R_L // R_{LC}) \qquad [3.15]$$

$$\text{Or: } A_{V_o} = -g_{m1} \frac{(1+g_{m1} r_{DS2}) r_{DS1} R_L}{R_L + r_{DS2} + (1+g_{m2} r_{DS2}) r_{DS1}} \qquad [3.16]$$

If $g_{m2}\, r_{DS2} >> 1$, we find: $A_{v0} \approx -g_{m1}\, (R_L // R_{LC})$

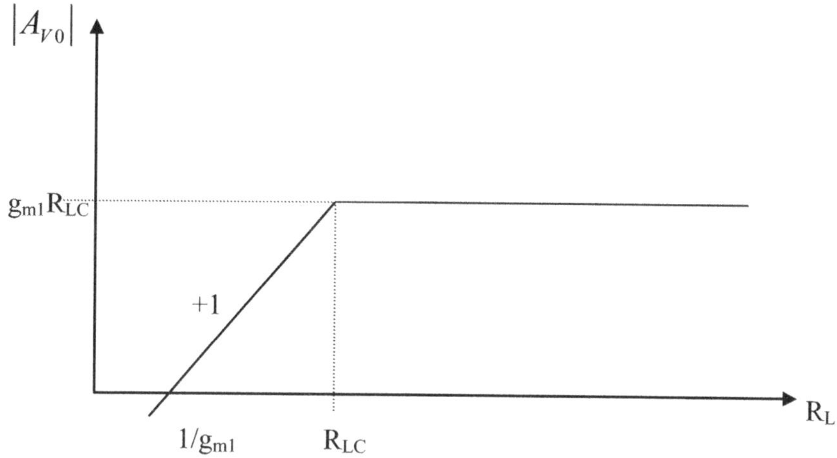

Figure 3.35. *Voltage gain: Bode diagram*

3.3.1.3. *Concluding remarks on the two mounts*

In order to qualitatively compare the LF voltage gain of the cascode assembly with that of a common source, with drive transistors (g_{m1} and r_{DS1}) that are identical and depending on the load resistance R_L, we calculate the ratio:

$$\frac{A_{V0cas}}{A_{V0SC}} \approx \frac{(R_L // R_{LC})}{(R_L // r_{DS1})} = \frac{1 + \dfrac{R_L}{r_{DS1}}}{1 + \dfrac{R_L}{R_{LC}}}$$

The asymptotic log-log diagram (Figure 3.36) of this expression (with $r_{DS1} < R_{LC}$) is:

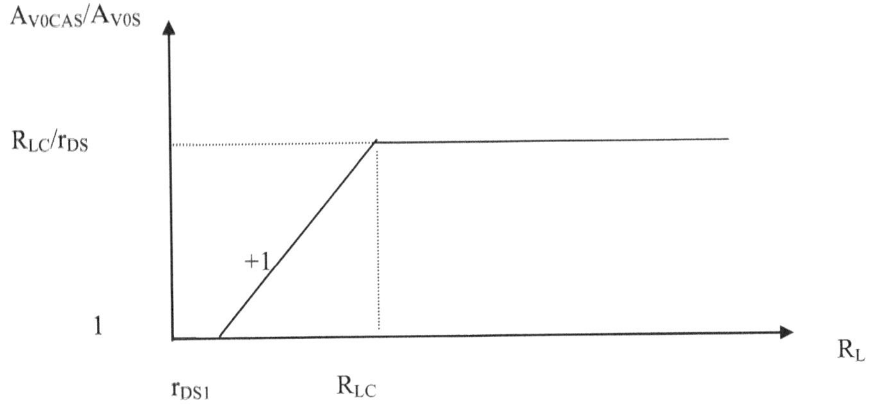

Figure 3.36. *Gain comparison: with or without "cascode"*

For a low R_L load in ohms ($R_L < r_{DS1}$), the introduction of a transistor (cascode) does not improve the low-frequency voltage gain of a common source circuit. On the other hand, if the R_L load is greater than the impedance (cascode) R_{LC}, the gain is enhanced by $\frac{R_{LC}}{r_{DS1}}$, that is, $g_{m2}\ r_{DS2}$ (which corresponds to the increase in output impedance seen toward the drive transistor M_1).

3.3.2. *Reminder on the general structure of the operational amplifier*

The amplifier is the basic analog circuit. We mainly focus on the input stage and its study in common mode (the differential mode is systematically treated in the academic literature). The goal here is to understand the operation of a circuit and to understand the operating regimes of the components it comprises.

3.3.2.1. *Characteristics and performances of Op-Amp*

The operational amplifier is defined as a voltage source controlled by a differential input voltage. The ratio between the variation of the output voltage and the variation of the differential input voltage defines the differential gain: AVD, the main characteristic of the Op-Amp. In order to achieve the desired function, Op-Amps are mostly used in a closed loop system, allowing the output voltage to be controlled according to the input voltage.

The voltage gain in open loop – in other words, the transfer function of the Op-Amp – must then be as high as possible so that the gain in the closed loop depends only on the feedback or on the gain in the open loop as little as possible. Under stability conditions and for certain physical reasons – for which the system has a transfer function (with a slope of –1) varying from the frequency of the dominant pole to the frequency of unity gain – the open- and closed-loop gain band products are equal.

The output voltage of the Op-Amp should depend on the input voltage as little as possible. We define the A_{VMC} common mode gain, the ratio of the variation of the output voltage to the variation of the common mode input voltage, and the common mode rejection ratio (CMRR).

Op-Amps are generally used in the linear regime and with feedback. In order to limit the operation in these regions, we consider some important concepts such as: maximum speed of variation of the output voltage (the slew rate), the harmonic distortion to the output signal introduced by the Op-Amp, the operating dynamic range for the common mode input voltage, CMR (common mode range) and the output dynamic range, the extreme value that the output voltage can take. If the CMR of an Op-Amp varies from negative V_{ss} to positive V_{DD}, we say that this Op-Amp has a rail-to-rail input, the same for the output if the output dynamic ranges from V_{ss} to V_{DD}.

An efficient Op-Amp is characterized mainly by a high differential gain. It is also characterized by the highest possible unit gain frequency while maintaining a slope of order –1 for the transfer function between the frequency of the dominant pole and the frequency at unit gain: a high common-mode rejection ratio over the entire frequency range of use, a significant slew rate, low output harmonic distortion for a sinusoidal input signal, the widest possible dynamic range of use in input and output (ideally from V_{SS} to V_{DD}, i.e. rail-to-rail mode). In addition, to approach the characteristics of an ideal controlled voltage source, an Op-Amp must have the highest input impedance and the lowest output impedance. All of these characteristics must be ensured with low power consumption.

3.3.2.2. *Op-Amp structure*

The block diagram of the structure of the Op-Amps is shown in Figure 3.37, with three essential functions illustrated:

A two-stage differential amplifier is called an operational amplifier; the input stage and possibly the gain stage contribute to the overall voltage gain of the Op-Amp The input stage consists of a differential amplifier (DA) providing the assembly with the inverted and non-inverted inputs, the large CMRR and the high input impedance. The second stage is a very high gain voltage amplifier, most needed for all open loop Op-Amps. To develop a large gain, the output impedance of the previous stage is very strong. Therefore, it is necessary to have a third and final stage – the output stage – giving low output impedance of the Op-Amp while ensuring high input impedance for the gain stage. Most DAs intended for high-frequency amplifications actually have three stages with voltage gain.

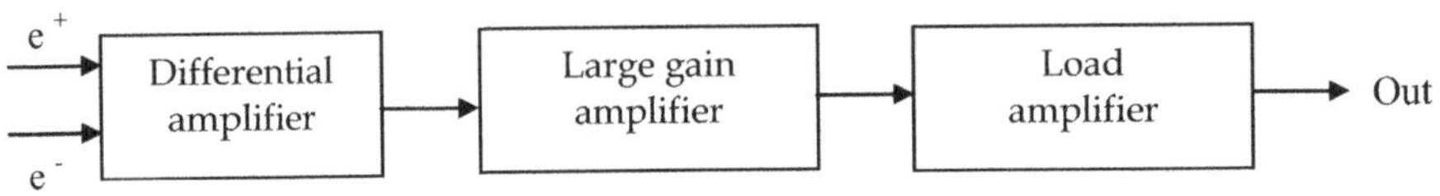

Figure 3.37. *Block diagram of the A.O*

The overall gain, provided by these three stages, therefore makes it possible to reduce the gain of each, which has the advantage of improving the slew rate (by reducing the transconductance of the input stage, for example) or of increasing the frequency at unity gain since each stage has a high bandwidth.

This way, we can choose a low noise input stage. Evidently, there are countless variations on the internal structure of Op-Amps (number of stages, complexity of compensation networks, completely differential operational amplifiers, common mode feedback circuit, etc., but the basic principles remain the same as mentioned above).

3.3.2.3. *Rail-to-rail input stage*

3.3.2.3.1. Definition

We want to design a differential input stage that works for common mode transistors ranging from negative V_{SS} power to positive V_{DD} (rail-to-rail) power. The CMR common mode dynamic range is then centered on the half sum of the supply voltages which is, by definition, the analog ground. The main performances are a high differential gain and a high bandwidth, which will be as constant as possible over the whole CMR, and a low common mode gain.

3.3.2.3.2. Study of the dynamic range of common mode input voltage

The operation of the differential stages of the NMOS type and the PMOS type (Figure 3.38) depends on the polarization of the transistors they comprise. This bias itself depends on the input voltages and on the common mode voltage, in particular.

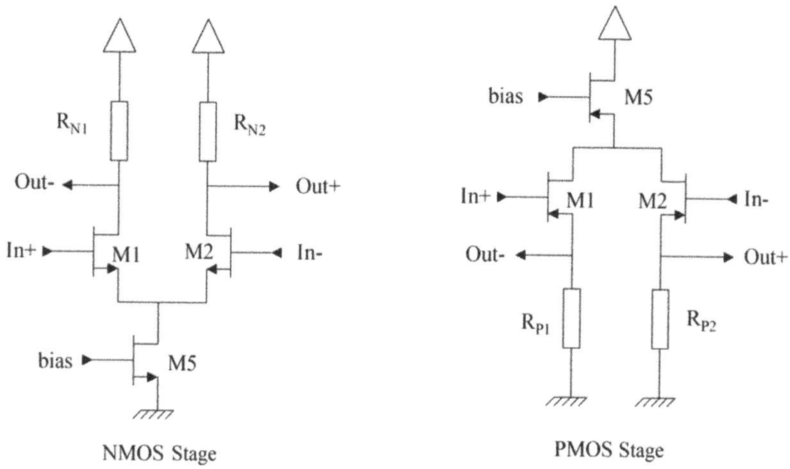

Figure 3.38. *Differential amplifiers. M_5: current source, M_1, M_2: differential pair, R_1 and R_2: active load*

Let us study the evolution of the amplitude of the variation of current i, for the same differential excitation V_{diff}, at the input (i.e. of constant amplitude whatever the common mode voltage), as a function of the common mode voltage V_{MC}. The latter will be between V_{SS} and V_{DD}. Based on the equivalent differential pair diagram (Figure 3.39), the low frequency differential gain of the input stage is calculated.

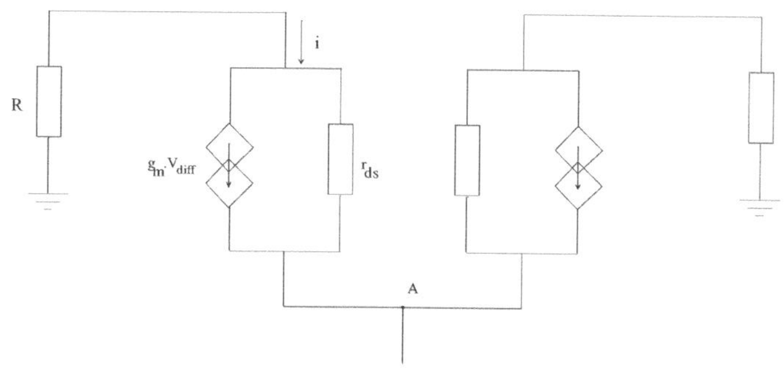

Figure 3.39. *Small signal equivalent differential pair*

We find: $i = g_m \cdot \left(\frac{r_{DS}}{r_{DS} + R_N} \right) \cdot v_{diff}$

3.3.2.3.3. Operating mode of an NMOS differential stage

The amplifier is used in common mode. We can distinguish four operating zones (Figure 3.40) for the diagram shown in Figure 3.38.

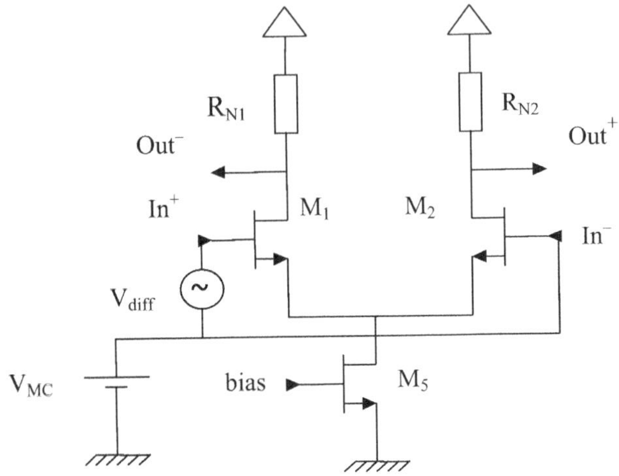

Figure 3.40 *Common mode differential amplifier*

Common mode voltage: V_{SS} to $V_{SS} + V_{th}(M_1)$

When the common mode voltage is less than V_{th} (M_1), transistors M_1 and M_2 are blocked and no current flows through the differential stage. The drain potentials are then V_{DD} and the source potential is V_{SS}.

Common mode voltage between $V_{SS} + V_{th}(M_1)$ and $V_{SS} + V_{th}(M_1) + V_{dsat}(M_5)$

When the common mode voltage is between $V_{SS} + V_{th}$ (M_1) and $V_{SS} + V_{th}$ (M_1) + V_{dsat} (M_1), the voltage V_{DS} of transistor M_5 is lower than V_{dsat} and M_5 is in the ohmic region. Then the direct currents increase with the common mode voltage (see characteristic I_D (V_{DS}) of the MOS). The voltages V_{GS} of the transistors M_1 and M_2 increase with the drain currents while their drain potentials decrease (the current increases in the transistorized loads R_1 and R_2). In the ohmic region for M_5, the transconductance of the input transistors is therefore not constant since the drain current varies. It follows that the current variation ($g_m.v_{diff}$), due to the same variation of the input differential voltage, increases (in absolute value) when the common mode voltage varies from $V_{SS} + V_{th}$ (M_1) and $V_{SS} + V_{th}$ (M_1) + V_{dsat} (M_5).

Common mode voltage between $V_{SS} + V_{th}(M_1)+V_{dsat}(M_5)$ and V_{DD}-$[V_{dsat}(M_1)+R.I_0 - V_{GS0}(M_1)]$

In order for the all of the transistors of the differential stage to have reached their saturation region, the common mode voltage at the input must have exceeded the value $V_{SS} + V_{th}(M_1) + V_{dsat}(M_5)$. Under these conditions, the currents in the transistors M_1 and M_2 are constant and equal to I_0. Then the voltages V_{GS} of M_1 and M_2 are constant (V_{GS0}), they adapt to let the current I_0 pass through the channel. Therefore, the source potential follows the variations of the gate common mode potential. The drain potentials of these input transistors are constant and are equal to V_{DD}-R.I_0, R being the transistorized resistance for the current I_0. In a saturation region, at constant current and at constant V_{GS} voltage, the transconductance gm of the input transistors is constant. Consequently, the variation of current i due to the same variation of the input differential voltage is constant when the common mode voltage varies between $V_{SS} + V_{th}(M_1) + V_{dsat}(M_5)$ and $V_{DD} - [V_{dsat}(M_1) + R.I_0 - V_{GS0}(M_1)]$.

Common mode voltage between $V_{DD} - [V_{dsat}(M_1)+R.I_0 - V_{GS0}(M_1)]$ and V_{DD}

The currents of the transistors remain constant, imposed by M_5, but the V_{DS} voltages of the input transistors become lower than V_{dsat} since the drain voltages remain constant and the common source potential increases. The transistors M_1 and M_2, from the common mode voltage equal to V_{DD} - $[V_{dsat}(M_1) + R.I_0 - V_{GS0}(M_1)]$, then enter the ohmic region at constant current, where, to compensate for a decrease in V_{DS}, V_{GS} must increase. As a result, the common source potential of M_1 and M_2 increases less quickly than that of the gate common mode and it stabilizes. In the ohmic region, at constant current, the transconductance g_m of the input transistors decreases when the common mode voltage varies between $V_{DD} - [V_{dsat}(M_1) + R.I_0 - V_{GS0}(M_1)]$ and V_{DD}. The amplitude of the variation in current i decreases faster than the transconductance gm because the r_{DS} resistance of transistor M_1 also decreases and $i = \left(\frac{r_{DS}}{r_{DS}+R_N}\right) g_m V_{diff}$.

In order for the all of the transistors of the differential stage to have reached their saturation region, the common mode voltage at the input must be noted. Note that this last operating zone only exists if $V_{dsat}(M_1) - R.I_0 > V_{GS0}(M_1)$.

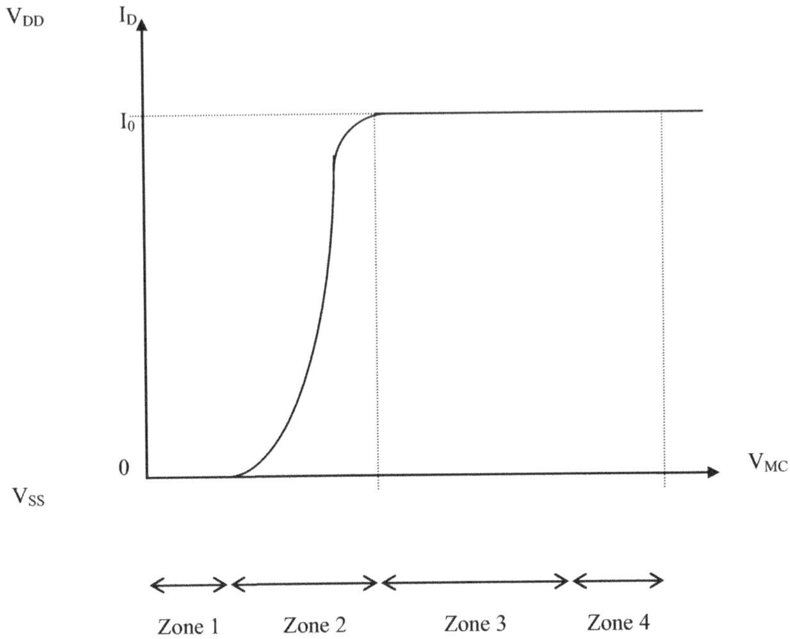

Figure 3.41. *Common mode voltage*

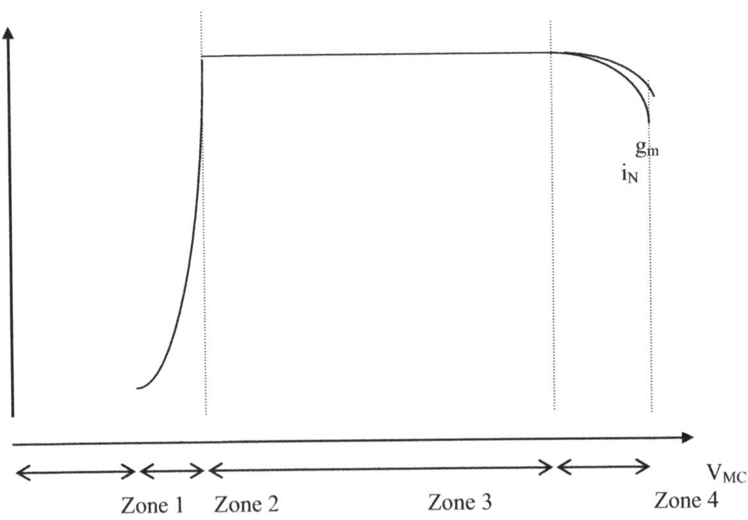

Figure 3.42. *I_N and g_m*

3.3.2.4. *Total rail-to-rail differential structure*

In CMOS technology, it is customary to replace resistive loads R_1 and R_2 with PMOS transistors mounted as M_3 and M_4 diodes. It should be noted that, in this case, the output voltage is not equal to V_{DD} when the common mode voltage at the input is between V_{SS} and V_{SS} + V_{th} (M_1); this is because the loads are PMOS transistors mounted as a diode. As a result, the nominal low frequency voltage gain is lower with "diode" transistors as loads, because the value of the equivalent resistance in small signals (1/g) is lower than the V/I ratio (R: DC resistance) polarization (Figure 3.43).

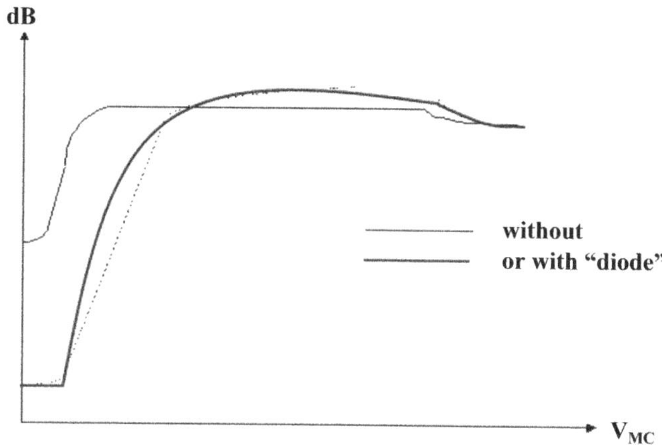

Figure 3.43. *Nominal low frequency voltage gain, without or with "diode" transistors as loads. For a color version of this figure, see www.iste.co.uk/abouelatta/smartpower.zip*

On the other hand, at low current, i.e. for a common mode voltage close to V_{SS}, the transconductance g_m of M_3 and M_4 (without transistors mounted as a diode), becomes greater than the fixed value of the resistors R_{Ni} (with transistors mounted as a diode), and the differential gain is then greater. Therefore the loss of gain over the entire input dynamic range is less. The same study, applied to a PMOS differential stage, brings us to the same results as the previous one in NMOS. This leads to the observation that there is a common mode voltage zone for which the amplitude of the current variation i_N (or i_P) is canceled out.

By carrying on the same graph for these variations of current amplitude i_N and i_P as a function of the common mode voltage, it is interesting to note that when one of these amplitudes is canceled out, the other is not. Therefore the sum of i_N and i_P is never zero (Figure 3.44).

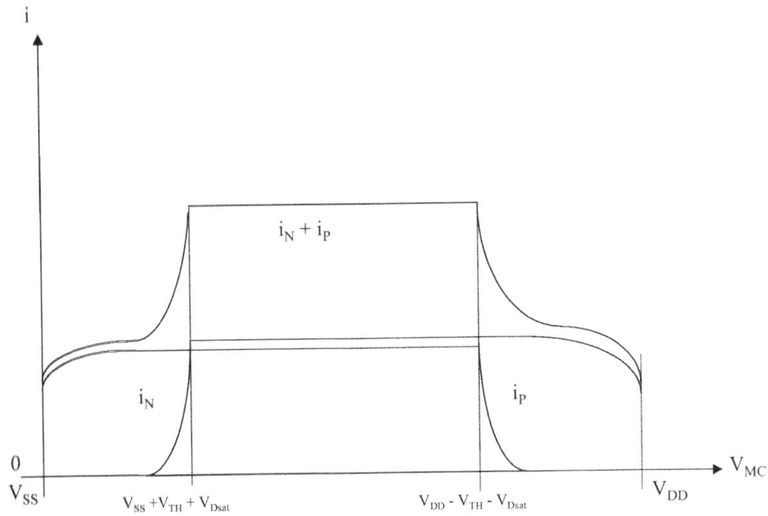

Figure 3.44. *I_N, I_p, I_N+I_P*

3.3.3. *Digital*

3.3.3.1. *The ASICs world*

Figure 3.45 illustrates the wide variety of products that can be found under the generic name "ASIC" (application specific integrated circuit): from small programmable circuits to integrated "full customized" circuits.

3.3.4. *The notion of codesign*

In general, codesign is a design that uses mixed resources (Figure 3.46):

– software (micro-processor, DSP, etc.); the source code is usually C, C++ (or assembler);

– hardware (programmable logic circuit, ASIC, etc.); the code describing the function is usually an HDL (hardware description language) code such as VHDL (very high speed integrated circuit HDL).

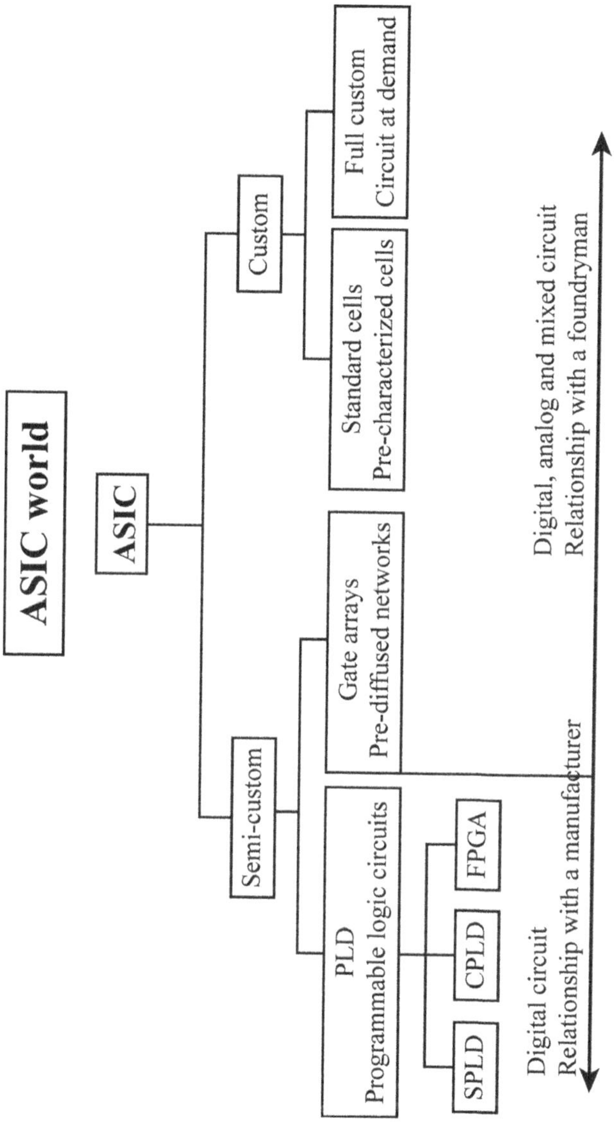

Figure 3.45. *ASIC*

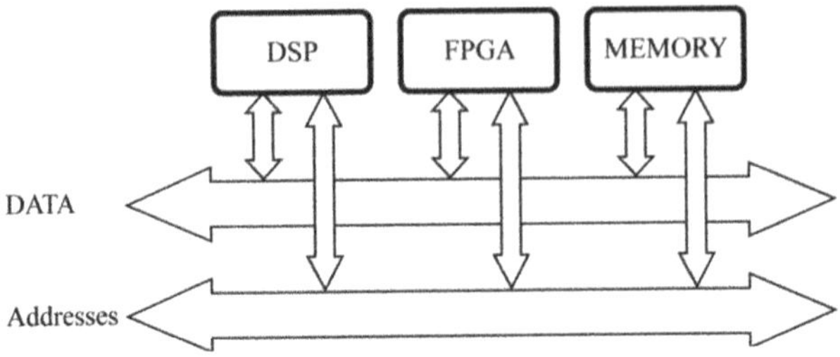

Figure 3.46. *Mixed resources: hardware and software*

The optimum (in terms of performance) the designer is looking for is somewhere in-between the software and the hardware. To find this optimum, we can follow the synoptic in Figure 3.47.

In addition, we denote the constant evolution of the specifications of a system during its development. For a product to be powerful, the technological target tends to move from software to hardware. Figures 3.47 and 3.48 illustrate this finding.

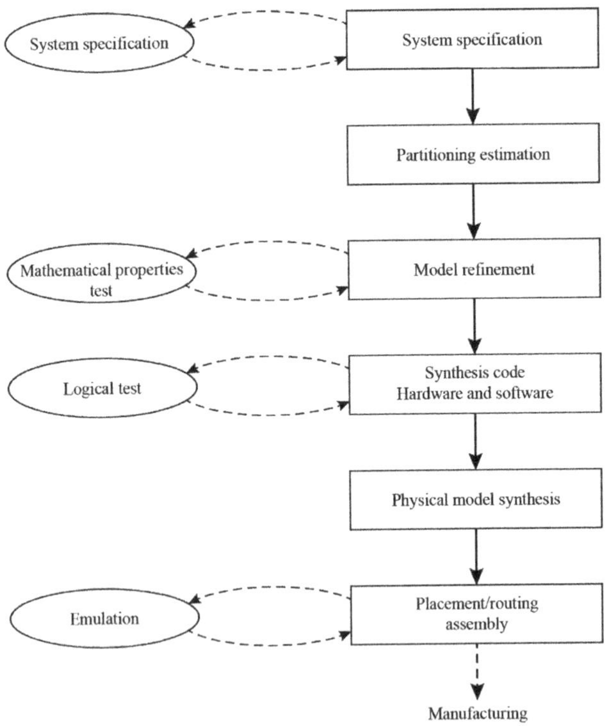

Figure 3.47. *Generic mixed design method*

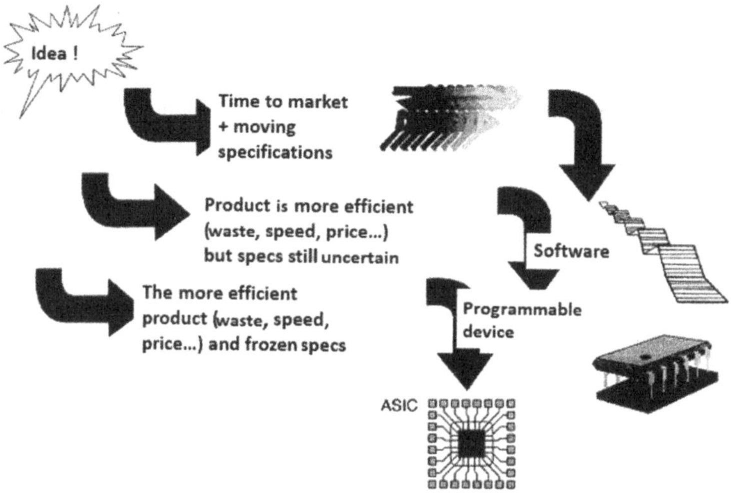

Figure 3.48. *From an idea to IC*

3.3.5. *Assessment*

Power integration is a developing field. If we want to go directly toward monolithic integration, it seems desirable and realistic to start from the power device and manufacture everything that is part of control, protection, etc., around it in a way that is compatible with its technology. Smart power integration will eventually have to make use of the same production lines as the integrated circuits while keeping a specificity on the insulation of the low-level part from the power stage. Functional integration requires a very specific technology (fine wafer, diffusion on both sides, etc.), making this path incompatible with a standard process. On the other hand, functional integration could quickly turn to the field of microsystems or sensors. Multipastilles integration (several chips in the same housing) will most likely compete with monolithic integration because it makes the association of chips of different natures possible while maintaining the technological processes adapted to each of them. With regard to hybridization, the approach seems to be most flexible. We can start from our own know-how. In this case, in our laboratory (AMPERE, Lyon), the algorithm and control methodology of electrical machines and the A/D and D/A converters have already been independently investigated. For example, we are currently focusing on the following issues:

– lightening the task of the processor (DSP or other) by developing programmable or non-programmable ASICs (this is the co-design);

– analog/digital conversion (and A/D), focusing on the problems of insulation between the different parts, noise generated and, more generally, to the problems of electromagnetic compatibility between the subsets of the system and the design of analog circuits;

– developing high-voltage devices that are compatible with standard low-level technology in part control, control, protection, etc.

In the future, know-how will shift from the system manufacturer to the semiconductor manufacturer. The system designer will become an architect by assembling subsystems and no longer by assembing devices. As for the manufacturer of power semiconductors, they will have to focus on integrating these subsystems.

4

Technology for Simulating Power Integrated Systems

This chapter presents an attempt to synthesize a design methodology for the system integration of an engine control chain. The latter incorporates the mixed design of control algorithms for electrical machines; the goal is to optimize integration, and a number of tools and algorithms are used.

4.1. Introduction

Power integration has been a developing field for the past 20 years. The system approach must prevail from the very early stages of design to ensure better security, less congestion and fewer costs. Simulation software has already demonstrated its ability to provide fast and accurate information in order to predict system behavior and robustness in various areas. This approach is developing in the field of power integration.

It is clear that, at present, the design of systems incorporating engine control is becoming more and more complex. This is inherent in the need to precisely manage the position and torque control, as well as energy efficiency, safety and even problems of phonic pollution.

Today, the control of electric actuators requires more and more efficient and inexpensive electronic resources. In addition, electronic engineers are looking for information regarding the state of the system (converter and/or electrical machine): this is referred to as diagnostics. This information – received at the level of digital electronic regulation, in addition to measured parameters such as currents, voltage, speed, torque, and so on – allows the control to be self-adapted in the event of a transient or permanent defect: diagnosis becomes active. In this context, our ultimate goal is to develop wide-application systems that can rapidly prototype control strategies and diagnosis of electrical actuators.

The development of dedicated specific integrated circuits therefore finds its value in an optimized and inexpensive solution (provided the production volume is large enough). It is not necessarily a question of rebuilding everything in a single ASIC (or in a programmable logic circuit) but rather of pairing it with a generic processor such as a microcontroller and a digital signal processor (DSP).

In practice, this can no longer be done using traditional means; simulation tools have become indispensable. Thus, we present a summary of an electric traction chain, integrating the development of a control solution for asynchronous electric motors. We present practical results and also consider the "device" aspect that may constitute the pilot circuits of the power converter.

4.2. Hardware and software design of engine control

The synoptic of an engine bench is described in Figure 4.1. The main elements are as follows:

– continuous power (300 V/10 A) to power the inverter (at IGBTs);

– the inverter, with its PWM interface (Pulse Width Modulation);

– instrumentation: three current measurements, a continuous voltage measurement and a measure of speed.

Co-design – that is, software and hardware – has become a means of simultaneously meeting the criteria of complexity and commercialization. Hardware and software co-design has changed the game from the traditional sequential simulation. More specifically, it is the unification of these two worlds that is a technological revolution. Design can begin before developing the architecture of the system, through its specification.

In practice, hardware and software designers must work closely with those in system architecture. Moreover, the level of library hierarchies depends on the functions they reflect. Those at the "system'" level are much more general than those "upstream", where devices are described at the physical level of the design of the layout/masks, depending on the technology effectively used. A basic idea is to be able to describe libraries at the lowest possible level, so that one is not dependent on technology at the level of the "hard" or "soft" implementation; they can move from one design to another.

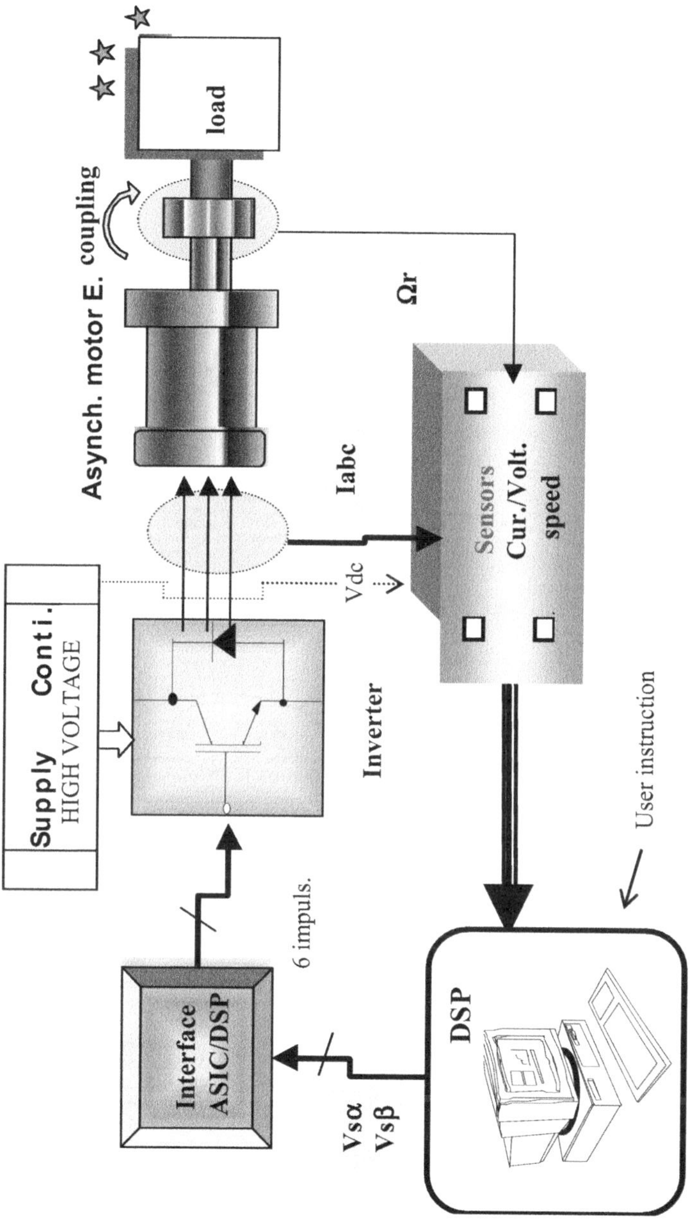

Figure 4.1. *Synoptic of the structure of the engine measure bench. For a color version of this figure, see www.iste.co.uk/abouelatta/smartpower.zip*

In addition, for the purpose of reliability and productivity, blocks (see "Intelligent Property") that have already been proven in previous designs will be sought.

For the integration of the control system, there are four main phases (see Figures 4.2(a) and (b)):

– functional specification;

– exploring the space of solutions;

– mixed synthesis;

– the functional test.

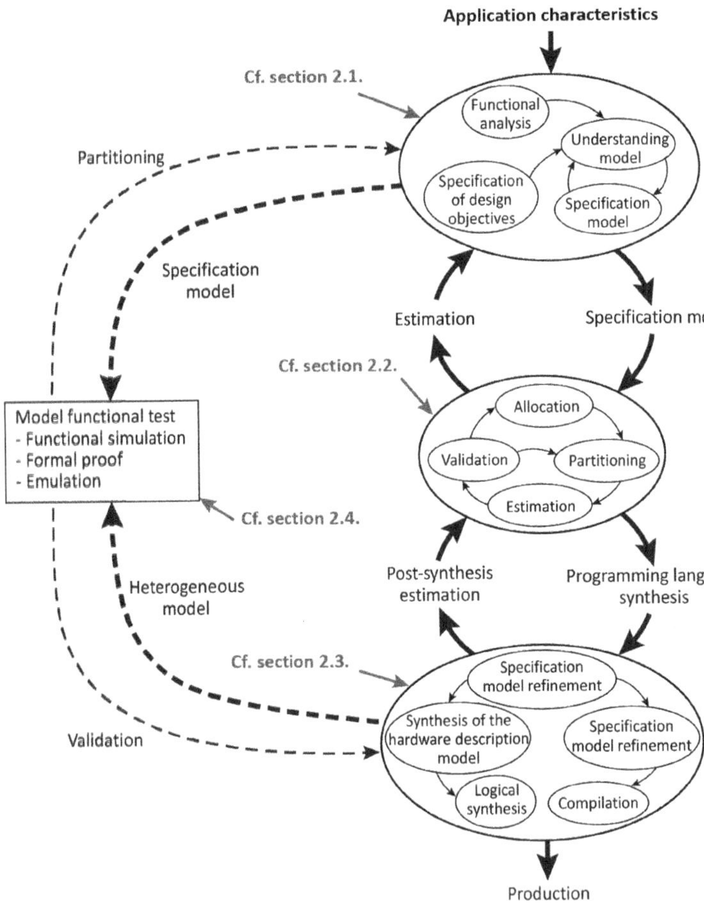

a)

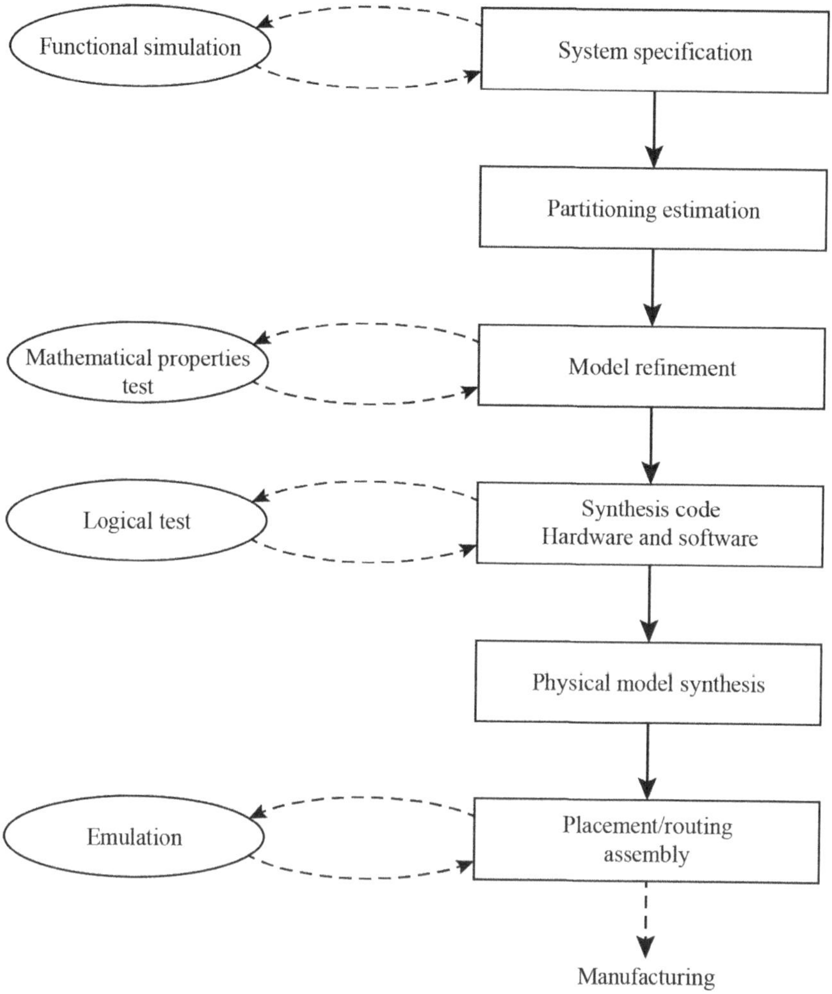

b)

Figure 4.2. *a) General methodology of mixed design. b) Our current approach to the mixed design stream*

4.2.1. *Functional specification*

It is obvious that errors need to be detected as soon as possible. The functional specification should be used to identify and define the system to be produced (communication protocol, algorithm, etc.). In practice, for data inputs, what are the outputs and how does the system calculate them?

The comprehension model

To fully describe the command, we need to represent several types of functions:

– the control algorithm in the form of a block diagram;

– protocols using a diagram of signals associated with a Petri network or GRAFCET.

The simulation of this comprehension model should allow the solution to be validated without knowing the target and the imperfections it engenders: calculation with real numbers, no notion of propagation time, delay, etc.

The specification model makes it possible to determine how the functions are actually performed, taking into account the specification of each block. The calculations in the blocks that are intended to be implanted in the fixed point format will be done, for example, in binary and with the appropriate encoding (number of bits, integer part and decimal part).

These steps (functional analysis, comprehension model and then specification) must be carried out in accordance with the specifications.

4.2.2. Exploring the space of solutions: the partitioned specification model

Exploration is a key concept in mixed design (hard/soft). Exploration of the space of solutions begins with allocation. Allocation involves architecture selection (chosen by design constraints): number of microprocessors, ASICs, memory, bus size, etc. Two strategies can be considered: freezing an architecture based on application considerations or limiting the choice of resources to a library of components.

Partitioning (mapping) is the action of allocating one or more objects of the lowest level of granularity (node) to each system resource and determining the order, setting priorities and execution of these nodes ("scheduling"). It aims to pair a function with a so-called "material" target for an ASIC, or "software" for a processor.

To carry out the partitioning, we need to know:

– the granularity (of course, to be adapted to the application and architecture of the target) that characterizes the smallest indivisible object considered during partitioning (also referred to as the "node");

– constraints (frequency, surface, consumption, etc.);

– models for optimizing each constraint and cost functions (weighting of constraints).

The execution of operations (nodes) depends on the chosen partition (parallel or sequential execution). There are two strategies:

– the static sequence, where the execution order is frozen in design;

– dynamic sequencing, when it is based on a priority allocation strategy in order to minimize wait times.

Estimating performance and optimization sizes is difficult because we have to get closer to physical implantation (synthesis and compilation of codes); this is very expensive in terms of computational time (not feasible for iterative exploration of a large number of solutions). Instead, we seek to find an order of magnitude by using a library of devices and functions for which the optimization sizes are known.

For specification models, as well as for the types of functions to be represented (data streams and reactive blocks), we use the GRAFCET standard. There is no “tool” to simulate a GRAFCET model; we therefore translate the GRAFCET model and algorithmic functions into a simulation language (MATLAB). Specification models (simulated by MATLAB) describe binary operations. The MATLAB sequence makes it possible to jointly simulate a binary model of the system (specification model) with an analog model of the engine (block schema). The choice of this tool (or PYTHON) is related to the algorithm’s comprehension stage and the functional verification of the system. SYNOPSYS is a set of tools used for VHDL simulation and synthesis; Altera/MAX+PLUS for compilation, routing placement and prototyping; and CADENCE for routing placement in semi-custom technology.

4.2.3. *Mixed synthesis, hardware and software code*

A refinement of the specification model is necessary; it will be completed by adding:

– memory management;

– the allocation of addressable space;

– interfacing between components;

– arbitration of access to shared resources.

We then look at the synthesis of the high-level code. With each node in the specification model assigned to a system resource, code synthesis involves generating hardware description language (HDL) code for hardware nodes and programming code (C, assembler, etc.) for software nodes. From the order of the execution of the nodes, the global sequencer is built.

The next phase is logical synthesis and compilation. Both synthesis and compilation work best when the processed code follows the instructions of the tool used. In terms of logical synthesis, we get better results for a circuit described by register transfer level (RTL) rather than at the behavioral level. Synthesis is the least developed link; industrial system synthesis tools are not as well developed as modeling and analysis tools.

Co-simulation requires process and communication precautions. Code generation is clearly a problem in itself. Specialized architectures such as digital signal processors (DSPs) or microcontrollers dominate the buried systems market because of their relatively low cost.

In any case, it is still the manual generation of codes that is commonly used. Advanced languages, such as C, C++, Java, PYTHON or VHDL, are still difficult to use for creating programs from abstract models. Furthermore, the line between "soft" and "hard" implementations moves according to technical advances or new constraints. To be truly efficient, an automatic code generator must be able to cover as wide a design space as possible. One way to get around the problem is to use predefined and set-up blocks of code in libraries as soon as possible.

4.2.3.1. *Synthesis of the hardware description model*

For hardware, we use high-level synthesis tools such as SPW and VHDL-AMS. This brings a considerable amount to co-simulation, if not yet solving the problems of optimization of memory, parallel and heterogeneous hardware architectures, synthesis and optimization, programmable hardware and optimization of communication protocols.

Currently, we are in the presence of a software patchwork so we need effective gateways. The introduction of artificial intelligence should eventually solve portability problems. For the moment, automation can only be done, at best, on portions of the design.

The generation of software code is simpler because it can be automatic for particular processor targets (floating point) thanks, in particular, to the MATLAB-Simulink-Dspace tool. Regarding our choice of DSP architecture (fixed point) and FPGA (programmable circuits), the generation of hardware and software code is manual, cumbersome and long. However, once again, it uses the notion of IP reuse, that is, the design re-uses old blocks that have already been validated (software or hardware blocks).

It is important to be aware that a system is more than the sum of its parts. With the advent of system on chip (SOC), a concept made possible by technological evolution, the role of a circuit designer has changed. The concept of a reusable intellectual property (IP) block was born from the complexity of these systems and their rapid evolution. This virtual component, or "re-use design," logically derives from library circuits, now used for the design of commonly-used specific circuits. The component designation also means that it will be reused without any modification (such as a semiconductor device, for example), which is not yet well-ingrained in the minds of all designers.

The benefit of the IP block is, of course, its portability, even if it is indirect, from one product to another and from one technology to another. This can be a hardware function or a software function (for example, an algorithm), or a "mixed" (hardware/ software) function.

The portability of "soft" IP written in high-level language from one technology to another is, theoretically, assured. However, for its implementation, it is necessary to use a library that is dependent on a particular technology. Moreover, an IP app is never fully characterized in terms of time analysis; the length of the connections (depending on the routing placement) is not fixed. However, we must be convinced that the opposition between software and hardware is less and less justified; the "hard" IP, whose design is in GDS format (but then this IP can no longer be changed; the timing and functionality are frozen), has many advantages and its use is all the more justified as the geometric dimensions of the circuits are reduced.

It is now possible to migrate a physical drawing from one technology to a finer one, taking into account, of course, the new rules of drawing. However, the larger the dimensions of the strokes, the more time the routing placement steps will take. In this case, the IP approach will be less valuable; problems of noise and electromagnetic interference will have to be carefully considered at the drawing stage, hence; once again, defining software/material partitioning early, when designing. We can conclude that IP is more a matter of methodology when designing, which is progress, but is still far from being plug and play.

We see the appearance of tools that generate hardware or software code such as CADENCE's SPW software. At this level, logical simulation is widely used to validate hardware code VHDL (our choice) constituting the high-level physical model. A time simulation is made from an estimated clock period.

In the last step before the physical realization of the system (layout or object code), the physical model describes the system at the level of its temporal execution. Time verification does not necessarily require a mixed simulation with the drive chain.

– The synthesis of physical models: this step is automatic due to the use of "manufacturer" software dedicated to a family of processors and synthesis tools linked to a library. In the end, we get an Assembler code (software part) and a netlist in VHDL, EDIF, etc. (hardware part).

– Placement/routing and assembly: this uses processor-related tools and routing placement software for the design of digital integrated circuits. In our purely hardware implementation on FPGA we used MAX+PLUS/Altera. The post-placement/routing physical model estimates the maximum clock frequency to re-inject this value into the time simulations of the physical model. Emulation is done on a dedicated specific map.

4.2.4. *Model functional testing*

Three audit levels can be considered. First, the simulation allows us to calculate the outputs of the system for a given configuration of the inputs, from a model of it; this is a functional audit. There are two problems:

– simulation times are often too long to simulate all combinations of input vectors;

– it is difficult to simulate systems defined by several models of behavior, that is, a heterogeneous system. The simulator must be able to co-simulate all the models.

Formal evidence can be used to prove the equivalence of circuits against a supposedly accurate reference model (which is useful when making changes to the circuit). It can prove the mathematical properties of a circuit.

Emulation allows us to prototype the application on a system, for example, on one or more programmable logical circuit(s).

The last step before the physical realization of the system (layout or object code): the physical model describes the system at the level of its temporal execution. Here, the choice of VHDL is essential for our 100% hardware integration.

The current problem is the heterogeneous simulation involving codes of different natures (VHDL, C, MATLAB script, etc.) in order to validate the different models in simulation. The validation of the physical model is particularly important because this model generates code for prototyping.

– Will the system work regardless of the input vector? The goal is to look for vectors for which the system could be faulted. The system must be validated with the partitioned specification model (which defines the sequence of operations).

– Is the binary accuracy of the results assured? This needs to be checked from the partitioned specification model.

4.2.5. *Synthesis of the approach and related tools of the functional model*

Table 4.1 summarizes the process and the tools associated with it.

Step	Realized function	Model	Tools
Functional specification	Macro functions Definition and "scheduling"	Understanding model (block diagram, GRAFCET, signal diagram)	MATLAB (real format)
Partitioning/ estimation	Choice of architecture. Distribution of nodes. Refinement of scheduling		Manual realization
Model refinement	Choice: protocol, bus and data format etc. Properties mathematical test (overflow, accuracy, etc.)	Specification model	MATLAB (parameterized blocks, fixed point calculation)
Code synthesis (hard and soft)	Code generation: VHDL and C Test logic (simulation)	High-level physical model	SYNOPSYS (simulation), SPW Dedicated tools for the processor (Dspace, SPW, etc.)
Physical model synthesis	Generation of machine code (assembler) Physical synthesis (hardware netlist)		SYNOPSYS (synthesis) Dedicated tool for the processor
Placement routing and assembling	Emulation	Physical model (post placement routing)	MAX+PLUS/Altera, Cadence (Silicon Ensemble) Emulation table

Table 4.1. *Design flow and associated tools*

4.3. Proposed design stream: related tools

We now consider the integration of the vector control of an asynchronous motor (with torque regulation for application in electric vehicles, for example). Using a prototype, we have almost validated the 100% hardware integration with a description using synthesizeable VHDL code (the calculation time is of the order of 25 µs).

By taking a sampling period that is longer than the cutting period the inverter can operate up to 16 kHz. The architecture is described in the rest of this chapter.

The current work (experimental measurements) aims to find an optimum on the internal coding of the data and on the characteristic times (sampling, etc.) for satisfactory operation of the traction chain.

We are confirming mixed type integration with a first manual sharing between the hardware (ASIC, FPGA) and software (processor) domains by a common sense rule starting from the observation that the system contains two recurrence frequencies:

– pulse width modulation (vector PWM) whose cutting frequency (inverter) is of the order of 10 kHz;

– regulation (calculation of voltages V_d and V_q) at the sampling frequency of about 1 kHz.

To relieve the unit of calculation of the quantities V_d and V_q from repetitive calculations of the commutating instants of the inverter switches, to us it seemed appropriate to dedicate an ASIC (programmable FPGA for prototyping) to carrying out this calculation and systematic work. Thus, the computing unit has more time to support a more sophisticated regulation algorithm or to process other information (observation of quantities for active diagnosis, for example).

At present, we have a hard library (synthesizable VHDL description) and a soft library (compilable C code). Co-simulation concerns the simultaneous simulation of models of different natures (digital blocks, analog, software scripts, etc.). We will therefore present two tools (MATLAB and SPW) as well as a standard (VHDL-AMS language) that will allow us to simulate the complete chain (algorithmic control part, inverter, engine, sensor). Indeed, the single system language is far from being developed.

4.3.1. *Accuracy*

Precise calculation is essential here. The idea is to conduct the calculations in 64/32-bit floating point format but the cost is often unjustified because many sources of errors on the measurements decrease the accuracy (measurements of torque, speed, position, current, etc.). We have adopted a fixed-point data coding such that the accuracy of the calculation is greater than the maximum physical precision (the implementation of the control must not introduce an error on the quantities ordered). In this case, the study of precision will define the width of the binary words (surface, size of the programming code).

The accuracy on the torque control results in a level of accuracy that must be achieved on the calculation of the switching moments of the inverter. This is difficult because there is no mathematical relationship between the switching moments and the controlled quantities (torque, speed, position). It is always possible to simulate a degraded model of the command to check some special cases. In addition, for physical reasons, the nonlinearities of the engine and inverter cause deviations from the theory. For an inverter between 5 and 20 kW at 20 kHz, we try to calculate the switching moments to the nearest 1 μs.

4.3.2. *Resources and system architecture*

Resource allocation is too complex if the architecture is not fixed. As a starting point, we therefore choose an architecture containing a DSP, an FPGA, a RAM, two buses (data, addresses) and a global clock.

Although DSP, FPGA and memories work synchronously internally, the overall system is asynchronous because the communication protocol is asynchronous. The comprehension model is simulated with an engine model. It makes it possible to check the control and servo functions. The MATLAB tool is particularly suitable because it is dedicated to automation and makes it easy to create blocks containing personal scripts. One of the problems is being able to detect design errors as early as possible. We then create an abstract system model, a simulated or formally verified co-specification for as many properties of this system as possible. There are some problems with the specifications to be executed; the types of computational algorithms can be very different, such as event or data streams. We use MATLAB and SPW tools.

MATLAB

MATLAB allows us to simulate the comprehension model (real calculation) or the specification model (calculation in parameterized binary format) in conjunction with an analog model of the engine (block diagram). The simulation results of the two models (comprehension and specification) are shown in Figures 4.3(a) and (b).

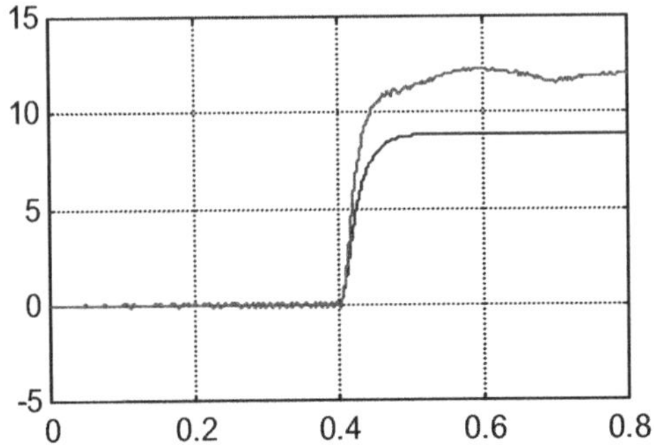

(i) Electromagnetic torque with its setpoint (N.m)

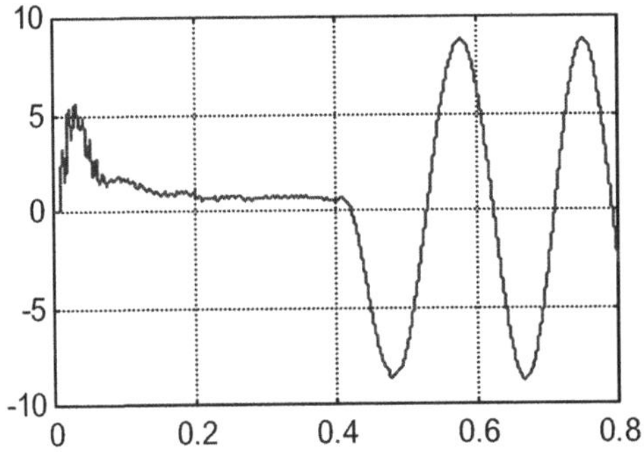

(ii) Phase current (I)

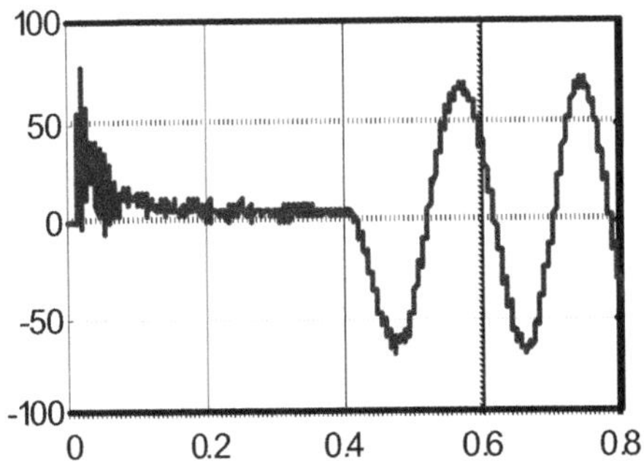

(iii) Phase voltage (V)

a)

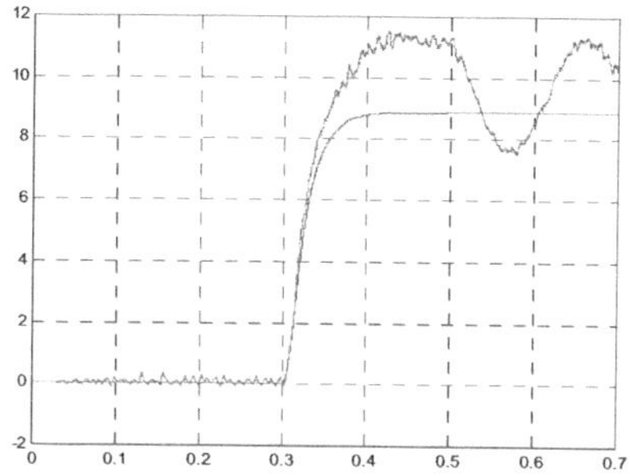

(i) Electromagnetic torque with its setpoint (N.m)

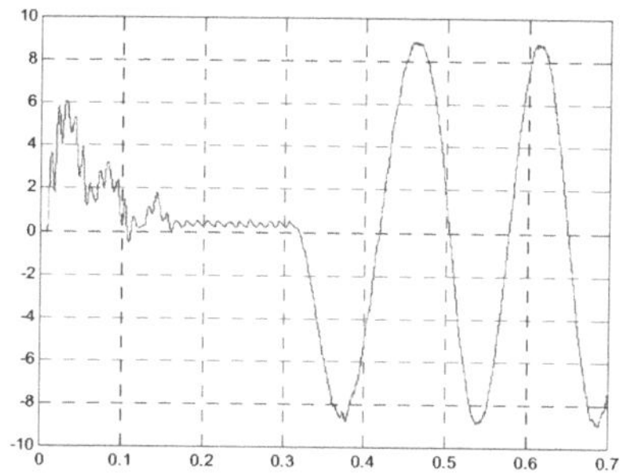

(ii) Phase current (I)

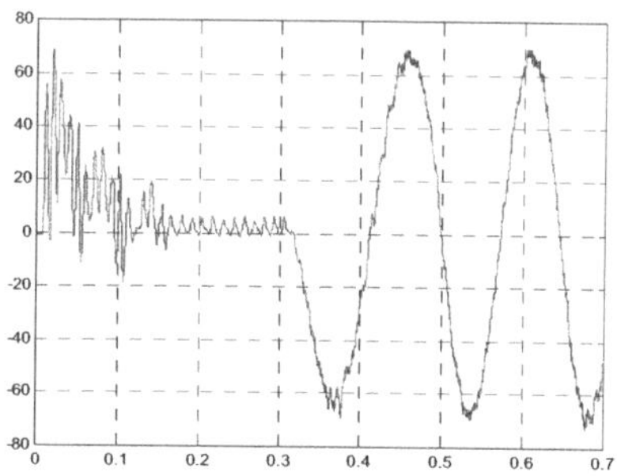

(iii) Phase voltage (V)

b)

Figure 4.3. *Simulation results of the two models in MATLAB. For a color version of this figure, see www.iste.co.uk/abouelatta/smartpower.zip*

The simulation times are very different for the start-up phase (estimated at 10 s). The simulation of the comprehension model takes a few minutes, while a second simulation with the specification model lasts several hours. These MATLAB simulations allow us to validate the algorithmic command on the principle and choose the appropriate binary encoding (sufficient accuracy, without overflow).

Alta SPW (Signal Processing Worksystem)

Alta SPW (Signal Processing Worksystem) is an environment proposed by the company CADENCE (see Figure 4.4). It makes it possible to:

– validate algorithmic solutions;

– work on a shared implementation hardware–software;

– generate VHDL or C code compilable to DSPs.

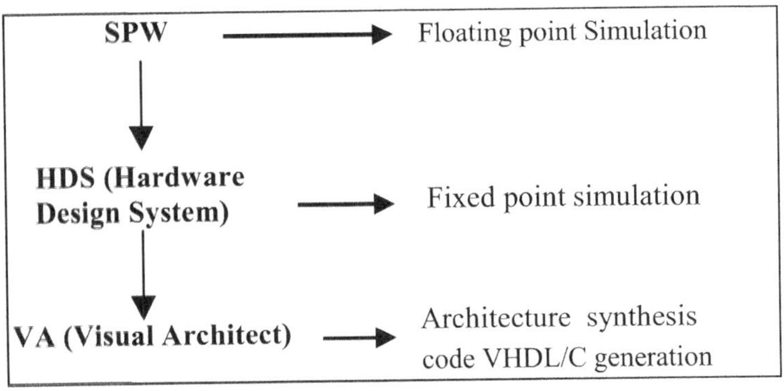

Figure 4.4. *Alta SPW: synoptic*

SPW is a data flow simulator, that is, it processes samples with a flow rate but does not use the notion of time. It can operate in three modes:

– Single rate: a re-entrant sample → an outgoing sample.

– Multirate: a re-entrant sample → N outgoing samples (known and fixed N).

– Dynamic: a re-entrant sample → N outgoing samples (N variable according to case analyses, for example, with a syntax of type "if ... then ... else").

Two models representing a complete engine bench have been developed: first, a model of an engine as well as the associated power inverter. Then, a coupling with the SYNOPSYS tool made it possible to generate a command module that could be simulated with SPW. The scalar command was chosen for this co-simulation for simplicity. This command, described in VHDL language, receives a frequency instruction to ensure the output generation of pulse signals to control the power inverter in input. In practice, we have effectively coupled SPW and SYNOPSYS software. Figure 4.5 shows the block diagram of the three-phase inverter under SPW and its internal structure (the inverter model of Figure 4.5(a) is depicted in detail in Figure 4.5(b)).

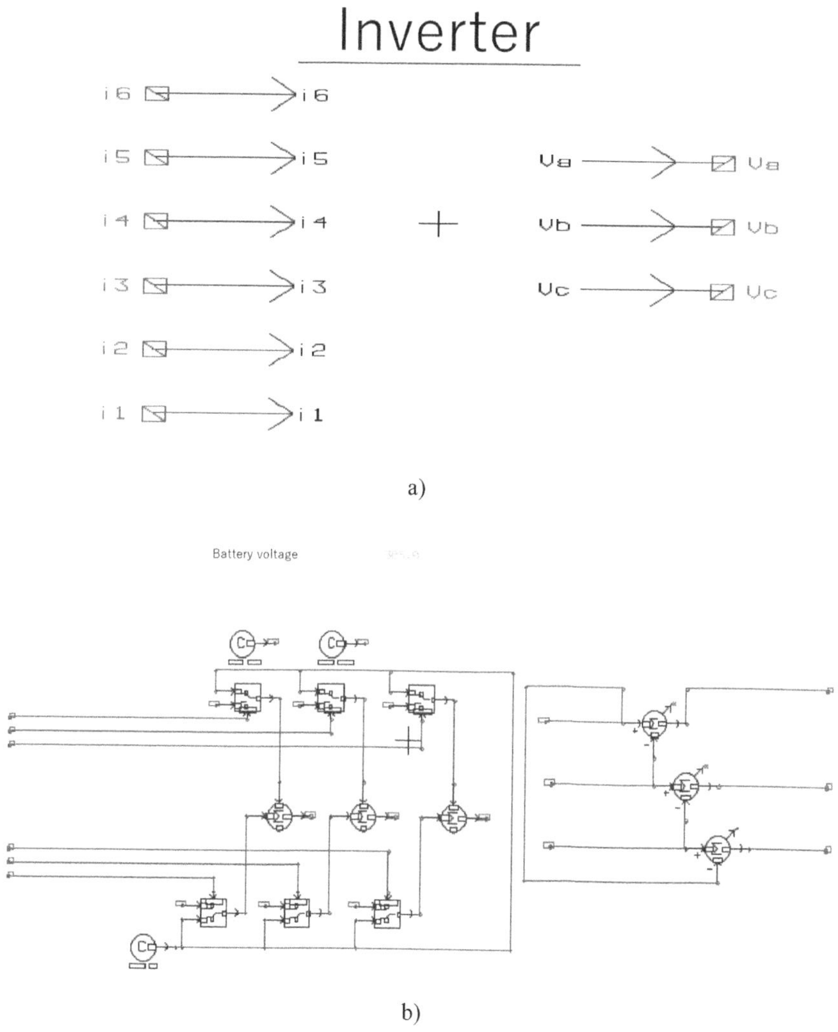

Figure 4.5. *a) Inverter and b) block diagram of the three-phase inverter under SPW and its internal structure. For a color version of this figure, see www.iste.co.uk/abouelatta/smartpower.zip*

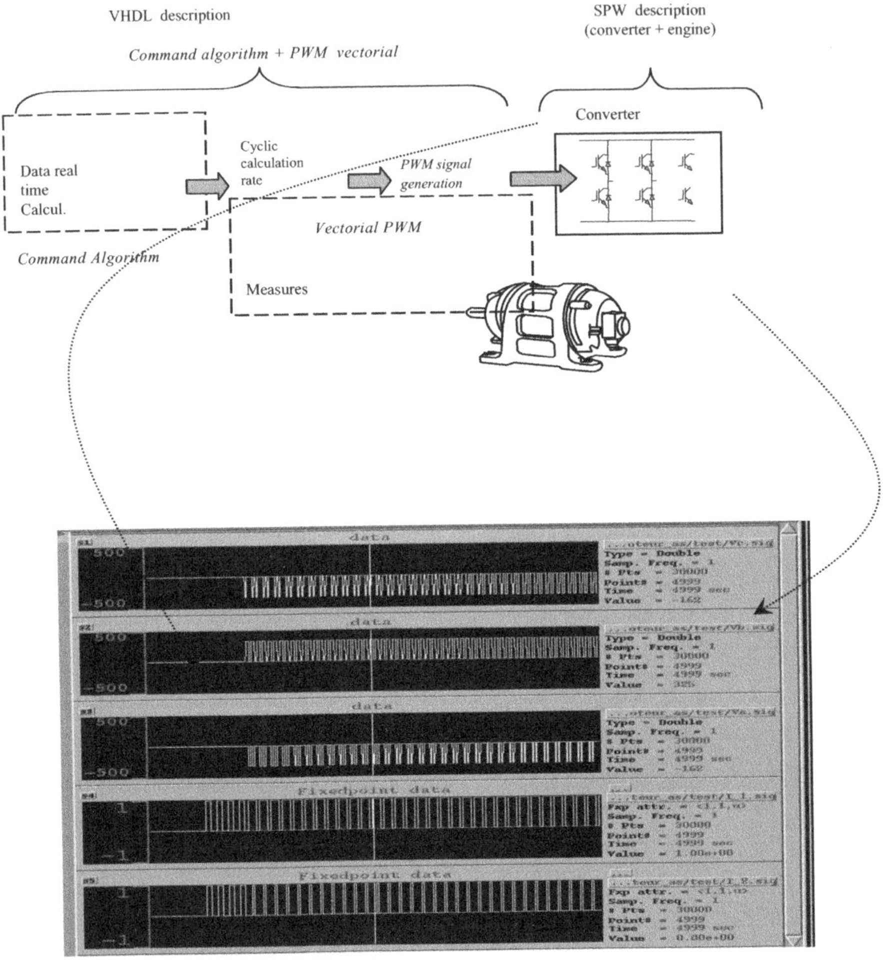

Figure 4.6. *Co-simulation results. The top three signals are the voltage slots output of the inverter and the other two are complementary PWM signals on one arm of the converter; they are generated by the VHDL block. For a color version of this figure, see www.iste.co.uk/abouelatta/smartpower.zip*

The VHDL language

The expansion to analog had become a need to allow analog as well as mixed digital/analog description and simulation. Fields other than the electrical are affordable by this language.

Because of this standard, the VHDL-AMS language allows the description and modeling of conservation as well as non-conservative, continuous and mixed continuous/discontinuous systems. Simulation algorithms are not part of the language; we can therefore describe mixed multi-technological models (thermal, hydraulic, mechanical, etc.). VHDL-AMS allows us to quickly develop models.

The two new types of objects appearing are as follows:

– Entity: this provides description of the outer view of the component (cf. black box). When writing an entity, we only declare the interface:

- generic, which groups the declaration of the parameters of the devices;

- port, which describes the interface with the external environment, via the external nodes.

– Architecture: this is a possible description (there may be more than one) of the model function. An architecture always refers to a single entity; it dictates the behavior of the black box. The equations of the function and the declarations of all the local variables belong to the architecture.

Asynchronous motor VHDL_AMS model

The verification of a complete system containing a digital processor such as the ASIC circuit requires the modeling of its application environment, which can be "non-electric", mechanical, for example. The laws describing this environment are the laws of physics that can have many analogies with electricity. The simulation of mixed platforms (analog models/digital models described with a high-level language), such as a drive chain with its control, is therefore made possible. In the case of machine controls, the control, described with a high-level language such as VHDL, will be co-simulated with an analog model of the motor. Figure 4.7 describes the principle of this co-simulation.

Figure 4.8 shows some simulations of an asynchronous motor model described with the VHDL-AMS language and simulated with the advanced MS tool.

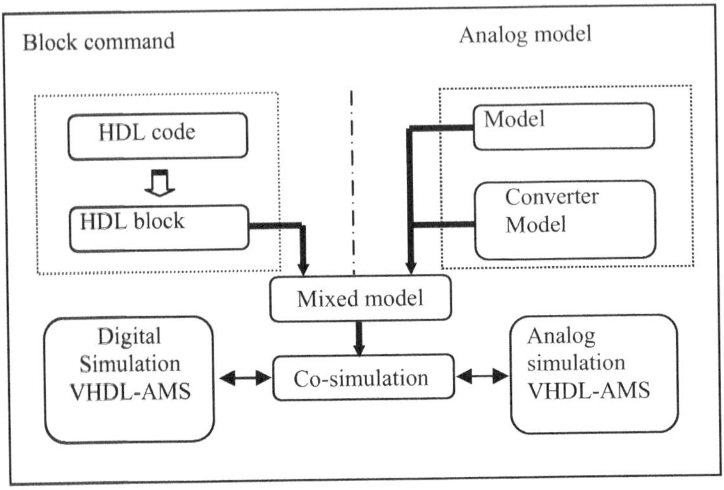

Figure 4.7. *Schematic diagram of a mixed simulation with VHDL-AMS*

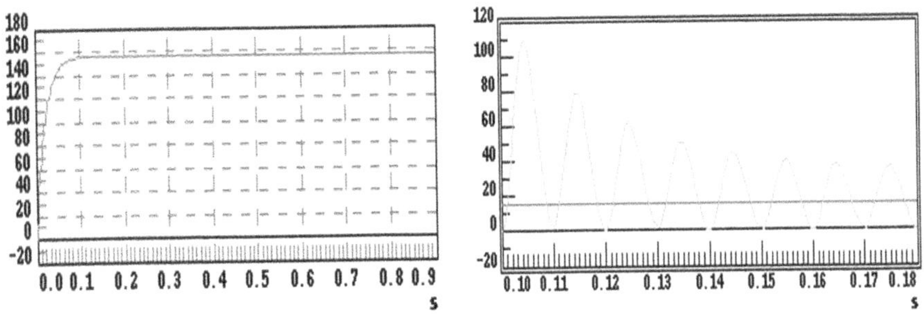

a) Starting speed (rd/s)

b) Torque stabilization around the setpoint (N·m)

Figure 4.8. *Simulation of starting a three-phase asynchronous motor described using VHDL-AMS. For a color version of this figure, see www.iste.co.uk/abouelatta/smartpower.zip*

4.3.3. *Realization*

Figure 4.9 presents the type of architecture that allows the control vector of an asynchronous motor to be received. This structure includes (1) a general sequencer (control unit); (2) a memory control unit; (3) a data acquisition unit of the converters; (4) two calculation units (ALU and CORDIC for rotation calculations); (5) a generating unit of PWM signals; and (6) four buses and various registers.

This structure is therefore described in VHDL, simulated and compiled with SYNOPSYS and then routed with Altera/MAX+PLUS. This architecture occupies 90% of the resources of an FPGA. The internal RAM (configured in ROM) is used to store the engine settings. This same architecture can receive simpler motor control (e.g. scalar control) primarily by modifying the general sequencer (state machine). This architecture also serves as the basis for current work developing an ASIC that only performs vector PWM.

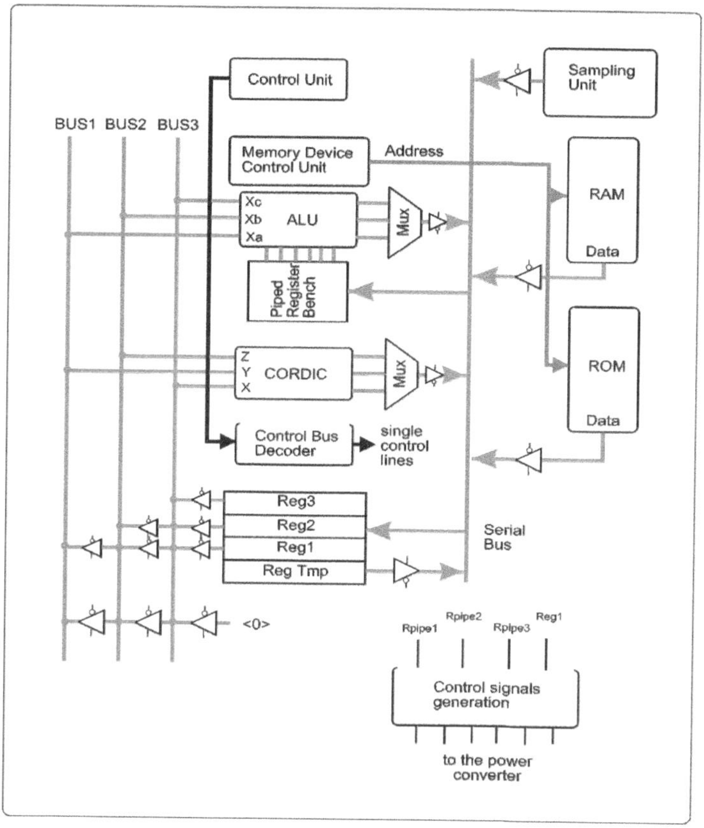

Figure 4.9. *Vectorial or scalar control ASIC architecture*

The choice of data bus size is guided by both the desired computational accuracy and the silicon surface of the circuit. A compromise is needed to satisfy both constraints. The binary simulations under MATLAB led us to fix the choice on a fixed point arithmetic with the size of 16 bits for all four buses of the architecture. Therefore we have developed mainly two systems: the first hardware (Figure 4.10) comprises a programmable logic circuit of large capacity, and the second contains a DSP Texas

Instrument fixed point and the same FPGA described previously (Figures 4.11(a) and (b)).

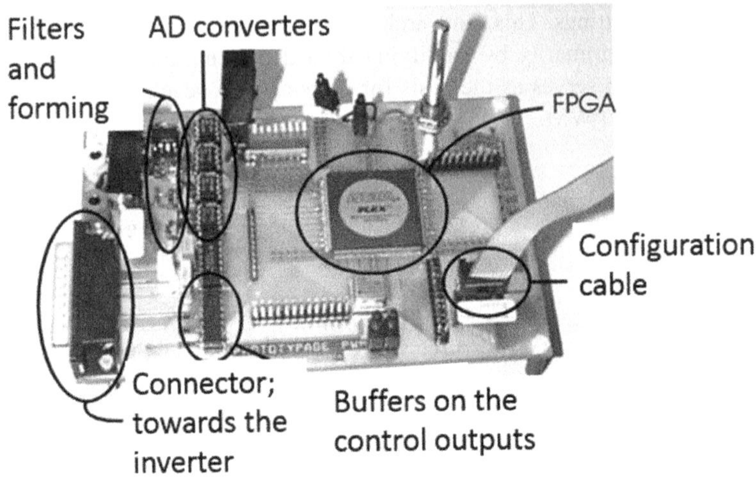

Figure 4.10. *Hardware prototyping board. For a color version of this figure, see www.iste.co.uk/abouelatta/smartpower.zip*

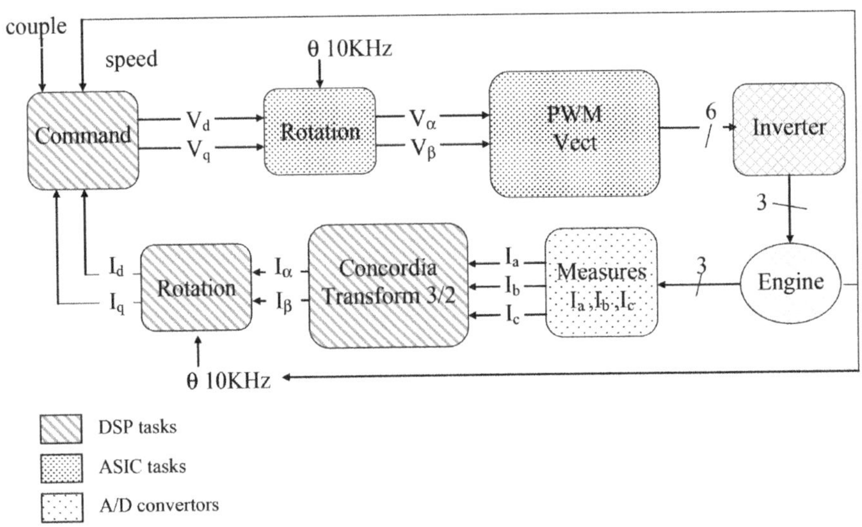

a)

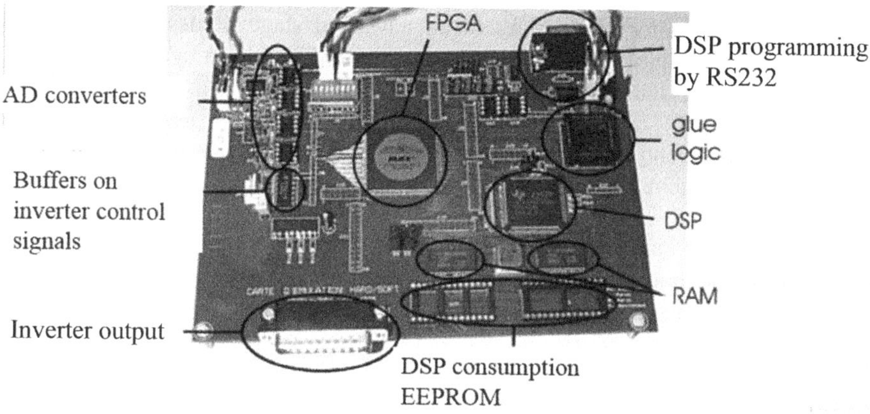

b)

Figure 4.11. *a) Distribution of the "hard" and "soft" tasks of the chain. b) Hard/soft card prototyping. For a color version of this figure, see www.iste.co.uk/abouelatta/smartpower.zip*

A third modular system (in the form of cards interconnected by a VME bus) is being developed. This system includes a DSP card, a programmable ASIC card, an AD/DA input–output card, as well as dedicated cards (speed acquisition for incremental encoder, absolute, etc.). These different prototyping boards make it possible to test motor control algorithm integrations either in pure hardware (with the FPGA) or in a mixed approach with a sharing between hardware (FPGA) and software (DSP).

4.4. Conclusion

This chapter focuses on the system integration of drive chains of electrical machines. This system includes both the control, of the engine itself, of course, but also of the power converters and their drivers. This is a collaborative work that presents an initial synthesis. We start with a reflection on a flow of control circuit design for electric actuators. The objective is to offer the designer of the command a methodology and tools facilitating the integration of this command; in practice, we have focused on the integration of a control vector. We:

– take into account the implementation constraints from the algorithmic design phase;

– postpone technological choices to an advanced stage of design in order to maintain relative technological independence, shorten the design cycle by means of automation, which is currently partial, of tasks.

We relied on techniques and tools used in microelectronics for the design of complex mixed systems combining hardware and software.

We show some simulation results and realizations of an engine bench with its close control. These tests make it possible to verify the effect of the control on the engine and to predict possible failures of the control. This saves a lot in terms of design time.

Automating most tasks is an ultimate goal in order to minimize the risk of error, increase the processing capacity of complex systems and reduce development times.

However, at present, there is no platform to solve the problems of mixed implementation of motorized control-command policies. The tools tested never fully meet the expectations of designers. For example, CADENCE-SPW allowed us to co-simulate the open-loop chain by associating SPW with the SYNOPSYS simulator for the part described in VHDL.

This is functional, but only after heavy work describing the analog power part. In addition, SPW is not a time simulator, so it is difficult to check the timing. On the other hand, standards appear, such as VHDL-AMS or Verilog-A, for the description and simulation of mixed systems. Here, we present the preliminary results with AdvanceMS (Mentor Graphics VHDL-AMS simulator). In practice, we could not perform a complete co-simulation (VHDL and power part: inverter and motor) because the version used did not fully support the VHDL standard. In summary, the difficulties remain numerous, and many steps must be carried out manually.

In conclusion, we briefly present three types of architecture: the first is fully hardware (FPGA circuit), the second is mixed (FPGA and DSP fixed point) and the last is modular (FPGA and floating-point DSP and any other card interconnected by the VME bus). This last configuration is more dedicated to development work to implement, test and compare different implementations or control-command algorithms. The first two solutions are much more interesting with a view to industrializing the product. Finally, we would like to remind you of an essential fact; it is important to have a very short design cycle because technological supports evolve very quickly. As a result, the solutions developed can quickly become obsolete. This reinforces our desire to develop an effective design method and not a solution that is optimal at a given moment and outdated the next.

5

3D Electrothermal Integration

5.1. Introduction

In recent years, many electrothermal (ET) co-analysis methods have been developed to design integrated circuits (ICs). To extract the distribution of temperatures, the heat equation is solved in 2D and 3D. The electrical resistance of the circuit will then change with the temperature and an appropriate overhaul of the electrical solution will be achieved. Such a process will be the starting point for updating thermal issues for heat sources. When the convergence reaches a fixed accuracy, the iteration process stops.

Currently, there are two main techniques depending on how the thermal problem is solved. The first method is to solve the thermal conduction equation using the finite element method (FEM), the finite volume method (FVM) or the finite difference method (FDM), as indicated in commercial multiphysics simulators, simulating 3D geometries and/or also including other phenomena such as micro-heat pipe cooling effects, with a very high calculation cost.

Second, the thermal problem is solved by using an equivalent circuit model that both thermal circuits can simulate with an electric solver, such as the SPICE type (Advanced Design System (ADS), ELDO). This technique can significantly reduce computation costs and address high-speed electrothermal problems.

A number of ET methods have been studying chip or PCB in the power distribution network (PDN) analyses, but on device-level projection. The PDN should be studying in focusing various parties, such as Coplanar Wave Guide (CPW), vias, C4 bumps, etc.

Although each simple model can be brought to a single node, the total equivalent network size for PDNs can easily exceed several million nodes. The main challenge of simulations and PDNs is how to reduce the cost of computation. Traditional model reduction techniques are more suitable for systems with a large number of poles; however, a PDN is usually composed of a large number of nodes with a small number of poles. Therefore, reducing the model for PDN is not so obvious. Recently, new reduction methods have been studied for power grids, which are only based on the concept of node reduction. Once the reduction methods are adapted to the electrical and thermal study, the reduction strategies can be used directly, with predefined accuracy, which decreases the size of the network.

This chapter deals with the electrothermal modeling method for substrates, 3D circuits and multi-cores systems. To give a brief introduction to these phenomena, some simple models are presented. We use the resolution method in TSV level simulation, that is, the modified nodal analysis (MNA) method. The PDN will be a very interesting topic for future studies, but in this chapter we only use the simulation method. We are only interested in the low voltage study.

5.2. Electrothermal modeling of substrate

Coupling substrate noise in ICs has become an important objective in circuit design. Nowadays, micro(nano)-technology and the development of semiconductor technology allow designers to integrate multiple systems into a single chip, both in 2D (planar) and 3D (volume). This design technology reduces costs, improves performance and makes a system-on-a-chip possible.

3D integration seems to be the true paradigm to compete with Moore's law (more than Moore vs. Moore). However, it is still difficult to go in this direction. In 2010, the question was: Why 3D? Now, questions surround how and when 3D is used. Stacking 3D chips are known for overwhelming problems with conventional 2D ICs, using deep contacts or specific vias passing through silicon for the transmission of the signal.

The type of silicon substrate that the 3D interconnects are made from is a key parameter and has a clear impact on the electrical behavior of the 3D interconnect. Electrical modeling methods depend on the type of substrate of the application. Two different classes of P silicon substrates are essentially used in CMOS/BiCMOS processes: uniform lightly doped substrates (>1 Ω · cm) and heavily doped substrates (<1 Ω · cm) with a lightly thinned doped (>1 Ω · cm) epitaxial layer. Other semiconductor bulks are used in micro-/nanotechnologies and could integrate 3D interconnections (silicon on insulation technologies, micro-electro-mechanical system technologies, sensor technologies, etc.).

Here, we try to propose an analysis of studies of dynamic electrothermal phenomena in circuits in tri-dimensional layers. First, we resolve the Poisson equation using Green's function. The latter mathematical method is an analogue of the transmission line matrix (TLM), built on the transmission/reflection of voltage and current through layers, adapted to buried contacts, of any thickness or shape. We then solve the equation of 3D heat, in transient regime. In this regard, we use the algorithms developed for electrical modeling, using the concept of impedance spreading. These electrical and thermal spreading impedances must be considered as entangled. This approach allows us to start tentatively modeling the more general problem of noise in 3D ICs using a new paradigm: the so-called transfer impedance.

5.2.1. *Brief introduction to mathematical tools*

Here, Green kernels are applied to homogeneous stacked layers for the extraction of substrate circuit models, as compared to numerical methods; the resolution speed of the first method is much faster. Some reminders and basic concepts are introduced. The use of discrete cosine transformations (DTCs), applying a fast Fourier transform (FFT) twice, consecutively, does accelerate the computational speed. Then an improved model, which can be applied to the substrate with buried contacts (or via), is shown; contacts in different layers can be treated. As a test, an FEM is used to examine the validity of the calculation presented.

5.2.1.1. *Green kernels*

The high-density integration and high frequency of the systems make substrate noise coupling one of the most important design considerations due to their significant impact on the performance of ICs. The purpose of this 3D substrate analysis is to efficiently extract the parameters of impedance Z between contacts (electrode or via) positioned on or in the silicon substrate. An efficient impedance extraction tool for any two contacts can help the designer to speed up the design and optimize the final layout. As with flat technologies, 3D interconnects can be built as an electrical model equivalent to resistance, inductance, capacity and conductance (RLCG), with a Π or T network. Typically, a simple compact model is built by modeling the substrate as a simple node.

Nevertheless, this hypothesis is only viable when the substrate is highly conductive, in low and medium frequency domains, and is not suitable for the multilayer substrates. For this reason, in this chapter we develop a substrate extraction method based on the TLM method over multilayered substrates and/or Green kernels, to model such effects in the volume.

5.2.1.2. *Analysis of the substrate*

Below, we present a 3D substrate paradigm with surface or buried contacts. The novelty that we consider and apply here is therefore to perform all of the calculations in the reciprocal domain, from the very beginning of the calculation to its end, in order to extract the temperature or voltage.

The *Z* parameters can be set as follows:

$$Z_{mn} = \frac{V_m}{I_n} | I_{k \neq n} \qquad [5.1]$$

In practice, by injecting a unit current as a source of excitation, I, at the *n*th point, and a unit excitation source of current, -I (equivalent to a sink current I) at a *m*th point and calculating the resulting voltage at the surveyed point, we can obtain the direct impedance Zmn between the contacts m and n.

Under quasi-static conditions, the potential on the substrate satisfies the Laplace equation:

$$\nabla^2 \varphi(x, y, z) = 0 \qquad [5.2]$$

It can also be written in Cartesian coordinates as:

$$\left(\frac{\partial^2}{\partial x^2} + \frac{\partial^2}{\partial y^2} + \frac{\partial^2}{\partial z^2}\right) \varphi(x, y, z) = 0 \qquad [5.3]$$

Hereafter, concerning the heat diffusion or Poisson equations, the analytical method is, *ab initio*, based on the algorithm of variable separation, solving an eigen problem.

By putting φ = X(x, x') · Y (y, y') · Z(z, z'), [5.3] can be rewritten as:

$$\mathrm{Z}(\mathrm{z}, \mathrm{z}') = \mathrm{Z}'(\mathrm{z}, \mathrm{z}') \cdot \cos\left(\frac{\mathrm{m} \cdot \pi \cdot \mathrm{x}'}{\mathrm{a}}\right) \cdot \cos\left(\frac{\mathrm{n} \cdot \pi \cdot \mathrm{y}'}{\mathrm{b}}\right) \qquad [5.4]$$

Then, we get a simple equation, for voltage induced by a point charge.

$$\frac{\mathrm{ab}}{4} \cdot \left[\frac{\mathrm{d}^2 \mathrm{Z}}{\mathrm{dz}^2} - \gamma_{\mathrm{mn}}{}^2 \cdot \mathrm{Z}\right] = -\frac{\delta(\mathrm{z} - \mathrm{z}')}{\varepsilon_{\mathrm{N}}} \text{ with } \gamma_{\mathrm{mn}} = \sqrt{\left(\frac{m \cdot \pi}{a}\right)^2 + \left(\frac{n \cdot \pi}{b}\right)^2} \qquad [5.5]$$

For z ≠ z', δ(z - z') = 0, equation [5.5] has a well-known general solution:

$$Z' = A \cdot e^{-\gamma_{mn} \cdot (d+z)} + B \cdot e^{\gamma_{mn} \cdot (d+z)} \qquad [5.6]$$

This equation also invokes a reflected wave and a transmitted one.

For example, in the case where the point charge and the observation point are in the same dielectric layer on the surface with z = z' = 0, Green's kernel is transformed into:

$$\emptyset(x, y; x', y')_{z=z'=0}\,(G_0)_{z=z'=0} = \left(\sum_{m=0}^{\infty}\sum_{n=0}^{\infty} f_{mn}C_{mn}\cos\left(\frac{m\pi x}{a}\right)\cos\left(\frac{m\pi x'}{a}\right)\cos\left(\frac{n\pi y}{b}\right)\cos\left(\frac{n\pi y'}{b}\right)\right) \quad [5.7]$$

where $C_{mn} = 0$ for (m = n = 0), $C_{mn} = 2$ for m = 0 or n = 0, but m ≠ n and $C_{mn} = 4$ for any other m and n (m > 0 and n > 0).

f_{mn} has the form:

$$f_{mn} = \frac{1}{ab\gamma\varepsilon_N}\frac{\beta_N\tanh(\gamma_{mn}d)+\Gamma_N}{\beta_N+\Gamma_N\tanh(\gamma_{mn}d)} \quad [5.8]$$

β_N and Γ_N can be computed recursively from:

$$\begin{bmatrix}\beta_k\\ \Gamma_k\end{bmatrix} = \frac{1}{2}\cdot\begin{bmatrix}1+\frac{\varepsilon_{k-1}}{\varepsilon_k} & (1-\frac{\varepsilon_{k-1}}{\varepsilon_k})e^{2\theta_k}\\ (1-\frac{\varepsilon_{k-1}}{\varepsilon_k})e^{-2\theta_k} & 1+\frac{\varepsilon_{k-1}}{\varepsilon_k}\end{bmatrix}\times\begin{bmatrix}\beta_{k-1}\\ \Gamma_{k-1}\end{bmatrix} \quad [5.9]$$

where $\theta_k = \gamma_{mn}(d-d_k), \beta_0 = 1, \Gamma_0 = 1$ [5.10]

Then, after deriving some formal mathematical solutions of the Poisson equation, taking into account boundary conditions, we describe a more accessible approach to the problem, using a TLM method with some brief reminders. Note again that, from the beginning of our algorithm, we have worked on the reciprocal space. Specifically, this means that we also take the FFT of the currents injected into the discretized contacts.

In practice, therefore, we can choose any shape for these buried contacts, while in the few other works dedicated to this, to the best of our knowledge, all of the contacts are rectangular. This last point seems crucial to us, and we should consider silicon vias (TSV) (cylindrical, truncated, etc.) in our 3D circuits.

After a two-dimensional spatial Fourier transformation in the x and y directions, [5.5] becomes:

$$\frac{\partial^2\Phi}{\partial z^2} - k_x^2\Phi - k_y^2\Phi = 0 \quad [5.11]$$

where Φ is the potential in the frequency range and k_x, k_y (m^{-1}) are the spatial frequency variables.

The current density: $J_Z = \sigma^* E_Z$ (σ* is the complex conductivity) and $E_Z = -\dfrac{\partial \varphi}{\partial z}$;

then:

$$J_z = -\sigma^* \frac{\partial \varphi}{\partial z} \quad [5.12]$$

In the spatial frequency domain, we get:

$$J_z(k_x, k_y, z) = -(\sigma + j\omega\varepsilon)\frac{\partial \varphi(k_x, k_y, z)}{\partial z} \quad [5.13]$$

After derivative calculations of the above equation, we get:

$$\begin{cases} \dfrac{\partial \varphi}{\partial z} = -\dfrac{J_z}{(\sigma + j\omega\varepsilon)} \\ \dfrac{\partial J_z}{\partial z} = -(\sigma + j\omega\varepsilon)(k_x^2 + k_y^2)\Phi \end{cases} \quad [5.14]$$

These equations show the relationship between the potential distribution and the current density in the z-direction in the spatial frequency domain.

5.2.1.3. *Transmission line analogy for multilayered media*

In its simplest form, a transmission line is a pair of conductors connecting two electrical systems (source and charge, for example) with forward and return paths; for cases where the return path is floating, a third (or more) conductor is introduced as a grounding shield. For microwaves, these are waveguides. In our case, the circulation of electromagnetic waves and their interference, through the silicon substrate, is one of the most serious obstacles in the constant trend toward the integration of current microelectronics. In fact, the transmission line matrix (TLM) method has been established in some cases concerning substrate surfaces; TLM can be seen as a more physical interpretation of the mathematical derivations presented above.

The main strength of this method seems to be well adapted to buried contacts, regardless of their number and shape. Consider a plane wave through a multilayer medium, its plane of incidence being parallel to the <x, y = ““‘> plane, providing the same boundary conditions as those mentioned above. Next, we consider the general case in which the impedance of the line is not the same as that of the load. A wavefront A hits the Z_L charge: part of the energy is absorbed by Z_L and the remaining energy is reflected; in this case, the voltage and current wavefronts are not in phase. This reflected wave may encounter another front of incident B.

The direction of the current flow depends on the polarity of the waveform at the time of observation; if two positive directed waveforms (one forward and one reflected) meet, the current waveforms subtract, but the voltage waveforms add. Similarly, if a positive directed waveform meets a negative directed waveform, the current will add and the voltage will subtract.

The expression for the impedance is given as:

$$Z = \frac{V_{total}}{i_{t\,otal}} = \frac{V_F \pm V_R}{i_F \mp i_R} \qquad [5.15]$$

The general solution for voltage and current is given as:

$$V(z) = V_f e^{\gamma z} + V_r e^{-\gamma z} \qquad [5.16a]$$

$$I(z) = Y_c (V_F e^{-\gamma z} - V_r e^{\gamma z}) \qquad [5.16b]$$

where the propagation coefficient γ is the square root of the product Z by Y, the linear impedance and admittance, respectively,

$\gamma = \alpha + j\beta$; α: attenuation, $\beta = 360°/\lambda_{si}$

By directly deriving from the chain matrix, we get:

$$\begin{bmatrix} V_1 \\ I_1 \end{bmatrix} = \begin{bmatrix} \cosh \gamma L & Z_c \sinh \gamma L \\ Y_c \sinh \gamma L & \cosh \gamma L \end{bmatrix} \begin{bmatrix} V_2 \\ -I_2 \end{bmatrix} \qquad [5.17]$$

And from the output branch $V_2 = -Z_L I_2$, we get:

$$\begin{bmatrix} V_1 \\ I_1 \end{bmatrix} = -I_2 \begin{bmatrix} Z_L \cosh \gamma L & Z_c \sinh \gamma L \\ Z_L Y_c \sinh \gamma L & \cosh \gamma L \end{bmatrix} \begin{bmatrix} 1 \\ 1 \end{bmatrix} \qquad [5.17b]$$

where $Z_c = 1/Y_c$ (square root of the Z and Y ratio) is the characteristic impedance (electric and magnetic fields' modulus ratio), Z_L is the charge impedance and L is the distance from the load. We can extract an input impedance, Z_l ($=V_1/I_1$), function of the load impedance Z_{l+1} (l is the layer number), as:

$$Z_l = Z_c \frac{Z_{l+1} + Z_c \tanh \gamma L}{Z_c + Z_{l+1} \tanh \gamma L} \qquad [5.18]$$

This well-known (TLM) equation Z_l is directly related to [5.8], where $Z_c = \beta_N$ and $Z_L = \Gamma_N$.

In fact, beginning again from [5.7] and taking into account the boundary conditions – continuity of potential and discontinuity of the electric field if there is a surface charge at the layer interface – we program this iterative solution along the substrate depth of [5.7]. In fact, these calculations are made only once at each frequency.

5.2.2. *Simulation results by using Green/TLM*

In this work, as a test, a Bi-CMOS compatible technology 0.35 µm is first considered.

We approach the actual profile using stacked layers of uniformly doped given thicknesses, as shown in Figure 5.1(b).

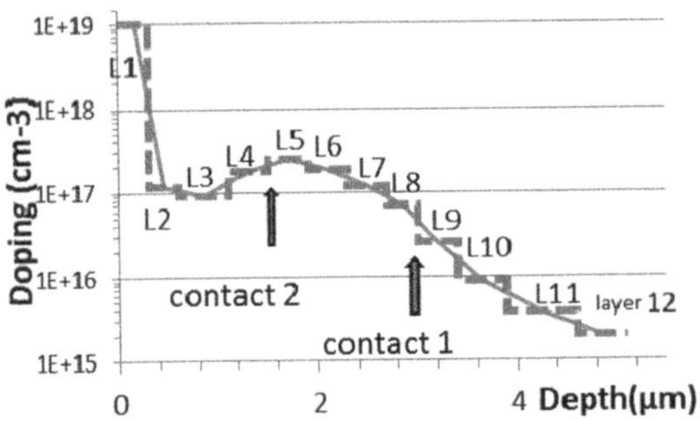

a)

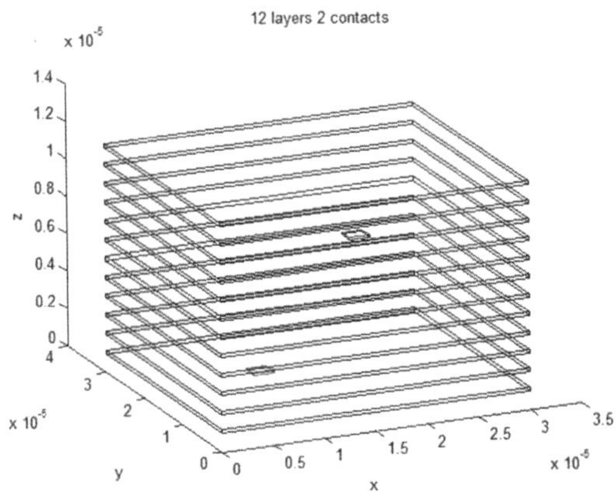

b)

Figure 5.1. *a) Specific depth profile of a 35-µm technology (p^+/p^- region); b) 3D schematics of the structure. For a color version of this figure, see www.iste.co.uk/abouelatta/smartpower.zip*

First, we carried out a number of digital experiments (not presented here) using COMSOL software, a well-known multiphysics simulator (electrical, thermal and mechanical couplings), to test its robustness and accuracy. It is also a tool dedicated to "full-wave" electromagnetic analysis. It mainly uses Galerkin-type algorithms; as a general rule, a 3D simulation can take 10 minutes or more than an hour.

Figure 5.2 synoptically shows the general algorithm for extracting the impedance of the substrate. Figure 5.3 shows the extracted voltage and impedance.

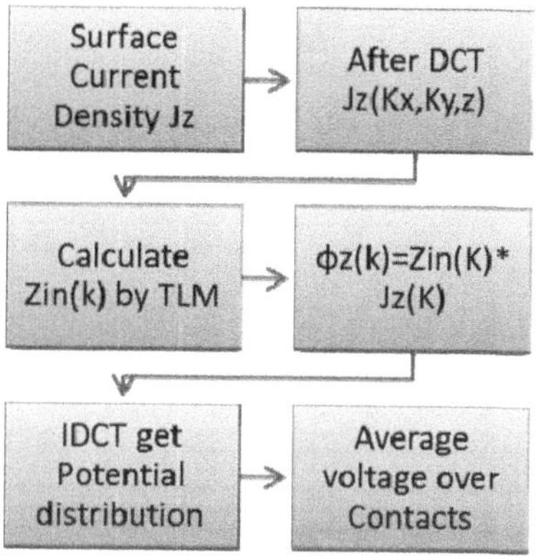

Figure 5.2. *General steps of 3D-TLE algorithm (transmission line extractor)*

– Embedded contacts

Consider the general case where the contacts (previous works exist for surface contacts) are totally buried. Then, the substrate can be considered as a real parallel connection of two separate parts (courant bypass). The current is injected on a contact of one *i*th layer so that the substrate is divided into two from this layer.

Now, it seems easy to calculate any transfer impedance by TLM. Optionally integrated into the substrate, the contacts can be introduced into any layer; they can be real (e.g. metal-like) or virtual of any form.

As mentioned previously, Figures 5.3(a) and (b) present the voltage map and the impedance module, respectively, between some Ci and Cj contacts, at the interfaces L3/L4 and L7/L8. It is typical of a Lorentzian curve (R//C, with $R \cdot C \equiv \rho\varepsilon$). From the above results, it can be seen that the comparison between TLM/Green and COMSOL is very satisfactory. The calculated results of the proposed method are consistent with those of the COMSOL simulation, while our method uses relatively less time (typically: 1/50–1/100 ratio).

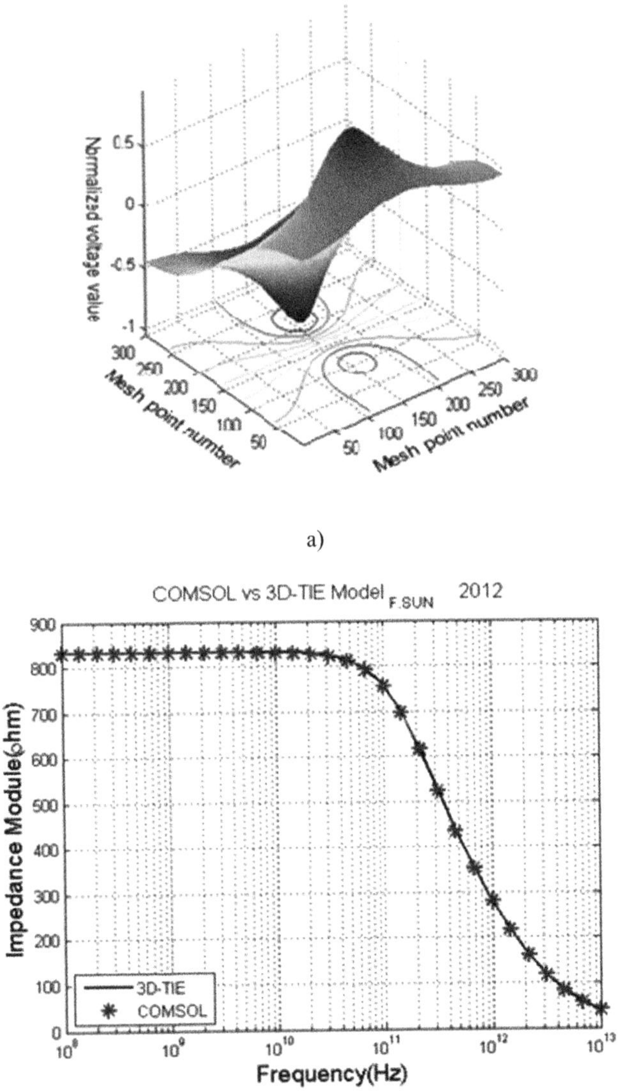

Figure 5.3. *a) Voltage map (layer interfaces: L3/L4). b) Corresponding impednce module. For a color version of this figure, see www.iste.co.uk/abouelatta/smartpower.zip*

It can be seen that the surface potential is not uniform (high voltage region). The non-uniformity is caused by an injection of uniform current density when calculating the impedance of two resistive regions, rather than two metal contacts, for instance, oxide-isolated conductive regions (e.g. pads and coplanar waveguide) and junction-isolated well regions (e.g. regions of n wells on a p-type substrate). In these cases, the substrate plates could not be considered equipotential.

The calculations obtained with TLM/Green implemented in MATLAB and compared with COMSOL were found to be virtually equal. Indeed, in MATLAB we choose layers like perfect parallelepipeds (layers of uniform thickness). We are fully aware that the quasi-electrostatic modeling framework cannot work as perfectly up to 10 THz (compared to experiments), which implies a wavelength of about a few tens of micrometers in silicon. This wavelength is 10 orders smaller compared to the contact–contact distance (3 μm) used; in this work, we use a quasi-static hypothesis, not full wave Maxwell's equations.

Next, Figure 5.4 shows a wobulation frequency of up to 10 THz, as we want to know how to simulate parasitic effects at very high frequency; the typical RC model still works for the substrate in our case with validation of FEM results, but which are no longer physically valid; we should use a full wave method for very high frequency (>10 GHz). However, there is still a long way to go to compare the FEM method (COMSOL) and our calculations. In addition, for cases of TSV, we need to be more careful; we want to establish a valid compact model for a frequency >10 GHz. That is why we do the optimization for the compact TSV model, because at high frequency (>10 GHz), if we keep the same RLCG model, adjusting their values, we can always adapt to the measurement data (up to 10 GHz).

We have not explicitly taken into account permeability because it is not realistic to go beyond 20 GHz in this analysis (which is not so bad). However, we think it is interesting to compare the FEM, in order to test the strength of our algorithm. In our simulator, using MATLAB, we use a matrix formalism, extending the impedance (Z) and current transmission (T) from one layer to its adjacent layer, from a k layer

$$\begin{bmatrix} Z_{k,l+1} \\ T_{k,l+1} \end{bmatrix} = \begin{bmatrix} A_k & -B_k \\ C_k & -D_k \end{bmatrix} \begin{bmatrix} Z_{k,l} \\ T_{k,,l} \end{bmatrix} \qquad [5.19]$$

where T deals with the diversion of the current (between two parallel regions). Based on the transmission line method, a substrate extractor is developed, as illustrated in Figure 5.4.

$$\mathrm{V} = \frac{\sum_{n=1}^{N} j\omega c \varphi_n}{\sum_{n=1}^{N} j\omega c} = \frac{1}{N} \sum_{n=1}^{N} \varphi_n \qquad [5.20]$$

where φ_n is the voltage of the substrate at a node and N is the number of contact sub-areas. Therefore, in our modified algorithm, the contact voltage is then calculated by averaging.

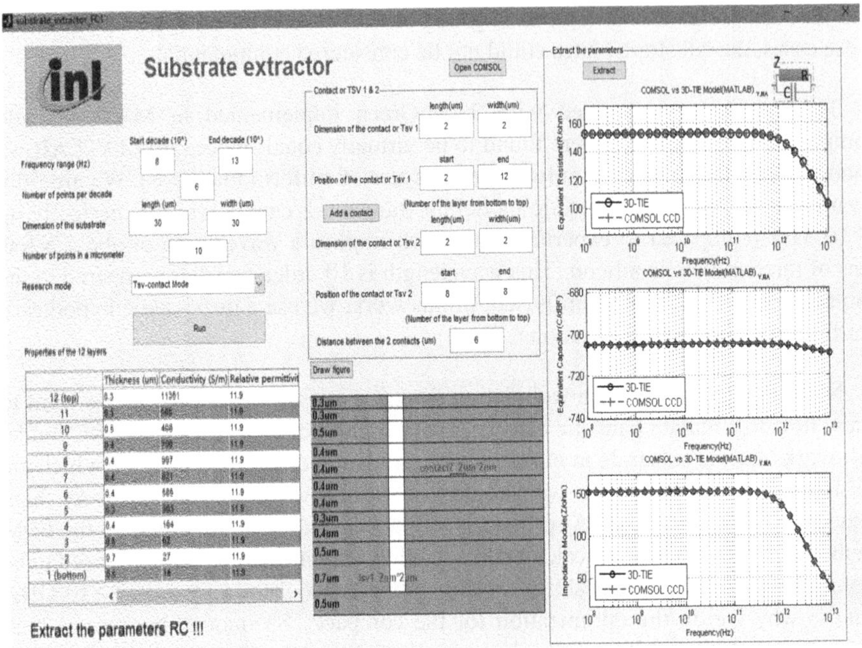

Figure 5.4. *3D-TLE substrate extractor, e.g. Impedance Z17, capacity and resistance C17 and R17 extracted between a contact and a TSV (valid for a frequency below 20 GHz and quasi-static hypothesis). For a color version of this figure, see www.iste.co.uk/abouelatta/smartpower.zip*

5.2.2.1. *Simulation and optimization of two TSVs U-model: a key piece*

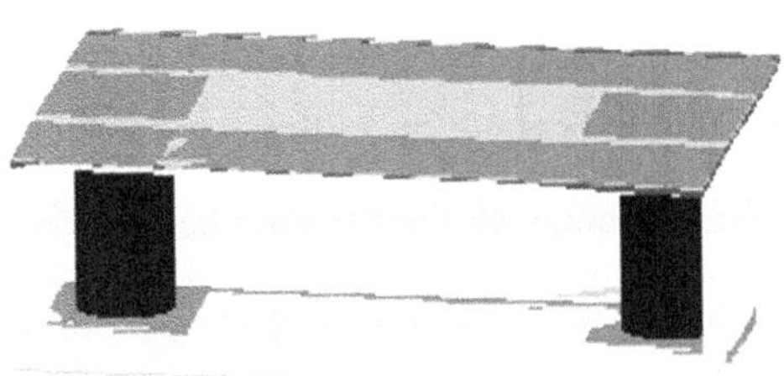

Figure 5.5. *Model geometry of two TSVs. For a color version of this figure, see www.iste.co.uk/abouelatta/smartpower.zip*

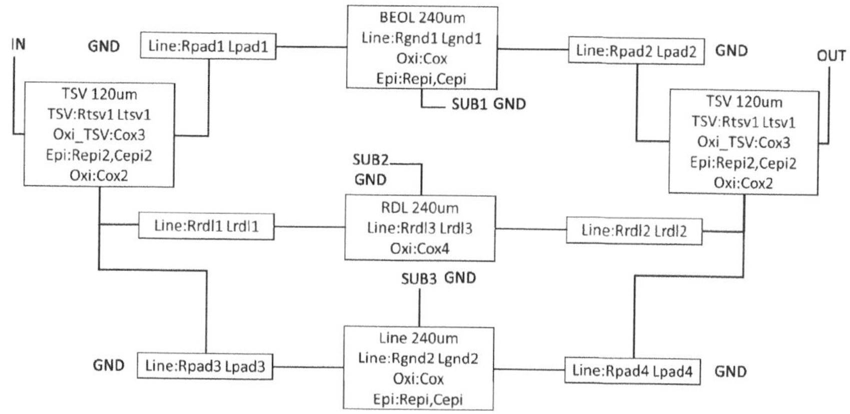

Figure 5.6. *Input script description*

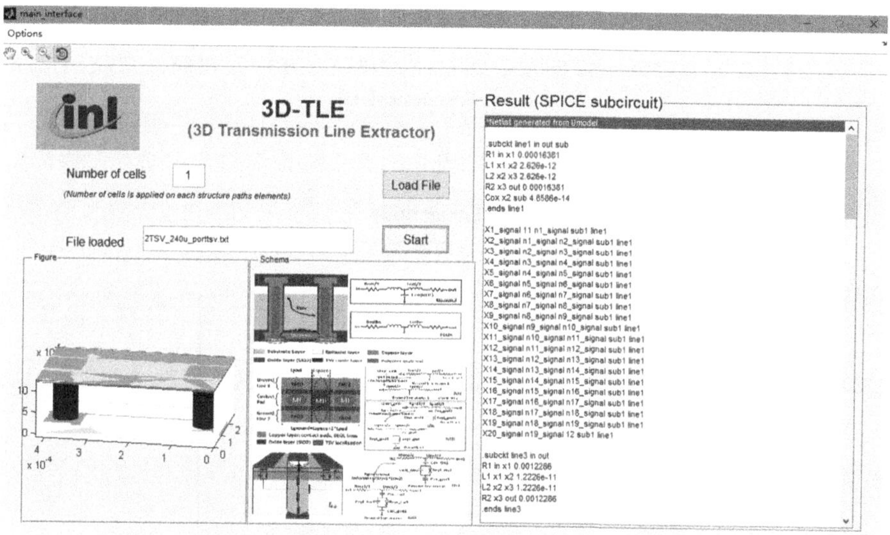

Figure 5.7. *3D-TLE extractor setting. For a color version of this figure, see www.iste.co.uk/abouelatta/smartpower.zip*

The parasitic parameters of the substrate can be expressed by a parallel of capacity and conductance. And their expressions are indicated as follows:

$$C_{Si\_sub} = \frac{\pi \times \varepsilon_0 \varepsilon_{r,si}}{cosh^{-1}(\frac{p_{TSV}}{d_{TSV}})} \times h_{TSV} [F] \quad [5.21]$$

$$G_{Si\_sub} = \frac{\pi \times \sigma_{si}}{cosh^{-1}(\frac{p_{TSV}}{d_{TSV}})} \times h_{TSV}[S] \qquad [5.22]$$

Here, when we extract the measurements, we must apply the de-coating method (or peeling, de-embedding); Figures 5.8(a) and (b) show the difference with and without de-coating.

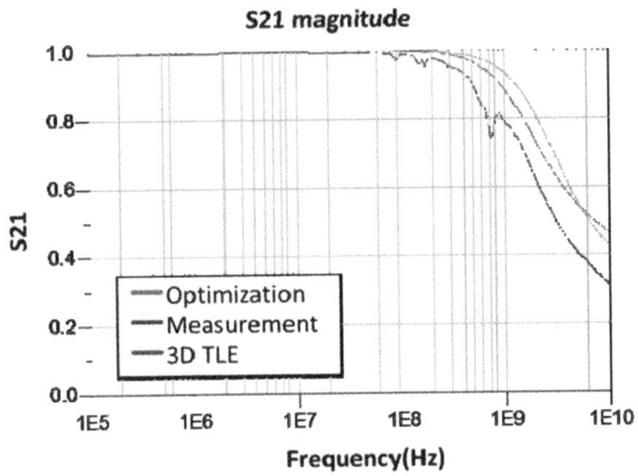

a)

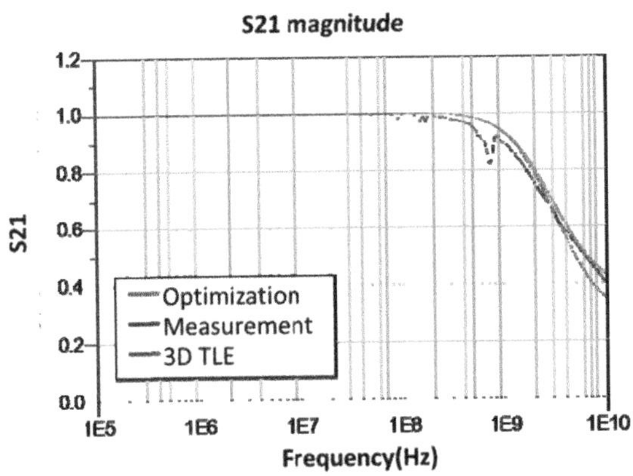

b)

Figure 5.8. *2xTSVs model S parameters: a) S21 magnitude without de-embedding; b) S21 magnitude with de-embedding. For a color version of this figure, see www.iste.co.uk/abouelatta/smartpower.zip*

As we can see in both figures, the result with de-coating is closer to our simulated circuits.

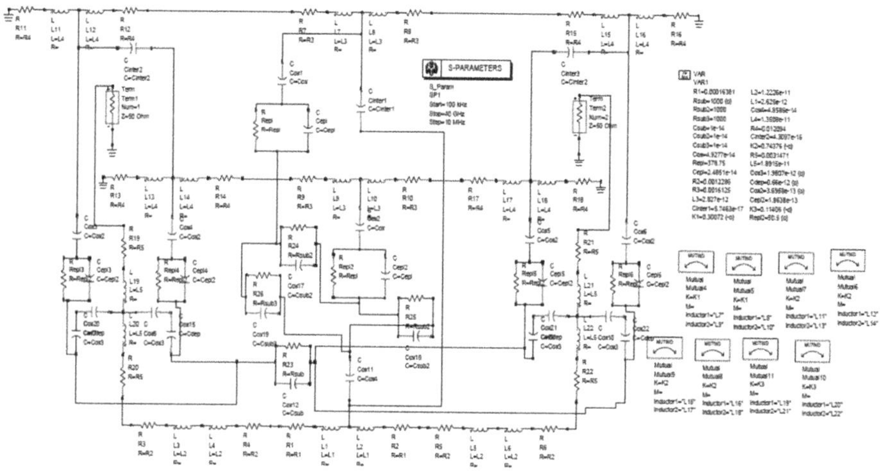

Figure 5.9. *S parameters study for the 2xTSVs (ADS) model*

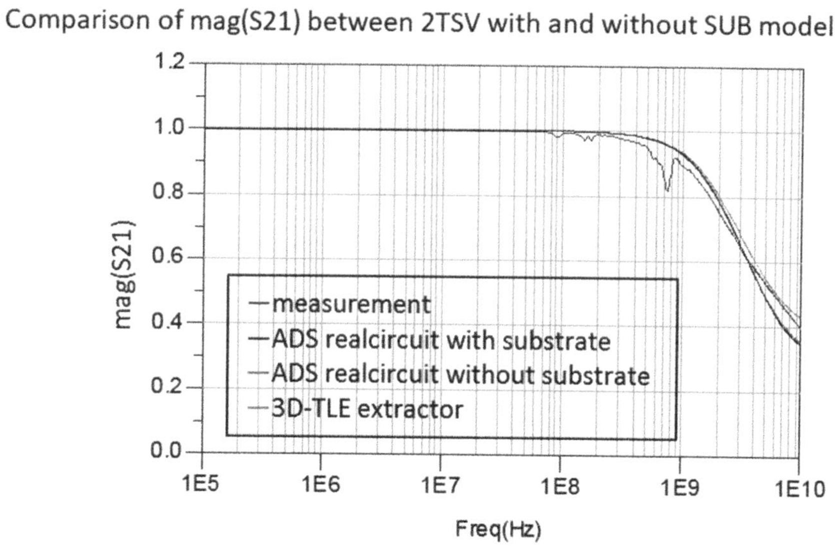

Figure 5.10. *Comparison of S_{12} between 2TSV with and without SUB model: measurement (blue), optimized circuit result without SUB resistance (black), optimized circuit result with SUB resistance (red) and 3D-TLE result (pink). For a color version of this figure, see www.iste.co.uk/abouelatta/smartpower.zip*

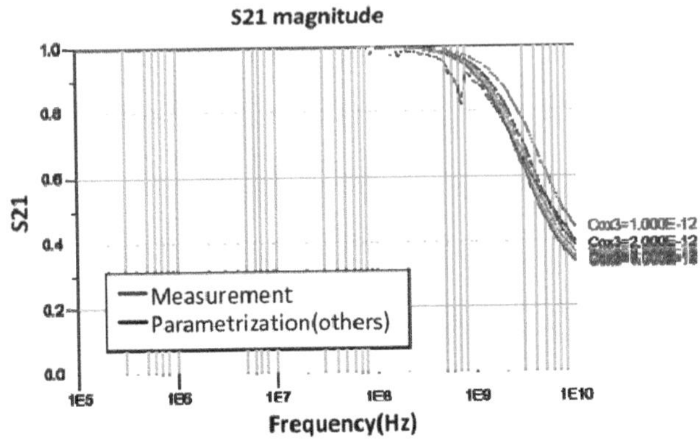

a) Change Cox3(TSV oxide)

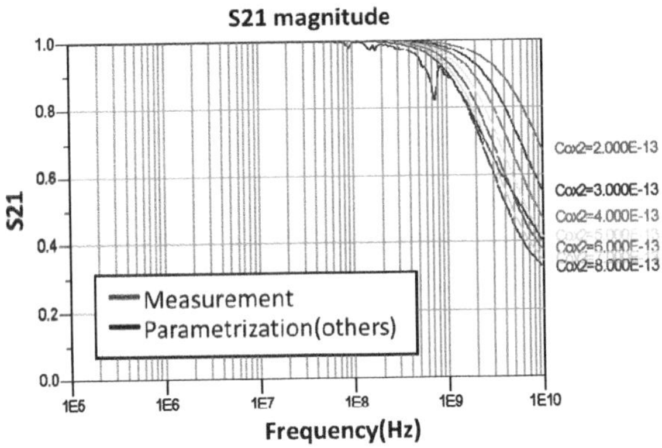

b) Change Cox2(oxide layer capacitance close to TSV)

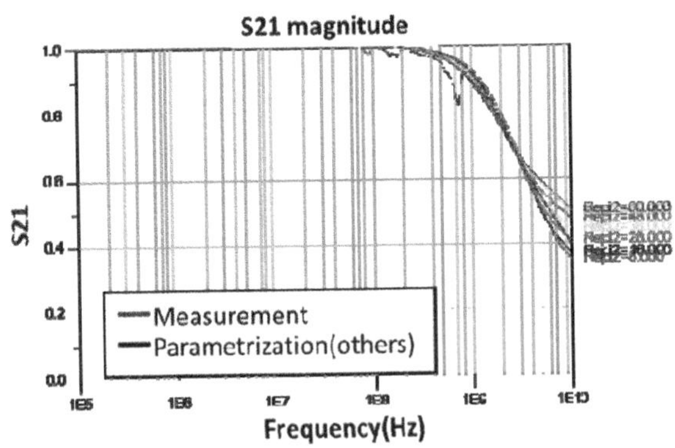

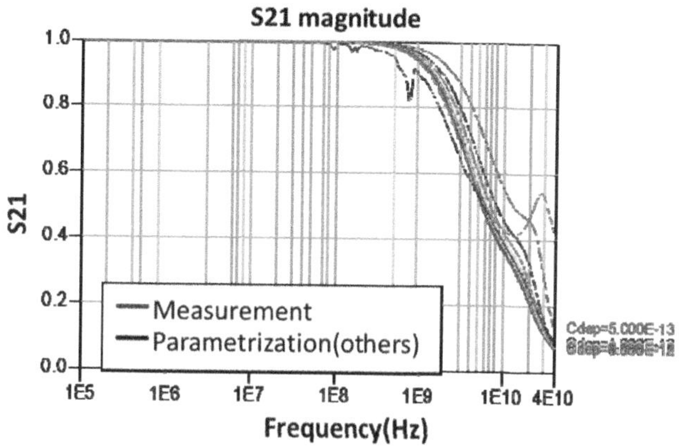

Figure 5.11. *S_{21} collision/distribution/scattering parameters. For a color version of this figure, see www.iste.co.uk/abouelatta/smartpower.zip*

5.2.2.2. *Heat equation*

A common problem encountered in the thermal analysis of electronic packaging, or in 3D ICs, is that of thermal resistance diffusion. Thermal spreading resistance occurs when heat flows by conduction between a source and a tank with different cross-sections. An example is a multilayer chip. Another example is that of a module housing with a thermal blubber path between a chip and the encapsulation of the module. In both cases, the heat flows from the chip through a given area in the substrate or encapsulation and spreads to another area. This case is that of a (non)stable temperature due to one or more heat sources, in the latter case using the principle of superposition. A thermal spreading resistance (always in the reciprocal range) can be extracted from a discrete heat source.

$$\frac{d^2T}{dx^2} + \frac{d^2T}{dy^2} + \frac{d^2T}{dz^2} - \frac{1}{\alpha}\frac{dT}{dt} = 0 \qquad [5.23]$$

where $\alpha = \frac{k}{\rho c}$ is thermal diffusivity, k is the effective isotropic thermal conductivity and c is the isobaric mass heat. For the equilibrium state, the heat equation reduces to the Laplace equation, similar to the Poisson equation when no electric charge is present. In the same way as for the Poisson equation, we extract an associated Green kernel, that is, if we assume:

$$T = X(x, x') \cdot Y(y, y') \cdot Z(z, z')\tau(t, t')$$

[5.23] gives:

$$\frac{1}{X}\cdot\frac{d^2X}{dx^2}+\frac{1}{Y}\cdot\frac{d^2Y}{dy^2}+\frac{1}{Z}\cdot\frac{d^2Z}{dz^2}=-A^2\frac{\delta(x-x\prime)\cdot\delta(y-y\prime)\cdot\delta(z-z\prime)\cdot\delta(t-t\prime)}{\alpha} \quad [5.24]$$

Then:

$$\frac{d^2X}{dx^2}+\lambda^2X=\frac{d^2Y}{dy^2}+\mu^2Y=\frac{d^2Z}{dz^2}-\gamma^2Z=\frac{d\tau}{dt}-\alpha(\lambda^2+\mu^2-\gamma^2)\,\tau=0 \quad [5.25]$$

where λ, μ and γ are the eigenvalues from the Green kernels, with:

$$\lambda^2+\mu^2-\gamma^2+\frac{A^2}{\alpha}=0 \quad [5.26]$$

The last equation [5.26] gets an evanescent solution as a function of time; therefore, for a layer, we can write:

$$\tau_i=\tau_{i0}\cdot\exp\left(-(\lambda_{ix}^2+\mu_{iy}^2-\gamma_{iz}^2)\cdot\alpha\cdot t\right) \quad [5.27]$$

In the reciprocal domain (therefore, we can also take advantage of data compression, inherent to the DCT), [5.21] becomes:

$$\frac{d^2X}{dx^2}+\frac{d^2Y}{dy^2}+\frac{d^2Z}{dz^2}-\frac{1}{\tau}\cdot\frac{d\tau}{dt}=-\lambda^2X-\mu^2Y+\gamma^2Z-\left(\frac{j}{\alpha}\right)\omega\tau=0\ (j^2=-1) \quad [5.28]$$

We can define the Fourier coefficients of a spread impedance, transcendental, as the following ratio:

$$R_{FCSi}=A_i/B_i \quad [5.29]$$

where Ai and Bi are the Fourier coefficients of the *i*th general solution for temperature, in the direction z perpendicular to the stacked layers:

$$FCS_i=A_i\cdot\cosh(\gamma z)+B_i\cdot\sinh(\gamma z) \quad [5.30]$$

With regard to boundary conditions, they can be formulated as follows: the continuity of thermal conductivity at the level of the interface layers leads to

$$k_i\,\frac{dT_i}{dz_i}=k_{i+1}\frac{dT_{i+1}}{dz_{i+1}}\ \text{(Neumann condition)} \quad [5.31]$$

At an interface

$$X_i(t)=X_{i+1}(t), Y_i(t)=Y_{i+1}(t)\ \text{(Dirichlet condition)} \quad [5.32]$$

$$k_i \frac{T_{i,z}}{T_{i+1,z}} = k_{i+1} \frac{R_{FCDS_{i,z}}}{R_{FCDS_{i+1,z}}} \quad [5.33]$$

And

$$\varepsilon_i E_i = \varepsilon_{i+1} E_{i+1} \text{ (E: Electrical field)} \quad [5.34]$$

If [5.31] and [5.33] are used, we get:

$$R_{FCS_{i,z}} \cdot Y_{i(y)} \cdot e^{-(-\gamma_{i,z}^2 + \mu_{i,y}^2 + \lambda_{i,x}^2)t} = -B_{i,z} \cdot R_{FCS_{i+1,z}} \cdot Y_{i+1(y)} \cdot e^{-(-\gamma_{i+1,z}^2 + \mu_{i+1,y}^2 + \lambda_{i+1,x}^2)t} \quad [5.35]$$

$$k_{i,z} \cdot R_{FCS_{i,z}} \cdot Y_{i(y)} \cdot e^{-(-\gamma_{i,z}^2 + \mu_{i,y}^2 + \lambda_{i,x}^2)t} = -k_{i+1,z} \cdot B_{i+1,z} \cdot \gamma\lambda R_{FCS_{i+1,z}} \cdot Y_{i+1(y)} \cdot e^{-(-\gamma_{i+1,z}^2 + \mu_{i+1,y}^2 + \lambda_{i+1,x}^2)t} \quad [5.36]$$

We have the same set – [5.34] and [5.35] – to replace Y by X. And [5.35] and [5.36] are verified if:

$$\frac{B_{i,z}}{B_{i+1,z}} = \frac{R_{FCS_{i+1,z}}}{R_{FCS_{i,z}}} \quad [5.37]$$

$$\frac{k_{i,z}}{k_{i+1,z}} = \frac{R_{FCS_{i,z}}}{R_{FCS_{i+1,z}}} \quad [5.38]$$

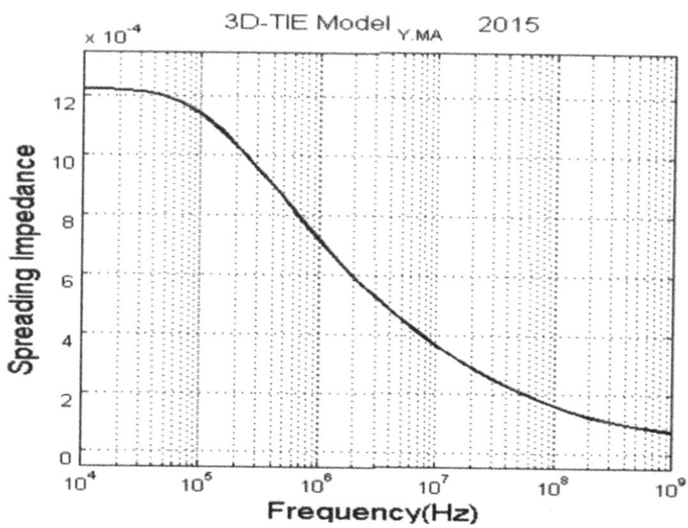

Figure 5.12. *Spreading impedance of the temperature*

$$\gamma_{i+1,z} = \left[\frac{\alpha_i}{\alpha_{i+1}} \cdot \gamma_{i,z}^2 + \left(\frac{\alpha_i}{\alpha_{i+1}} - 1\right) \cdot (\lambda_{i,x}^2 + \mu_{i,y}^2)\right]^{\frac{1}{2}} \qquad [5.39]$$

Then, we obtain:

$$A_{iz} = B_{iz} \cdot R_{FCDS_{i,z}} \cdot (\gamma_{i,z}) \qquad [5.40]$$

$$R_{FCS_{i,z}}(\gamma_n) = \frac{k_{i+1}R_{FCS_{i+1,z}}(\gamma_n)+k_i \cdot tanh(\gamma_{i,z} \cdot t_i)}{k_i \cdot R_{FCS_{i+1,z}}(\gamma_n) \cdot tanh(\gamma_{i,z} \cdot t_i)} \qquad [5.41]$$

γ_n is associated with the eigenvalues of [5.26]; t_i is the thickness of the ith layer.

Using this low-pass filter, setting a maximum frequency should allow us to examine the behavior of the heat flow, for example, through the interface of each bilayer for a frequency range. Furthermore, by taking the inverse Fourier transforms, we could compare with the direct calculation of the time-dependent thermal profile (see Figure 5.13).

Finally, the real solution of [5.21] can take the following form:

$$\Delta T\ (x, y, x', y', z, z') = T - T_\infty = T_0 \cdot exp\left(-\left(\lambda_{i,x}^2 + \mu_{i,y}^2 - \gamma_{i,z}^2\right) \cdot \alpha \cdot t\right) \cdot \sum_{m=1}^{\infty}\sum_{n=1}^{\infty}\left[\frac{2\left(A_N e^{-\gamma_{mn}(d+z_l)} + B_N e^{\gamma_{mn}(d+z_l)}\right)\left(e^{-\gamma_{mn}z_h} + e^{\gamma_{mn}z_h}\right)}{ab_N\gamma_{mn}\left(B_N e^{d} - A_N e^{-d}\right)}\right] \cos\left(\frac{mx}{a}\right) \cos\left(\frac{ny}{b}\right) \qquad [5.42]$$

where T_∞ is the stationary temperature.

We then built our simulator, taking into account the analogies between the electric and thermal models in the first instance. Specifically, if the calculation of all pertinent variables (voltages/temperature, current/heat flow density, noise power/current fluctuations) is performed in the reciprocal space and frequency domain, in the main program, we develop a single special subroutine to calculate the impedance (electric or thermal, or transfer impedance) for noise extraction (between any two points in the volume). In this work, we take 300 points × 300 points, and 12 layers. The first results are accessible for the temperature from both the FEM algorithm and our method, where the temperature source profile is applied to one or two contacts is shown in Figure 5.13(a) and the temperature map is shown in Figure 5.13(b) (FEM results (left) and Green's results (right)).

We believe these modeling technologies can be effective for the relevant industry, for example, 3D electrical simulations. We can also look for hot spots, involving drastic transient regimes, using the thermal approach. We should work on the placement routine of chillers via studies on heat propagation in three dimensions. *Ab initio* electrothermal coupling can be initiated by a simple linear formula such as:

$$R = R_0(1 + \alpha_0 \Delta T) \qquad [5.43]$$

Finally, we should start studying noise phenomena, fundamental for the nanoscale, for example, thermal noise, 4kRT, electric and thermal field coupling, or generation-recombination (GR), as well as their coupling.

Figure 5.13 presents a modeling of thermal phenomena for 3D circuits.

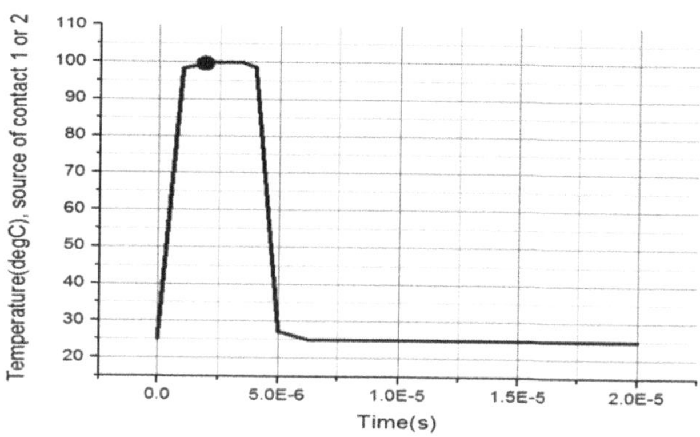

a) Imput temperature pulse

COMSOL (FEM)

MATLAB (Green Kernels/TLM)

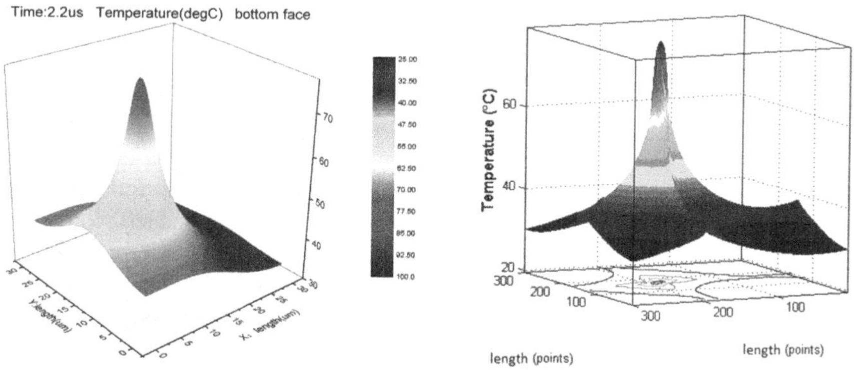

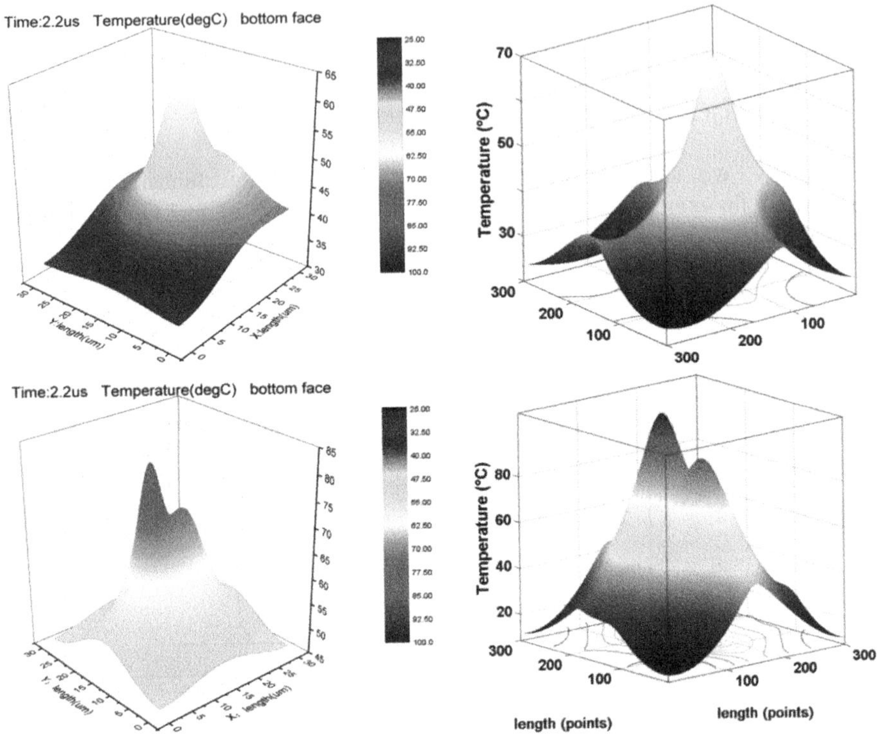

b)

Figure 5.13. *a) Temperature command profile; simulation date: 2.2 µs. b) Temperature maps: (left) FEM results and (right) "Green results"; t = 2.3 µs (see a)). For a color version of this figure, see www.iste.co.uk/abouelatta/smartpower.zip*

5.2.3. *Thermal management in a 3D-integrated figure*

Improvements in IC technology are driven by the need to increase performance and functionality, while reducing weight, size, energy consumption and manufacturing costs. As Gordon Moore forecast in his inspiring paper, reducing the size of the feature also reduces the surface area of the chip, thereby improving production while also reducing the cost per function. Scaling laws have shown that improved device and processor speed can be achieved through dimensional scaling. However, all paradigma ultimately have limits, and Moore's law is no exception. The limits of scaling Moore's law came simultaneously from several directions. Lithographic limitations made integration extremely difficult: more features on a semiconductor chip and the most advanced lithographic techniques needed to scale are becoming prohibitively expensive for most fabrication lines.

A diagram of an optical lithography system is shown in Figure 5.14. The optical projection systems used have very complex multi-element lenses that correct virtually all common aberrations and operate at the diffraction limit. The resolution of a lithography system is usually stated in terms of wavelength and numerical aperture (NA) as follows:

$$\text{Resolution} = k_1 \frac{\lambda}{NA}$$

where the constant k_1 depends on the process used. In the manufacturing of ICs, typical values of k_1 range from 0.5 to 0.8, with a higher number reflecting a less rigorous process. The NA of optical lithography tools ranges from about 0.5 to 0.6. Thus, the typical rule of thumb is that the smallest features that can be printed are roughly equal to the wavelength of the light used. Historically, improvements in the resolution of IC lithography have been driven by decreases in print wavelength, as shown in Figure 5.15. Lighting sources were initially based on mercury arc lamps filtered for different spectral lines. Figure 5.14 shows the progression from line G at 435 nm to line I at 365 nm. This was followed by a switch to excimer laser sources with KrF (krypton fluoride) at 248 nm and, more recently, ArF (argon fluoride) at 193 nm.

The most advanced manufacturing of ICs currently uses KrF technology with the introduction of ArF tools from 2001. It can also be seen from Figure 5.15 that the progression in the minimum size of ICs is on a much steeper slope than that of the lithography wavelength. Before the introduction of KrF lithography, the minimum sizes of printed features in practice were greater than the wavelength, with the crossing at the generation 250 nm and KrF. With the introduction of 180 nm technology in 1999, the most advanced manufacturing of ICs was carried out with characteristic sizes significantly below the wavelength (248 nm).

In addition, short-channel effects and random fluctuations make the geometries of conventional planar devices obsolete. Interconnection becomes a more important limiting factor for power dissipation and chip performance. The wires get closer to one other and the length of the interconnect increases due to the larger size of the matrix, as the size of the entities decreases. The capacity and resistance of the interconnect increases, while the parasites of the device decrease as the wires get closer to each other and the thickness of the wire does not decrease at the same scale as the size of the component decreases. Finally, the fact that scaling has taken place without appreciable voltage reduction over the past decade has amplified power densities to the precipice of cooling and limits of reliability.

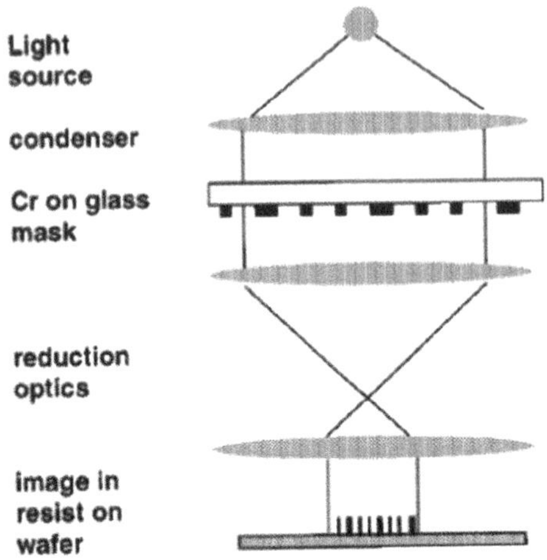

Figure 5.14. *Diagram of the optical lithography tool step by step and repeated. The pattern on the mask usually represents a level of a chip (or sometimes a few). The mask pattern is printed on the wafer; then the wafer is moved to a new location and the process is repeated until the entire wafer is exposed. Real optical projection systems have 25 or more lens elements for aberration compensation and field flattening to achieve diffraction-limited performance*

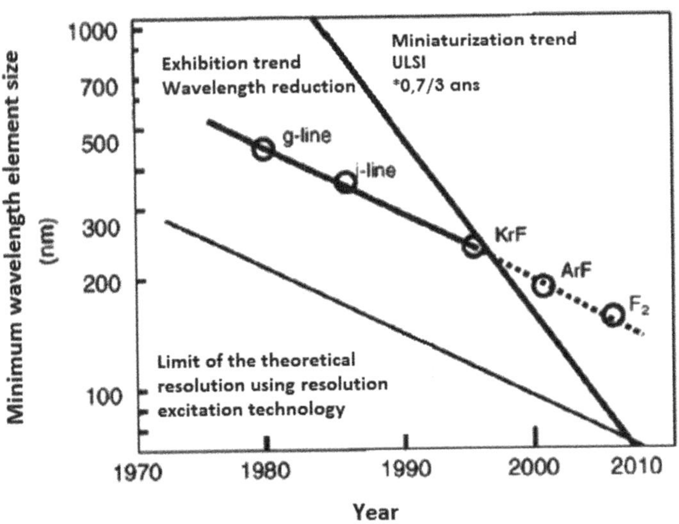

Figure 5.15. *Comparison of lithography wavelength trends with characteristic sizes of ICs elements*

Three-dimensional integration technology has emerged as an effective means of improving performance. Compared to conventional 2D schemes, the potential advantages of 3D integration are numerous. They include the following: (1) a shorter interconnection: reducing the length of the interconnect wire, which translates into improved performance and reduced power consumption; (2) heterogeneous integration: circuit layers can be built using different processes, or even on different types of wafers. This means devices can be optimized to a much higher degree than if they were built together on a single wafer. In addition, components whose manufacture is incompatible could be combined into a single 3D IC; (3) power: shorter wires also reduce power consumption by producing fewer parasitic capacities. Reduced energy budget results in reduced heat production, extended battery life and lower operating costs; and (4) higher encapsulation density and smaller footprint: due to the addition of a third dimension to the designs of conventional two-dimensional masks, and potentially results in a lower cost design.

However, one of the main concerns in the adoption of 3D technology is the increase in power densities that can result from placing one power-hungry block on top of another in the multilayer 3D stack. Since the increase in power density and the resulting thermal impact are already major concerns in 2D ICs, the switch to 3D ICs could accentuate the thermal problem due to the increase in power density, resulting in higher on-chip temperatures. A high temperature has negative effects on the performance of the circuit. The interconnection delay becomes greater while the driving force of a transistor decreases with increasing temperature. Increasing on-chip temperature can even lead to thermal runaways.

Therefore, it is very important to model thermal behaviors in 3D ICs and to study possible solutions to mitigate thermal problems, in order to take full advantage of the benefits offered by 3D technologies. Figure 5.16(a) shows a conventional planar IC modeled in five layers: metal layers (A), active silicon (B), bulk silicon (C), heat spreader (D) and heat sink (E). The heat sink is attached to bulk silicon with a thermal interface material. Figure 5.17(b) shows a two-die 3D IC constructed with two plane matrices stacked with their metal layers face to face (F2F). Figure 5.17(c) shows a five-die 3D IC that combines two 2-matrix F2F stacks with a B2B interface between the matrix pairs. With such a stack, the 3D IC stacked with two matrices (five dies) will have about half (quarter) of the footprint reliability.

Figure 5.17 displays a typical base processor. This baseline assumes a 70 nm technology with a supply voltage of 0.9 V and copper metallization with eight layers.

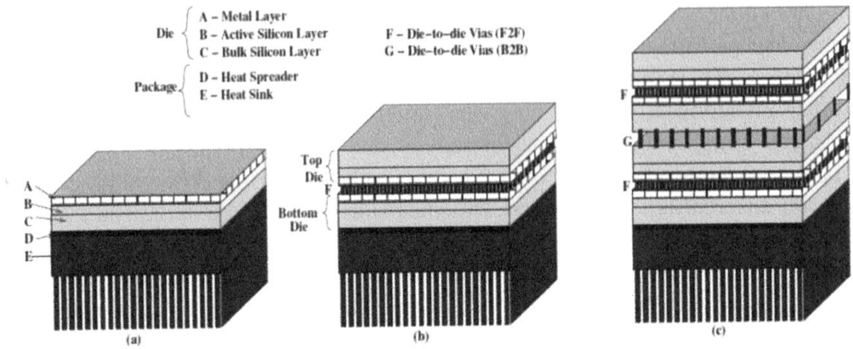

Figure 5.16. *a) A planar IC; b) a 3D IC with two dies; c) a 3D IC with five dies*

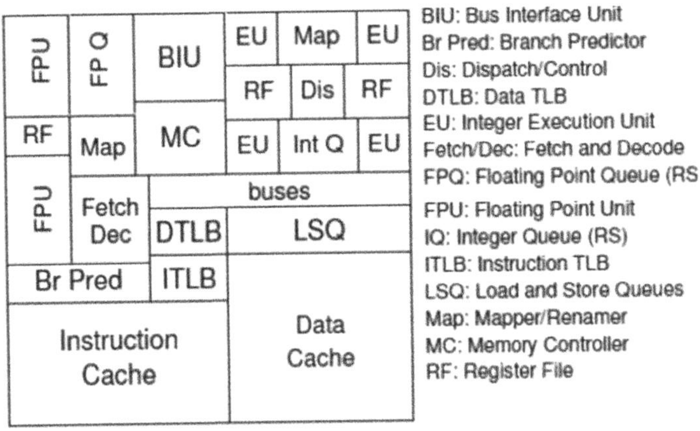

Figure 5.17. *Typical baseline processor*

Stacking functional blocks of the basic processor, as shown in Figure 5.17, results in high temperatures in each block as stacking matrices increase. Figure 5.18 shows the die temperatures of integrated planar and 3D circuits. The temperature equivalent to two matrices and five matrix 3D is 320.5 K (12.6 K increase over planar IC) and 335.1 B (increase of 26.2 K in IC planes).

Keep in mind that the rough power density ingestion experienced by 3D ICs only led to a moderate increase (12.6 K and 26.2 K for two-die and five-die ICs, respectively) in maximum temperatures on the chip.

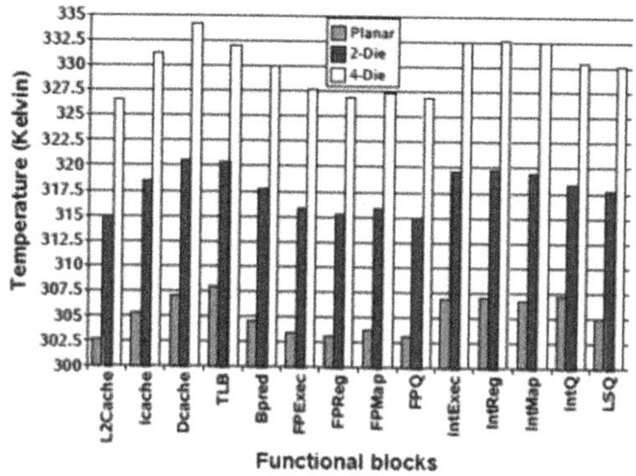

Figure 5.18. *Temperatures of functional blocks on planar and 3D ICs. For a color version of this figure, see www.iste.co.uk/abouelatta/smartpower.zip*

5.2.3.1. *Thermal modeling*

In order to properly analyze the thermal management of 3D ICs, thermal design guidelines would ideally be provided to address the limitations of 3D ICs without becoming thermally inaccessible. For instance, chip and encapsulation designers would benefit from knowing the location of the maximum temperature in a given multi-chip stack and its sensitivity to various design parameters. In the steady state, the thermal conduction in the substrate of the chip can be understood from the following differential equation:

$$K\left(\frac{\partial^2 T}{\partial x^2}+\frac{\partial^2 T}{\partial y^2}+\frac{\partial^2 T}{\partial z^2}\right)+Q(x,y,z)=0 \qquad [5.44]$$

where T is the temperature, K is the thermal conductivity and Q is the heat generated per unit volume. A single solution exists when convective, isothermal and/or insulating boundary conditions are correctly applied. The nature of the encapsulation and heat sink determines the boundary conditions. FEA or FDA methods were grasped for temperature calculations in these experiments.

5.2.3.2. *FEA background*

The FEM is one of the mathematical tools applied, used in numerical analysis; it is the most suitable for numerically solving partial differential equations (PDEs), and here it is used for complex geometries. First applied to structural mechanics problems, today it is used for most practical problems that are linear or nonlinear, stationary or time-dependent, defined in a one-, two- or three-dimensional domain, in order to represent the behavior of different physical, mechanical, thermodynamic, acoustic or electronic systems.

More specifically, it makes it possible, for example, to numerically calculate object behavior, even for complex ones, provided that they are "continuous" and described by a linear PDE: the motion of a cord shaken by one of its ends, the behavior of a rapidly moving fluid on an obstacle, the deformation of a metallic structure, etc.

Here, we are referring to a discrete mathematical algorithm that makes it possible to find an approximate solution for a PDE in a domain that has edge and/or internal conditions.

1) Discretize the continuum

In most cases, equations are impossible to solve on a large domain. Thus, we carry out an approximation to facilitate integral calculation.

The form of the mesh is a vital element in this method as it determines the points on which we calculate the function. If the mesh form is not adapted ("bad" form or not dense enough), we will not have an accurate solution. For example, if we try to discretize mountainous terrain to learn about it, but do not choose the mountain peaks as mesh nodes and the valleys as the lowest points, we will end up with an incorrect representation of this terrain and any study on this representation will be inaccurate. Thus, it is not necessary for the mesh to be regular and we tend to tighten the places of interest on the mesh (for instance, places where we think the solution will vary considerably).

Each element is made up of a number of nodes (Figure 5.19).

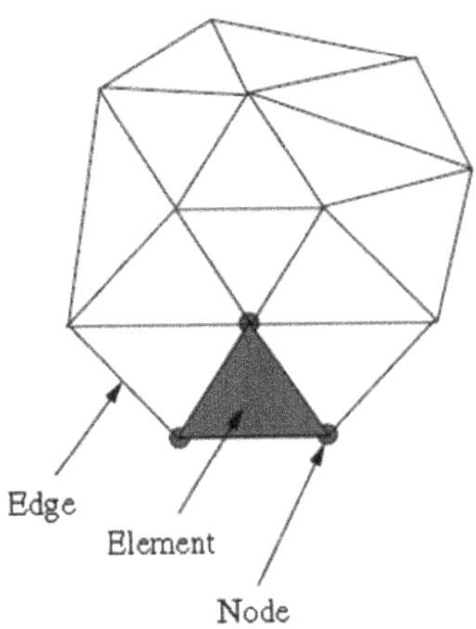

Figure 5.19. *Typical finite element mesh: elements, nodes and edges*

The number of nodes used to form an element depends on the type of element (or the interpolation function).

2) Select interpolation or shape functions

The next step is to select the type of nodes for an element; the nature and number of unknowns in each node determines the variation of a field variable in the element.

3) Form element equations (formulation)

Next, we must determine the matrix equations that discern the properties of each element by forming a left hand side (LHS) matrix. For example, a typical LHS matrix and a charge vector can be written as:

$$[\mathrm{k}]_{\mathrm{e}} = \frac{AK}{l}\begin{bmatrix} 1 & -1 \\ -1 & 1 \end{bmatrix} \quad [5.45]$$

$$\{\mathrm{f}\}_{\mathrm{e}} = \begin{Bmatrix} Q_i \\ Q_j \end{Bmatrix}$$

where the index e represents an element, Q is the total heat transferred, k is the thermal conductivity, l is thc lcngth of a one-dimensional linear element, and i and j represent the nodes forming an element. The unknowns are the temperature values on the nodes.

4) Assembling the element equations to obtain a system of simultaneous equations

When solving an equation system for all nodes, it is not sufficient to only consider one insulated element. In a finite element mesh, a node generally belongs to several elements and the load on each one of the elements results in the f value in the node.

The value of the function is calculated element by element; yet, as peaks are common to several elements, all local systems must be assembled into a one large system containing the equations of all nodes.

To construct this system, we will make use of the fact that the initial differential equation must be checked on each element, which means that we will have equations in each element peak. The equation on each peak depends on the values taken from the neighboring peaks and the partial derivative.

When we put the equations of each peak together, we get a system where the unknowns are values of the functions on mesh nodes (peaks) (generally, n unknowns and n-x equations, x being the size of E).

5) Solving the system of equations

Here, we apply boundary conditions that lead to one more new equation, making it possible to solve the system.

We determine the f_i, f values at each node, enabling us to trace the function using approximation.

When we have the f_i, all we have to do is carry out linear interpolation of these f_i to get f (approximate values) at each point.

If we do not tighten the mesh enough (in order to be able to manually calculate in a relatively less time), we will have huge gaps between the real solution and the approximated one. But if we only want values of the function at specific points in the domain, then this simple mesh will be enough. The power of computers now available makes it possible to use highly tightened meshes, which need expensive calculations and highly acceptable results (using some "Method error").

An important advantage of finite elements is the flexibility in mesh choices: element form, position and number of nodes, etc., as long as we tighten the mesh on specific points (or on peaks). Thus, a contour of some form can be approached from as close as we want, using a set of polygons. Using a non-uniform mesh is not a problem, given that integral calculations are carried out simultaneously on an element. So, we can tighten or loosen the mesh at desired places to increase precision or eliminate useless calculations. The only disadvantage of this method is the complex programming or computation involved. The time spent decrypting, creating or even adapting a program for finite element solving is huge, and must be adaptive to various boundary conditions in order to efficiently use it.

6) Calculate secondary quantities

From the nodal values, we can compute secondary quantities, for instance heat flows.

5.2.3.3. *FDA background*

The finite difference approach is very simple. The model, which is formed by writing the difference equations for a set of net points, can be improved by increasing the number of points. It has been adapted for rectangular domains as opposed to complex geometries, although many heat transfer problems can be solved with FDMs.

In the FDM, a theory of heat transfer is adopted, which constructs an equivalent thermal circuit through the thermal-electrical analogy. The equilibrium equation represents the network with thermal resistors connected between the nodes. The voltage and temperature at the nodes can be computed solving the system. The mass node is considered as a node of constant temperature, often the ambient one. Figure 5.20 shows, as an example, a simple thermal model for a chip.

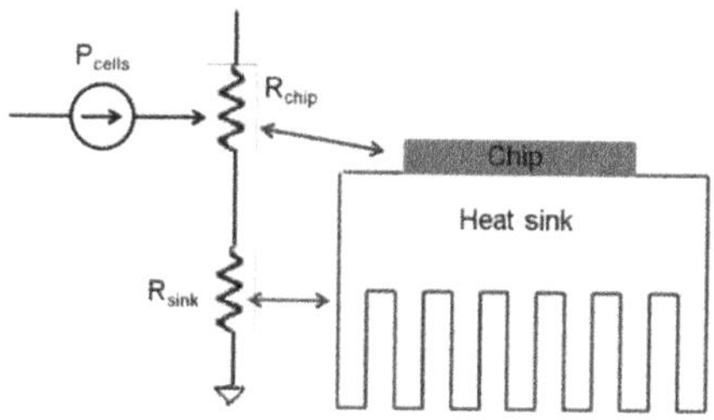

Figure 5.20. *A simple thermal model for 3D IC chip*

5.2.3.4. *Thermal physical design*

As one of the main concerns of 3D ICs is the increase in power density, due to the placement of one thermal critical block on top of another in the 3D stack, temperature-sensitive physical design methods should be developed to solve thermal problems.

5.2.3.5. *Thermal floor planning*

The 3D stage planner is used to place the blocks in layers. It partitions the blocks into appropriate layers, trying to minimize the difference in compressed area between the layers and minimize the total length of the wire using disturbance methods.

5.2.3.6. *Thermal placement*

The 3D thermal position problem is defined as follows: due to a circuit containing a set of cell instances and a set of networks, the number of device layers and placement area per layer, and the placement of cells in this area, look for a location for each cell so that the objective function of the weighted total wavelength is reduced under certain constraints, such as non-overlapping restrictions.

5.2.3.7. *Thermal routing*

The steering problem is defined given the design base, thermal conductivity of each physical layer, the result of a 3D mode or floor chart with a dedicated white area for delivery, a certain maximum temperature and circuit steering, so that the weighted cost of wire length and total number of TVs is minimized, taking into account the design rules.

5.2.3.8. *Thermal planning of layers*

The 3D scheduler is used to place the blocks in layers. It partitions the blocks into appropriate layers, trying to minimize the difference in the compressed area between the layers and the total length of the wire using all disturbance operations.

5.2.4. *Thermo-mechanical design*

Chips stacked on 3D architecture increase packaging density and thermal resistances, resulting in higher on-chip temperatures. Many studies have been carried out focusing on 3D thermal modeling, analysis and thermally sensitive design methodology. In addition, they used TSVs as thermal vias to build a vertical heat dissipation path, which leads to an increase in the heat load on the TSVs, as well as thermomechanical stresses, thus diminishing the reliability. Inadequate coefficient of thermal expansion (CTE) and thermal expansion directions can adversely affect TSV performance or even break the entire matrix, exacerbating the reliability of 3D ICs. To this end, 3D IC designs must carefully consider the aforementioned thermal trials at the time of design and execution.

– Implementation of TSVs that support thermal stress at design time to reduce TSV heat load and minimize thermomechanical stress for neighboring devices.

– Thermal management scheme at run time to analyze thermal cycle models and control thermal expansion, thus reducing mechanical stress on the chip and possibly achieving mechanical balance between each layer.

To minimize thermomechanical stress, TSVs must be placed intelligently at the time of design. Therefore, the corresponding analysis on the thermal stresses around the TSVs is fundamental for the solution. Several studies targeting thermal stress analysis have shown that the stress field in TSVs is uniform and can be represented by radial, circumferential and axial stresses. The constraints can be expressed as follows:

$$\sigma_\theta = \sigma_r = \frac{-E\,(\alpha_{tsv}-\alpha_{tsi})T_{sv}}{2-2v}, \ \sigma_z = 2\sigma_\theta \qquad [5.46]$$

where σ_r, σ_θ and σ_z are radial, circumferential and axial stresses, respectively, α_{tsv} is the CTE of TSVs, α_{si} represents the CDT of silicon, *TSV* is the heat load on TSV, *E* is the Young's modulus and υ is the Poisson ratio.

5.2.5. *Thermal modeling of the connectors*

After a thermal investigation on the substrate, we inspect the thermal behavior of the TSVs, the key points of the connections integrated in the substrate. The integration of the 3D IC can increase the integrity of the system, but as a side effect, the subsequent increase in the power density of the system can cause an important heat dissipation. This has induced a bottleneck for 3D IC. Then, how to remove heat from ICs and achieve efficient heat management becomes the aim of 3D IC technology.

In order to solve the problem of heat dissipation, researchers around the world have proposed heat management solutions, such as conductive TSVs, micro-pumps, using carbon nanotubes and graphene (which have better thermal conductivity than silicon) as a material. But these technologies are based on heat transfer analysis. The next part of this chapter will first give an accurate analytical model of heat transfer of cylindrical TSVs and will also study temperature characteristics. Next, a study of a single-chip model with four cores is processed in order to calculate the multi-core 3D IC temperature characteristics.

Here, we have not measured the change in the characteristics of the material depending on the temperature, or we can say that the temperature is relatively low, so the change in the characteristics of the material is small. We simply give a solution for each uniform layer in a low temperature design. In our study, we adopt a linear model; the overall effects of the substrate and connections can be divided. But, if the connections become very important, we cannot dissociate the substrate with TSV, RDL, etc. Correlation terms are involved; the system must be approached by nonlinear models.

5.3. Heat analysis for 3D ICs

5.3.1. *3D IC heat transfer compact model without TSVs*

As depicted in Figure 5.21(a), the 3D-IC is composed of n dies and the lower die is connected to a heat sink. Each die is connected by a layer of sticky coating, a Si substrate and an oxide layer; the heat is engendered by the heat source attached to each die, and the *n*th power dissipation of the die is assumed to be *Qn*.

Due to the low thermal conductivity between the oxide layer and the coating layer, thermal conduction between the chips becomes very difficult. Compared to the change in temperature of the dies, the temperature exchange between them can be disused. Then, we can use an average temperature to define the temperature of each layer. The 3D chip is composed of 2D chips of n layers with different temperatures. According to the analysis of the heat flow via "Fourier", using heat flow as an analogue of the current, temperature as an analogue of the voltage and resistance to heat as an analogue of the resistance, we can obtain the temperature of the *j*th layer.

$$T_j = \sum_{i=1}^{j}[R_i(\sum_{m=i}^{n} Q_m)] \qquad [5.47]$$

where n is the total layer number, and R_m and Q_m are, respectively, the mth layer's heat resistance and power dissipation.

Assuming that the power dissipation of each layer is the same, called Q, the thermal resistance of the first layer is R_1, and the thermal resistances of the other layers are the same, R, so that the n 3D IC layers, the temperature of the top layer, can be expressed as follows:

$$T_n = Q\left[\frac{R}{2}n^2 + (R_1 - \frac{R}{2})n\right] \qquad [5.48]$$

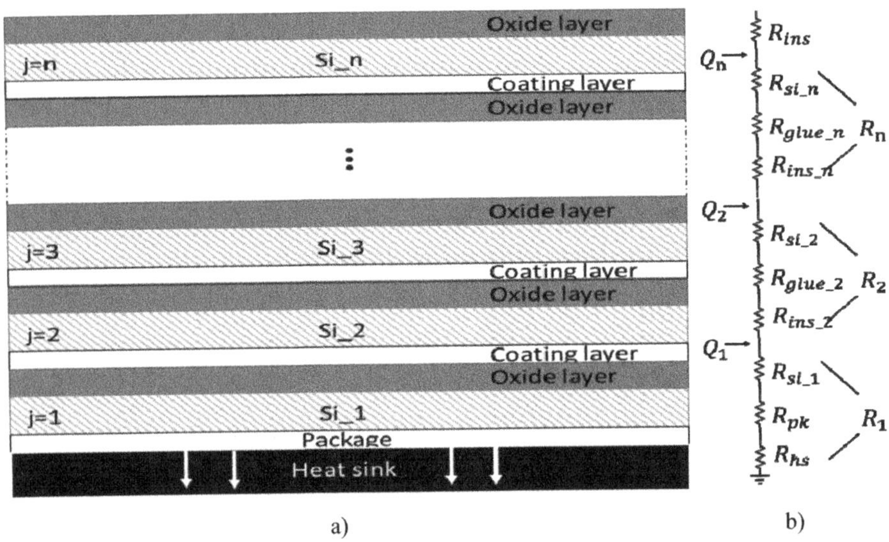

Figure 5.21. *a) Schematic of IC 3D structure (without vias); b) heat transfer model. For a color version of this figure, see www.iste.co.uk/abouelatta/smartpower.zip*

From Figure 5.21(b), we can observe that the thermal resistance of the first die R_1 includes the thermal resistance of the substrate S_{i_1}, package and heat well; the thermal resistance of the other dies R includes the thermal resistance of the Si substrate, coating layer and oxide layer. Thermal resistance is defined as: $R = l/(kS)$; l is the length of the heat flow, k is the thermal conductivity and S is the surface of the cross-section of the heat flow; Q represents power dissipation, and R_1 and R characterize, respectively

$$R_1 = R_{si} + R_{pk} + R_{hs} = \frac{l_{si}}{k_{si}S} + R_{pk} + R_{hs} \qquad [5.49]$$

$$R = R_{si} + R_{glue} + R_{ins} = \frac{l_{si}}{k_{si}S} + \frac{l_{glue}}{k_{glue}S} + \frac{l_{ins}}{k_{ins}S} \qquad [5.50]$$

where R_{si}, R_{pk}, R_{hs}, R_{glue} and R_{ins} represent, respectively, the thermal resistance of the silicon substrate of each chip, packaging, heat sink, coating layer and isolation layer; l_{si}, l_{glue}, and l_{ins} represent, respectively, the silicon substrate thickness, the coating layer and the insulation layer; k_{si}, k_{glue}, and k_{ins} represent, respectively, the silicon substrate thermal conductivity, coating and insulation layers; and S is the surface of the chip.

5.3.2. *IC model for analyzing the temperature of the chip of the top layer taking into account the TSVs*

In the 3D IC, the TSV is usually filled with copper, due to its high thermal conductivity and high heat dissipation power. If we neglect TSVs in the heat transfer model, this will lead to an overestimation of the temperature of the chip, so it is necessary to take into account parasitic TSVs for the analysis of 3D IC heat characteristics. Here, we adopt a 3D IC top layer analytical model, taking into account these TSVs. After adding TSVs in the 3D IC, the structure will be modified: the heat transfer model appears in Figures 5.22(a) and (b).

Assuming that the surface of each layer in the IC is S_{TSV}, the chip surface is S and the TSV surface ratio is r, we get:

$$r = \frac{S_{TSV}}{S} \qquad [5.51]$$

when examining TSVs, the thermal resistance of the layer will be paralleled between the thermal resistance of the TSVs and the original thermal resistance, as indicated in Figure 5.22(b). Assuming that the surface of the chip remains unchanged, by adding TSV, the surface area of the other parts of the layer will be reduced, and it will be $(1 - r)S$. Thus, the thermal resistance of the first layer will change from R_1 to R_1':

$$R_1' = \frac{l_{si}}{k_{si}(1-r)S}||R_{TSV} + R_{pk} + R_{hs} = \frac{l_{si}}{k_{si}(1-r)S}||\frac{l_{si}}{k_{TSV}rS} + R_{pk} + R_{hs} \quad [5.52]$$

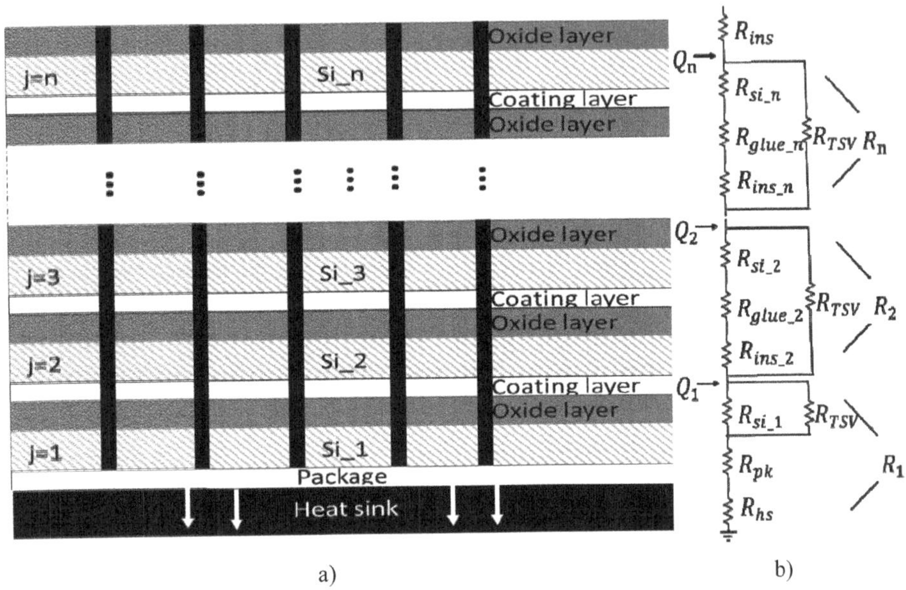

Figure 5.22. *a) Schematic of 3D IC structure with TSVs. b) Heat transfer model*

The thermal resistance of the other layers will change from R to R'

$$R' = \left(\frac{l_{si}}{k_{si}} + \frac{l_{glue}}{k_{glue}} + \frac{l_{ins}}{k_{ins}}\right)\frac{1}{(1-r)S}||R_{TSV1} =$$

$$\left(\frac{l_{si}}{k_{si}} + \frac{l_{glue}}{k_{glue}} + \frac{l_{ins}}{k_{ins}}\right)\frac{1}{(1-r)S}||\frac{l_{si}+l_{glue}+l_{ins}}{k_{TSV}rS} \quad [5.53]$$

Thus, after considering the TSV in the 3D IC, the analytical temperature model of the top layer can be given as:

$$T_n = Q[\frac{((\frac{l_{si}}{k_{si}}+\frac{l_{glue}}{k_{glue}}+\frac{l_{ins}}{k_{ins}})\frac{1}{(1-r)S}||\frac{l_{si}+l_{glue}+l_{ins}}{k_{TSV}rS})}{2}n^2 + ((\frac{l_{si}}{k_{si}(1-r)S}||\frac{l_{si}}{k_{TSV}rS} + R_{pk} + R_{hs}) -$$

$$(\frac{(\frac{l_{si}}{k_{si}}+\frac{l_{glue}}{k_{glue}}+\frac{l_{ins}}{k_{ins}})\frac{1}{(1-r)S}||\frac{l_{si}+l_{glue}+l_{ins}}{k_{TSV}rS}}{2}))n] \quad [5.54]$$

5.3.3. *3D IC thermal modeling result*

Here, we use an eight-die 3D IC model and choose MATLAB to do the simulation. The following table gives the model parameters.

Q (W)	Lsi (µm)	Lglue (µm)	Lins (µm)	ksi (W/mK)	kglue (W/mK)	kTSV (W/mK)	kins (W/mK)	Rhs (K/W)	Rpk (K/W)	S (mm^2)
2	50	2	8.8	150	0.25	390	0.07	2	20	10*10

Table 5.1. *Parameters of eight-die model: IC 3D*

Using MATLAB simulations, we can obtain the temperature change of the top layer of the TSV with the IC layer numbers, as shown in Figure 5.23. We can see that with the increase in the number of layers, the temperature of the upper surface of the chip increases linearly, the three colored lines represent r equal to 0.0001, 0.001, 0.01 and the black dotted line represents the dies without TSV. For the case without TSV, the upper surface temperature can reach 427K, but after considering the TSVs, the decrease in temperature is more obvious. The larger the r, the lower is the temperature. So, for an eight-layer 3D-IC, when r equals 0.0001, the temperature of the top layer decreases from 427 K to 421 K; when r = 0.001, it decreases to 392 K, and when r = 0.01, we have 360 K, because of the heat dissipation of TSVs.

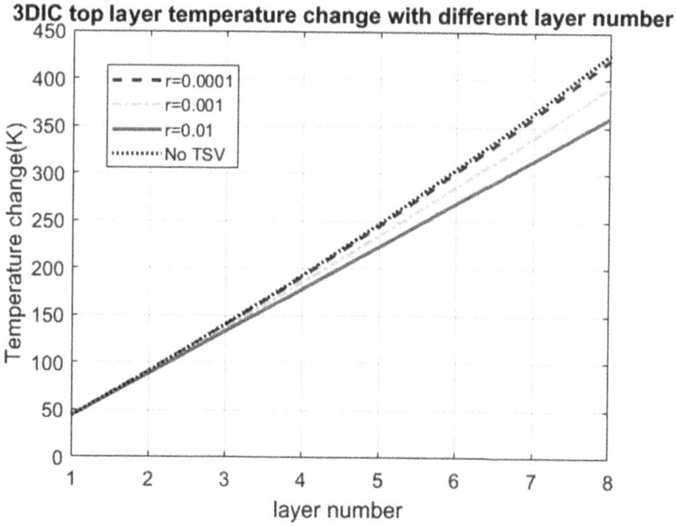

Figure 5.23. *Different layer numbers influence on the 3D IC top layer temperature. For a color version of this figure, see www.iste.co.uk/abouelatta/smartpower.zip*

As shown in Figure 5.24, the temperature of the top layer of the 3D IC changes with the surface ratio r of the TSV and the number n of dies. For the same r, with the increase in the number of dies n, r has an increasingly obvious influence on temperature. For a very high n, when one decreases r, at first, the temperature of the chip does not change, but although quite small, the temperature T increases sharply.

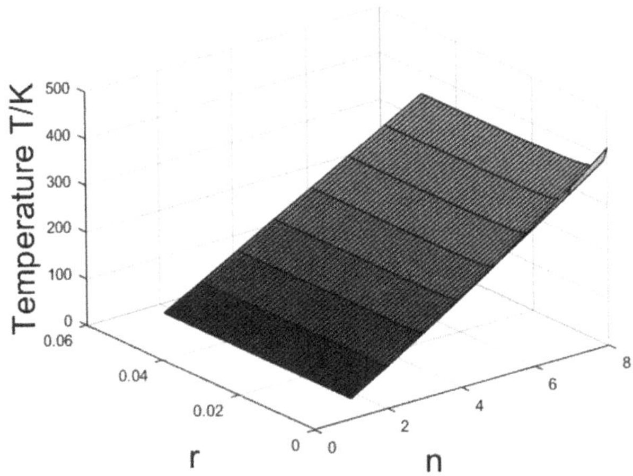

Figure 5.24. *Surface ratio r and dies' numbers influence on the 3D IC top layer temperature. For a color version of this figure, see www.iste.co.uk/abouelatta/smartpower.zip*

Figure 5.25 shows temperature changes when $n = 8$. When r is equal to 0.0001, the temperature of the upper surface IC 3D is 421K. As the r increases, the temperature drops. When r is equal to 0.005, the temperature drops to 366 K, and the decrease in value can reach 55 K, about 13%, when it is equal to 0.01. When r is equal to 0.005, the decrease is 6 K, about 1.6%, and when it is equal to 0.015, the temperature is 357 K. When $r = 0.01$, the decrease is only 3 K, about 0.8%, and with the augmentation of r, the temperature does not have a significant change.

Thus, in the 3D-IC design, for $n = 8$, the best surface ratio of TSVs is 0.5–1%. It can be adjusted according to requests. If the chip does not need as many interconnection signal lines, the other TSVs can be changed to thermal via and achieve better heat transfer. The designer of 3D ICs can read directly, the value of r as a function of temperature demand directly from Figure 5.25. When n is equal to another value, r can also be calculated this way.

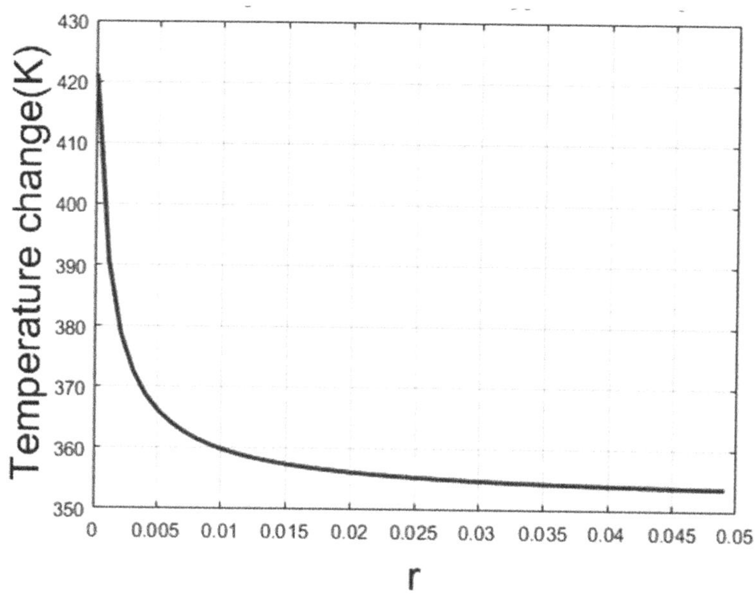

Figure 5.25. *Temperature change with surface ratio r (n = 8)*

However, if we want to maintain the temperature of each point of the 3D IC, another 2D heat transfer generator should be used to produce the exact 2D temperature distribution. Here, we use an FVM. The thermal diffusion equation is usually used to describe the heat pipe in a chip and calculate the temperature profile:

$$\rho c_p \frac{\partial T(\vec{r},t)}{\partial t} = \nabla \cdot [k(\vec{r},T)\nabla T(\vec{r},t)] + g(\vec{r},t) \quad [5.55]$$

with the general thermal conduction boundary condition.

$$k(\vec{r},T)\frac{\partial T(\vec{r},t)}{\partial n_i} = h_i(T_a - T(\vec{r},t)) \quad [5.56]$$

In the above equations, T is the temperature (K), k is the thermal conductivity [W/m/K], ρ is the density of the material (kg/m^3), cp is the specific heat [J/kg/K], g is the power density of the heat sources (W/m^3) and h_i is the heat transfer coefficient in the direction of the heat flow on the edge surface of the chip (W/m^2/K).

Figure 5.27 shows our total 3D thermal simulation work algorithm. In our case, we first get the surface ratio *r*, and calculate the thermal resistance dependent on *r*, such that R_{TSV}, R_{Si}, R_{glue}, R_{ins}.

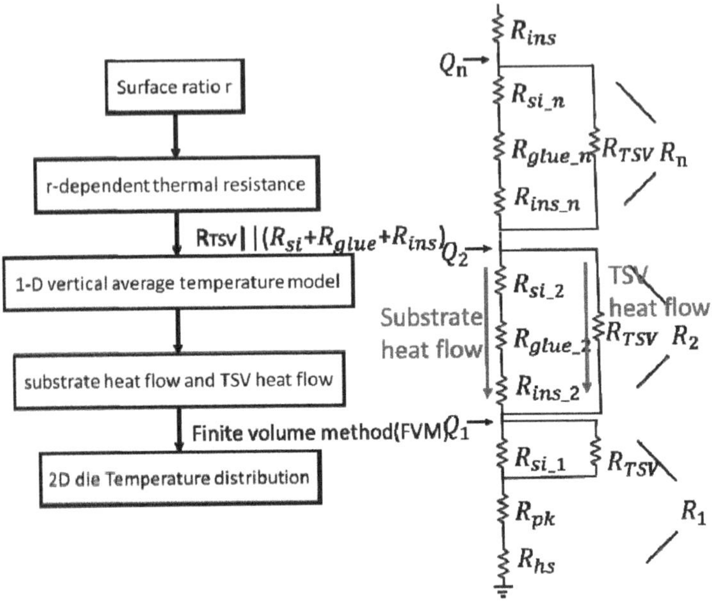

Figure 5.26. *3D thermal simulation work flow. For a color version of this figure, see www.iste.co.uk/abouelatta/smartpower.zip*

Using the analytical vertical model, the vertical temperature distribution can be obtained. With the temperature difference, which is analogous to the electric potential difference, we can also directly calculate the heat that passes through the substrate and the heat that comes through the TSVs, which can be computed by the following equation:

$$q_{TSV\_n} = \frac{(T_n - T_{n-1})}{R_{TSV}} \qquad [5.57]$$

$$q_{sub\_n} = \frac{(T_n - T_{n-1})}{R_{si\_n} + R_{glue\_n} + R_{ins\_n}} \qquad [5.57b]$$

Finally, we return to 2D thermal conduction with two different types of heat sources. Here, we adopt the FVM to solve the problem. Then, a 2D thermal distribution is generated. The following figures show the distribution of the thermal temperature of the lower and upper layer in the substrate of an eight-layer model with $r = 0.01$ and four TSVs. The TSV temperature of the lower layer is 344 K and the TSV temperature of the upper layer is 359.88 K.

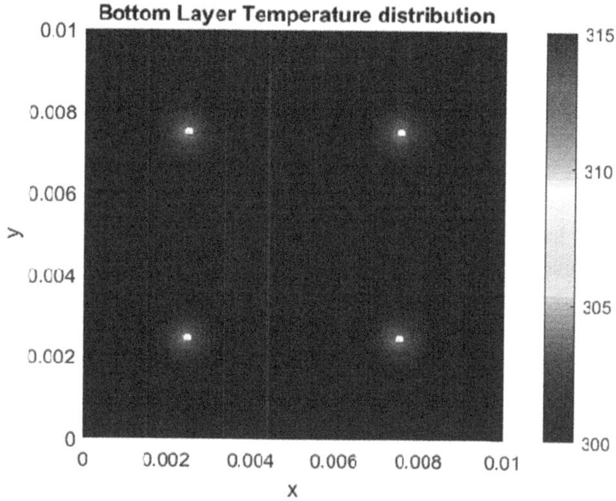

a)

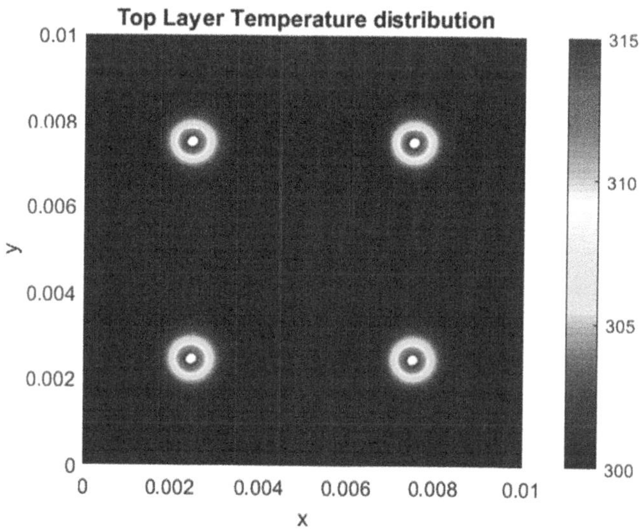

b)

Figure 5.27. *a) Bottom and b) top layer temperature distribution of four TSVs with surface ratio r = 0.01. For a color version of this figure, see www.iste.co.uk/abouelatta/smartpower.zip*

5.3.4. *Electrothermal (ET) modeling of very large scale circuits*

As stated in the IC roadmap, increasing power density requires decreasing the supply voltage, resulting in the need for a larger current supply through PDNs. Therefore, joule heating and power switching significantly increase the temperature of the circuit, which strongly influences the electrical performance, due to the strongly modified electrical temperature-dependent resistivity. Thus, an efficient and co-simulation PDN software is needed, for now, to calculate the effective IR drop, which is strongly related to electrical resistivity. In addition, ET co-simulations are inevitable in 3D IC design.

We use a standard structure for mass and power networks, as shown in Figure 5.28. The ground and power networks are separated by an insulation layer and connected by VDD and GND pins, respectively, due to a Z supply impedance series.

Electrical and thermal problems must be solved together in static and dynamic analysis. A device-level thermal and electrical modeling method will be provided, devoted to CPWs and TSVs. The main sources of heat come from the switching of the logic gates, as well as the "Joule" heating in the PDN interconnections. Thermal and electric transport are very analogous, with the heat flow corresponding to the electric current and the temperature difference playing the role of the difference in electrical potential. In any case, electrothermal co-simulation can be operated within the same solver of circuits.

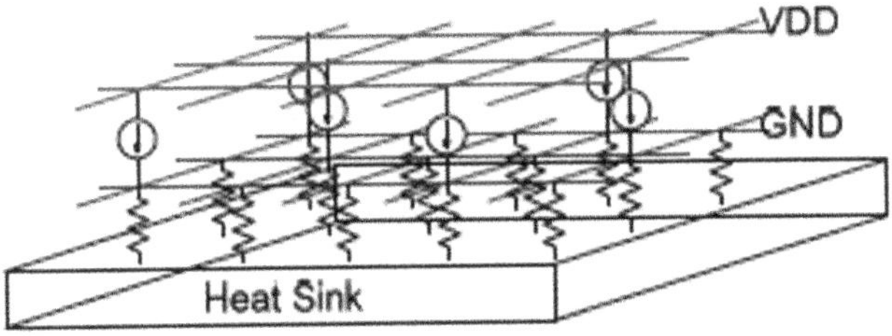

Figure 5.28. *Ground and power networks structures. For a color version of this figure, see www.iste.co.uk/abouelatta/smartpower.zip*

5.3.5. *Electrical modeling of very large scale*

The DC problem is modeled by two regular resistive grids and an internal node; temperature-dependent resistors R(T) connect neighboring nodes of the same layer. The power layer and the earth layer are connected to each other by an I0 current source. The solution of the electrical problem can be expressed as the voltage drop at node i:

$$V_d(i) = V_{DD} - (V_n(i) - V_g(i)) \quad [5.58]$$

where $V_n(i)$ and $V_g(i)$ are the node potentials of the power and ground planes, respectively.

A dynamic electrical model can also be developed when a dynamic, parasitic circuit has been defined for this structure. We use the first-order standard model for each physical cell, considering the parasitic inductive effect "L" between adjacent nodes, and a parallel capacitive "C" between neighboring nodes on the VDD and GND grids. In this way, we draw a dynamic elementary cell diagram as shown in Figure 5.29.

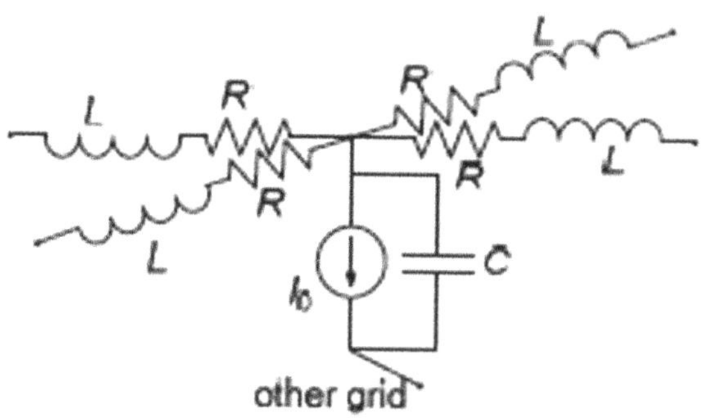

Figure 5.29. *Electrical dynamical element cell*

5.3.5.1. *Electrical modeling of the IC 3D*

– Electric modeling of segments, type 3xRLC

The model can be described through a SPICE-type script.

```
* Example RLC-3

Vin N_in1  0 PULSE 0 3 0.5m 0.5m  0.5m 4m  10m

X1 n_in1  n_in2  seg
X2 N_in2  N_in3  seg
X3 n_in3  n_out  seg
Cl n_out  0     10u

.SUBCKT seg n_in   n_out
Rseg  N_IN  N_1   3.5
Lseg  n_1   N_OUT  1.2m
Cseg  N_OUT  0     7.3u
.ENDS seg

.TRAN  0.004m 20m
.PROBE  V(n_out)
.print V(n_out)
.end
```

Figure 5.30. *3xRLC series model SPICE script*

The model describes a three-segment RLC circuit. The electrical circuit is shown in the following figures.

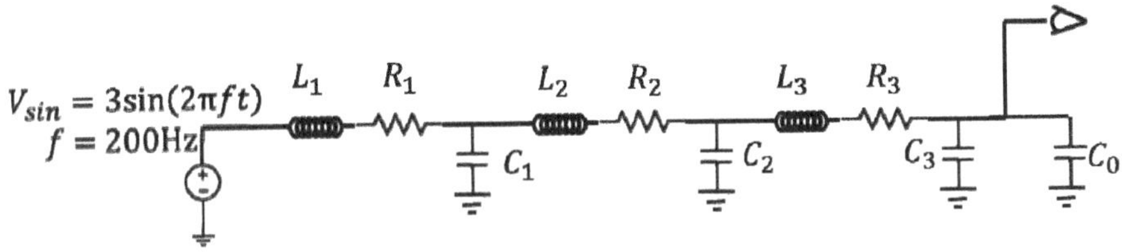

Figure 5.31. *"three-segment RLC" compact model*

The resistance of the segments is 3.5 Ω, the inductance is 1.2e-3 H, the capacity is 7.3e-6 F and the capacity added on the probe side is 10e-6 F. The voltage source is a sinusoidal signal, with an amplitude of 3 V, and a working frequency of 200 Hz. Here, we use the MNA method and the generalized conjugated residual solver (GCR), as well as the hollow matrix technique to solve and accelerate the calculation of the large-scale differential matrix. In order to verify the robustness of the method, the same simulation is performed with ADS by Keysight Technologies.

The signal is shown in Figure 5.32(a). The comparison of the output signal with ADS is shown in Figure 5.32(b). The latter shows a relative error of the MNA+GCR and ADS method. As can be seen in Figure 5.32(c), our MNA+GCR method has a very good accuracy compared to ADS; the maximum error is about 4%.

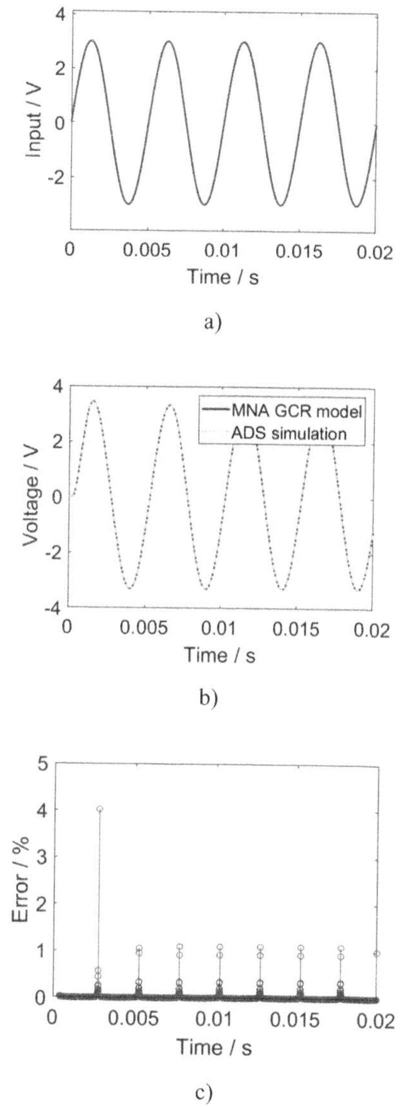

Figure 5.32. *a) Input voltage source; (b) output comparison between MNA and ADS, (c) relative error between MNA and ADS for the three-RLC segments model. For a color version of this figure, see www.iste.co.uk/abouelatta/smartpower.zip*

5.3.6. *Thermal modeling of very large scale circuits*

Due to the negligible heat diffusion in the dielectric, the thermal model can be assumed to be stable. Thus, the temperature distribution T(x) of metal interconnections can be solved by using the 1D heat diffusion equation:

$$\frac{d^2T(x)}{dx^2} = -\frac{g(x,T)}{k_m} \quad [5.59]$$

where k_m is the thermal conductivity of the metal [W/mK] and $g(x,T)$ is the heat sources term [W/m^3].

Assuming the defined network lengths are smaller than the characteristic length of heat, the network can be used to resolve electrical and thermal dissipation. Figure 5.33 is a typical node of the thermal network. It is composed of thermal resistors between neighboring nodes, and another thermal resistance related to the heat sink. The VDD and GND layers are connected by adding the heat sources P_s and P_J, P_s is the switching power, defined as: P_s (i) = I_0 Vd (i), related to the voltage drop at node "i" calculated by the electrical equation, and P_J is the Joule heating term from the power dissipation in PDN:

$$P_J(i) = \frac{1}{2}\sum_{k=1}^{4}\frac{v_{ik}^2}{R_{ik}}$$

where *k* is the four neighboring nodes' index.

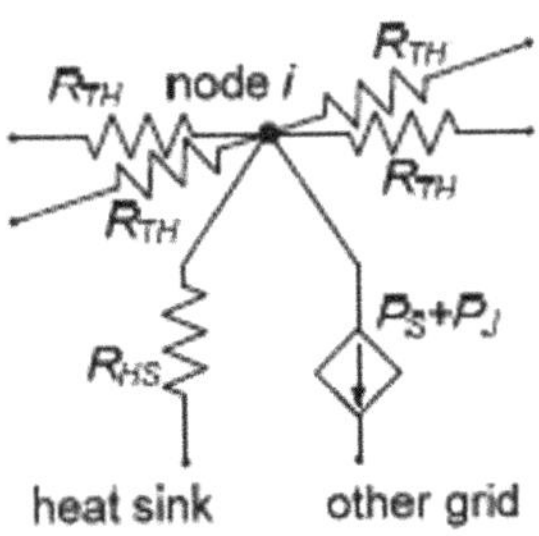

Figure 5.33. *Dynamic thermal elementary cell*

The problem of ET coupling is solved in a classical approach, by first solving the thermal network at room temperature, second, calculating the temperature-dependent electrical resistances R(T) and third, solving the electrical network, then obtaining the voltage. After computing the thermal heat sources, we can repeat the first three steps and so on. The final temperature and distributions of the voltage can be obtained when the relative error becomes less than the predefined error value. A more in-depth study of the dynamic analysis can be done from the previous static AND analysis by adding the following two steps.

The first step is to calculate the dynamic elements (capacitors, inductors, etc.) related to the voltage drop. The second step is to check whether the new thermal distribution is still compatible with the static electrothermal equilibrium. Thus, with a predefined impedance circuit, a frequency analysis can then be performed using the dynamic analysis workflow.

5.3.7. *Electrothermal modeling of very large scale circuits*

3D integration improves performance, increases functionality and improves implementation. Nevertheless, compared to 2D integration, due to the parasitic and large switching current and longer power distribution path, power and thermal integrity issues are still concerns of this 3D integration. In this section, we conduct an in-depth study of the power and thermal problems of 3D integration, specifically for TSVs. Our study is based on a three-layer system using technologies with TSVs. Starting from heat dissipation models, we present a power and thermal integrity analysis of 3D ICs.

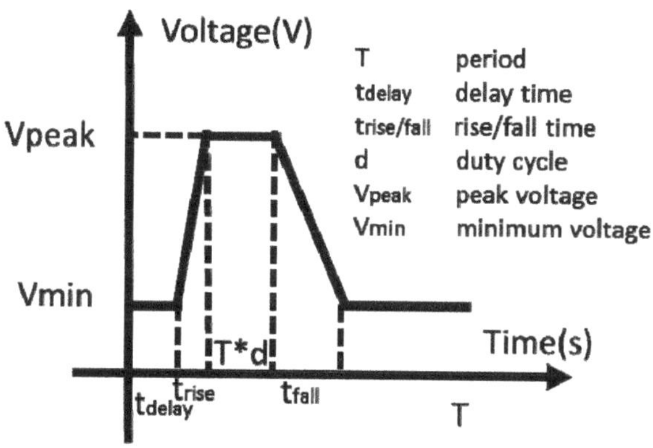

Figure 5.34. *Pulse signal definition*

Our study takes into account dynamic power analysis by combining the parasites of the 3D multilayer system. The extracted parasite deals with the coplanar line, the redistribution layer and TSVs. Coplanar line parasites are modeled by resistance, capacity and inductance. The redistribution layer is modeled by resistance, inductance and capacity. Time-dependent voltage sources can be represented by electrical characteristics: (1) switching frequency; (2) ascent and descent times; (3) the assignment cycle: (4) delay time and (5) peak voltage. Here, in our study, switching circuits are directly modeled by rectangular waveforms.

Voltage sources of the switching pulse type are expressed as:

$$V(t) = \begin{cases} V_{min}, \ 0 < t < t_{delay} \\ \left(t - t_{delay}\right) * \frac{V_{peak} - V_{min}}{t_{rise}}, \ t_{delay} \leq t < t_{delay} + t_{rise} \\ V_{peak}, \ t_{delay} + t_{rise} \leq t < t_{delay} + t_{rise} + T * d \\ V_{peak} - \left(t - t_{delay} - t_{rise} - T * d\right) * \frac{V_{peak} - V_{min}}{t_{fall}}, t_{delay} + t_{rise} + T * d \leq t < t_{delay} + t_{rise} + T * d + t_{fall} \\ V_{min}, \ t > t_{delay} + t_{rise} + T * d + t_{fall} \end{cases} \qquad [5.60]$$

The analysis is established on a 3D multilayer structure. The models are built from our 3D-TLE extractor that can extract and generate the SPICE-compatible interconnect model. The main challenge of this analysis is to simultaneously calculate the electrical analysis and the thermal analysis. Therefore, in the first approach, temperature-dependent resistivity is used in this analysis.

$$\rho = \rho_0[1 + \beta(T_{emperature} - T_0)] \qquad [5.61]$$

where ρ_0 (1.53265e-8 Ω m) is the electrical resistivity at T_0, 300 K, and β (0.0067) is the temperature coefficient of the electrical resistivity. The electrical resistivity of the conductors increases with temperature, and the power supply will be greatly influenced. In addition, the voltage drop and the current circulating on the transmission line will cause a "Joule" heating. After a sufficiently long time, the difference in temperature distribution can lead to electron migration (EM) problems. By using this method, electrical and thermal problems are coupled.

$$[G + pC]x(p) = I(p) \qquad [5.62]$$

To perform the electrical analysis of the 3D multilayer system, we use Kirchhoff's law solving linear equations. Matrix formulation follows the rules of the MNA method that is often used for power grid analysis. These matrices can be expressed as a Laplace transformation:

$$[G + sC]x(s) = I(s) \qquad [5.63]$$

where G is the n-by-n conductance matrix and C is the n-by-n capacity and inductance matrix, and I is the n-by-1 vector of current sources and voltage sources. $x(s)$ is the vector n by 1 that represents the voltage and current of the nodes. To obtain the voltage drop of the nodes and the current distributions, we solve the power supply network by reversing this matrix [G + sC]. For the analysis of large systems, we use a hollow matrix and a conjugate residue generalization (CRG) method to facilitate matrix calculations.

As a conductor of heat from one layer to another, TSVs can sometimes be very beneficial for dissipating heat away from a hot layer, but sometimes they can be harmful to nearby layers or devices. In this work, following the principle of electrical-thermal duality, the heat flow through a thermal resistance is equivalent to the current passing through the electrical resistance, and the temperature difference is equivalent to the voltage difference.

Heat is normally generated from the top layers and transferred by TSVs and other layers (e.g. isolation layers) to the silicon substrate. Insulation layers have much lower thermal conductivities than silicon; they therefore act as thermal barriers. Voltage changes in switching circuits will contribute to heat generation. TSVs, redistribution layers, coplanar lines and heat sink are modeled as thermal resistors in order to drive vertical and lateral heat flows into the 3D system. Due to the difference between the vertical thermal resistances of TSVs and the thermal resistances of the other layers of materials, the heat dissipates at different speeds. Equivalent thermal resistances are represented as the parallel connection of the resistors of the TSVs and the thermal resistances of other layers. The temperature of the heat sink is assumed to be the ambient temperature. Thanks to the principle of electrical-thermal duality, the thermal network can also be solved using the MNA method. The thermal linear equations of the nodes can be merged with the following equation:

$$G_{th}T_{emp} = Q \qquad [5.64]$$

where G_{th} is the thermal resistance matrix of size m-by-m and m is the number of nodes of the 3D system. T_{emp} is the m-par-1 vector of temperature nodes and Q is the m-by-1 vector for heat sources. We develop a thermal and electrical and analysis to compute the ET coupling effect of a 3D system. The proposed method is an iterative process, and Figure 5.35 shows the flow of thermal and electrical analyses.

Electrical analysis is transient; the voltage drop on the metal tracks will lead to a rise in temperatures. The temperature increase is then treated as a thermal source and is applied to the thermal analysis step. Thermal analysis provides a temperature distribution for the different layers and TSVs, where the modified temperatures will play on the resistivity of the conductors and then influence the electrical analysis. The iterative calculation of resistivity and temperature will converge when there are no more changes to the voltage drop and temperature distribution. Here, thermal analysis is a steady-state analysis that captures the temperature distribution from a given electrical grid. This method is limited to capture dynamic thermal distribution. However, fast and accurate capture of temperature distribution is the main objective.

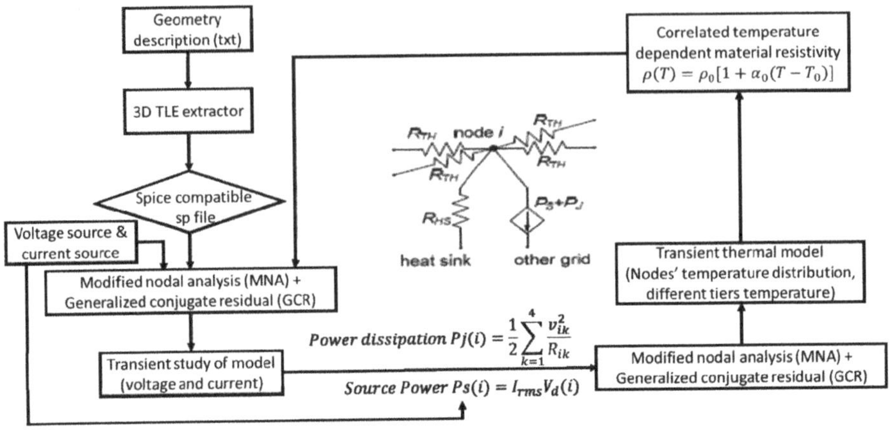

Figure 5.35. *3D IC electrothermal coupling flow*

– Case I: One-via model

The one-via model is shown in Figure 5.36(b). The chip is composed of a coating layer, an oxide layer and a silicon layer, the center is a copper TSV. The bottom is attached to a package, and a heat sink is used on the outside of the package. The chip size is 10 mm × 10 mm, the thickness of the silicon layer is 50 μm, and the coating layer and the insulation layer are 2 μm and 8.8 μm thick, respectively. The surface area ratio *r* of TSV is set at 0.01.

The electrical circuit and the thermal circuit are shown in Figures 5.36(a) and (c). We enter a pulse signal (defined in the previous section), and a parallel resistance of 50 Ω at the input and output, as well as a capacity of 150 fF at the output, in order to see the output signal. Here, the resistance of the TSV is about $0.0052e^{-4}$ Ω, so in comparison to 50 Ω, the resistance of the TSV is too small. Thus, the effective resistance of the parallel circuit depends on the resistance TSV.

We enter a pulse signal (as defined previously), and parallel, a resistance of 50 Ω at the input and output, and a capacity of 150 fF at the output, in order to see the output signal. Here, the resistance of the TSV is about 0.0052e-4 Ω, so in comparison with 50 Ω, the resistance of the TSV is too small. Thus, the effective resistance of the parallel circuit depends on the TSV resistor and the parallel resistance connection of 50 Ω will not change the output signal. The current will turn into heat through the resistance of TSV. The heat generated is dissipated in the 3D geometry.

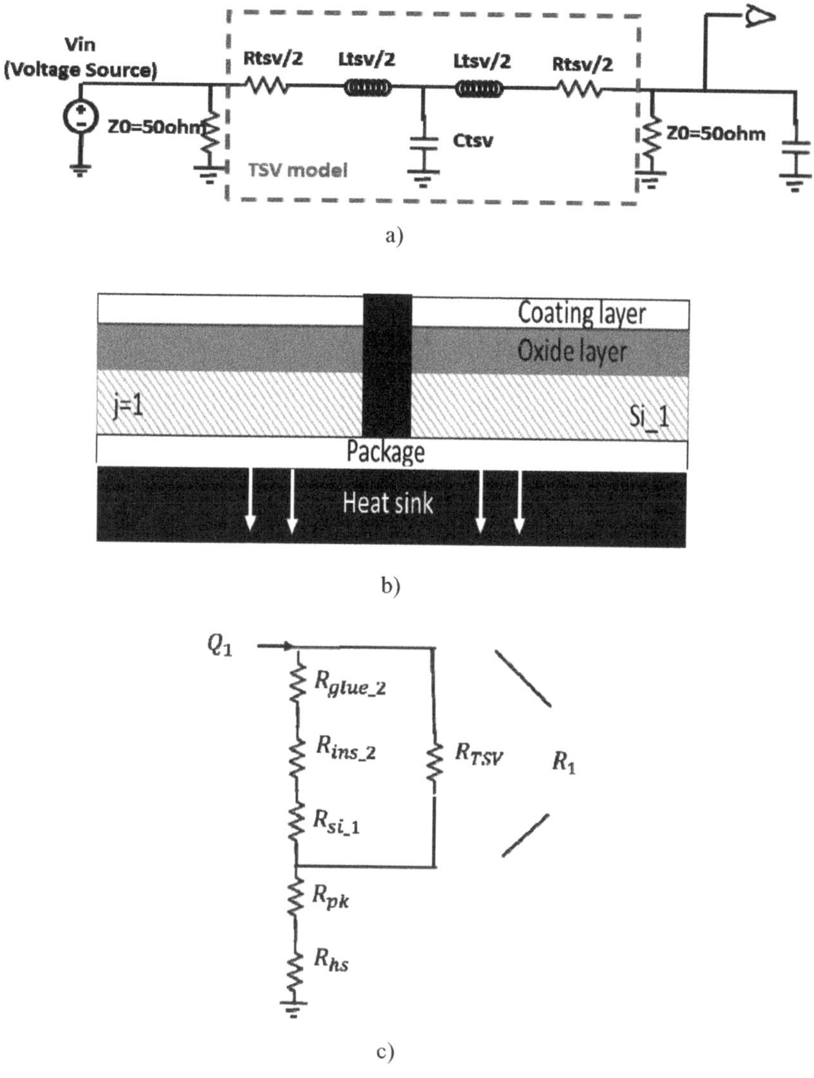

Figure 5.36. *a) Electrical compact mode; b) 2D geometry; and c) thermal compact model of one via model. For a color version of this figure, see www.iste.co.uk/abouelatta/smartpower.zip*

The thermal path is composed of TSV conduction and serial thermal conduction of the other three layers, the heat passing separately through the series of three layers and also through the TSV (due to the high thermal conductivity of copper). Then, the heat is collected at the packaging level, and carried through the packaging to the heat sink. The packaging and heat sink also have their own thermal conductivities, especially for the heat sink, it has a very high thermal conductivity and therefore a very low thermal resistance. The ambient temperature is about 300 K.

Geometry	
Silicon layer thickness (lsi)	50um
Glue layer thickness (lglue)	2um
Insulation layer thickness (lins)	8.8um
Length of chip (Lx)	10mm
Width of chip (Ly)	10mm
TSV number	1
TSV ratio (r)	0.01

Table 5.2. *Geometry parameters for the one-via model*

As shown in Figure 5.37, the S parameter is calculated using the ADS simulator.

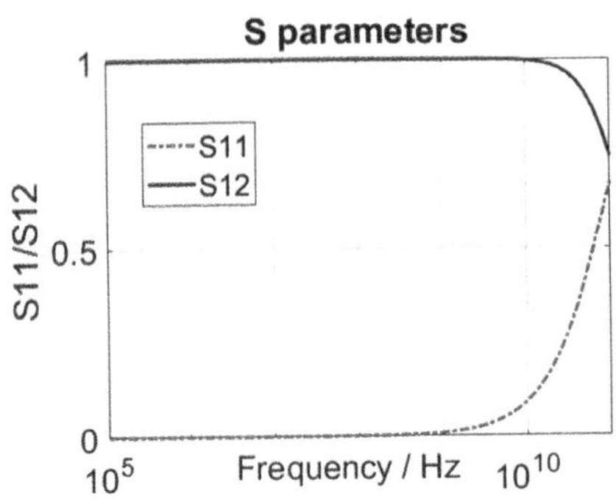

Figure 5.37. *S parameters of the ADS model 1-TSV (valid for the frequency below 20 GHz and the quasi-static hypothesis). For a color version of this figure, see www.iste.co.uk/abouelatta/smartpower.zip*

The output voltage error Err can be defined as:

$$Err = \frac{\max(abs(V1-V2))}{Vmax\_in} \quad [5.65]$$

where $Vmax\_in$ is the input maximum voltage, $V1$ is the first transient voltage and $V2$ is the second transient voltage.

Geometry	
Silicon layer thickness (lsi)	50um
Glue layer thickness (lglue)	2um
Insulation layer thickness (lins)	8.8um
Length of chip (Lx)	10mm
Width of chip (Ly)	10mm
TSV number	1
TSV ratio (r)	0.01

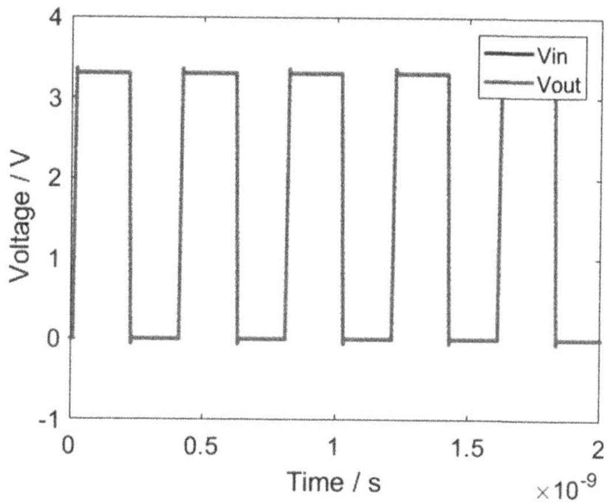

a)

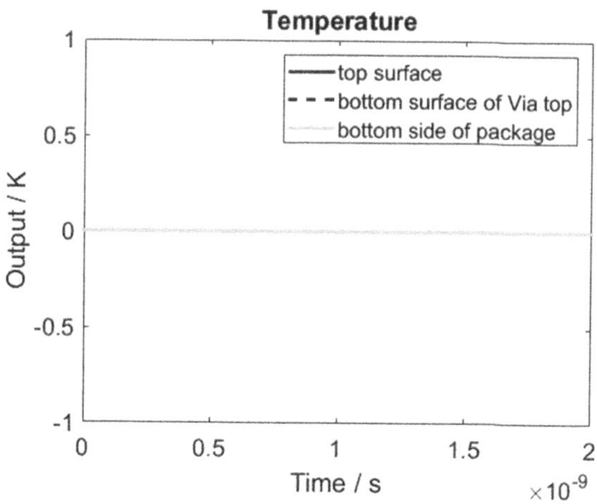

b)

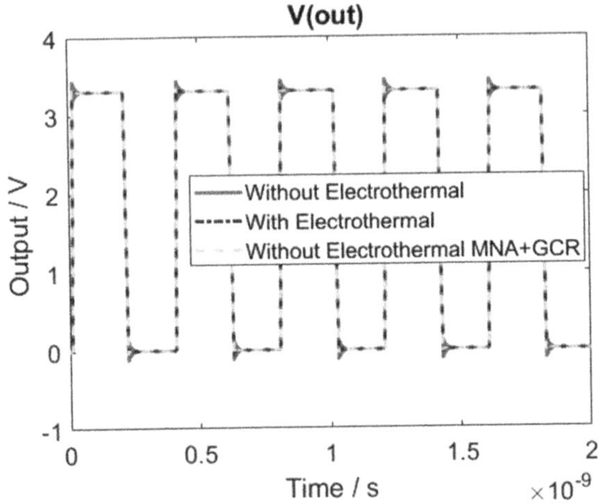

c)

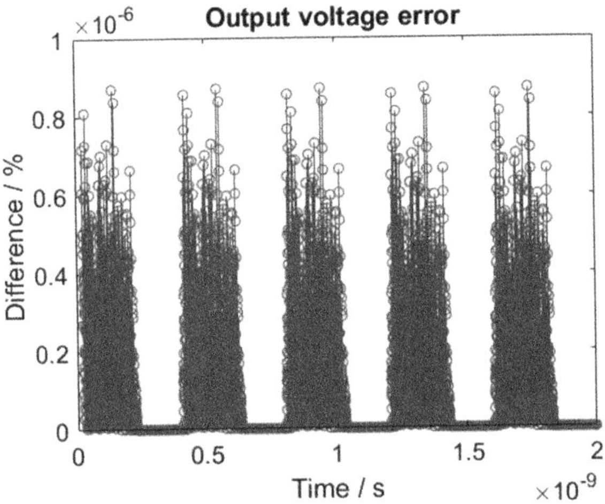

d)

Figure 5.38. *Signal transient integrity analysis for the 1-TSV 2.5 GHz model. a) Input signal relative to output. b) Maximum temperature rise. c) Comparison between ADS (without electrothermal consideration), MNA method (with electrothermal consideration) and MNA (without electrothermal consideration). d) Output voltage error (without electrothermal consideration). For a color version of this figure, see www.iste.co.uk/abouelatta/smartpower.zip*

Figure 5.38 shows the transmission effects of the 2.5 GHz signal. Graph (a) shows the comparison of the input and output signal. Graph (b) shows the top of the coating layer, the bottom of the TSV and the lower side of the package temperature, respectively. Graph (c) shows the change in voltage transmission when examining the electrothermal coupling effects between ADS (without electrothermal consideration), MNA method (with electrothermal consideration) and MNA (without electrothermal consideration). Graph (d) shows the output voltage error without taking into account the electrothermal effect. As we can see in the results of the 2.5 GHz study, the output signal has almost the same waveform as the input signal, and the temperature change in the circuit is 0 K.

In the above results, the unique TSV model does not have an electrothermal effect at very low working frequency. During a certain period of frequencies, the output frequency increases linearly, and this will generate a considerable amount of heat and increase the temperature of the via. Therefore, we discover that the electrothermal effect has a relationship with the tendency of the parameters S. The S parameters remain stable at low frequency and very high frequency, but for a frequency band, this changes radically. We can overlay the parameters S with the error curves and the maximum temperature curve.

Figure 5.39 displays the result. But due to the nonlinear system, from 20 GHz to 10 THz, the results are now expected to use full-wave models for electrical study. The current result is not valid in this frequency band.

As expected, S parameters really have a good trend with electrothermal effects. Strong electrothermal effects appear while the S parameters change.

– Case II: A TSV model with RC effect on substrate

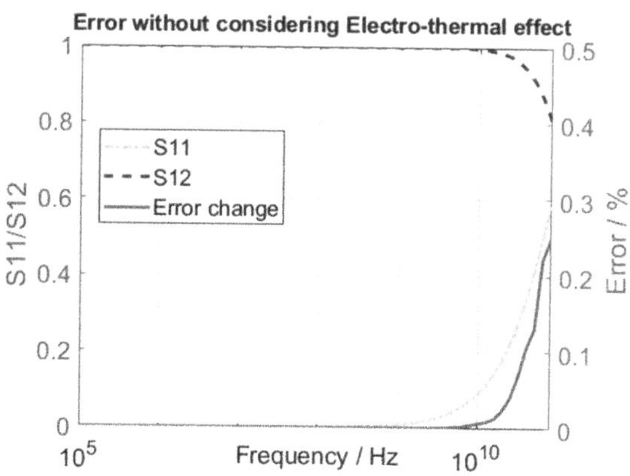

a)

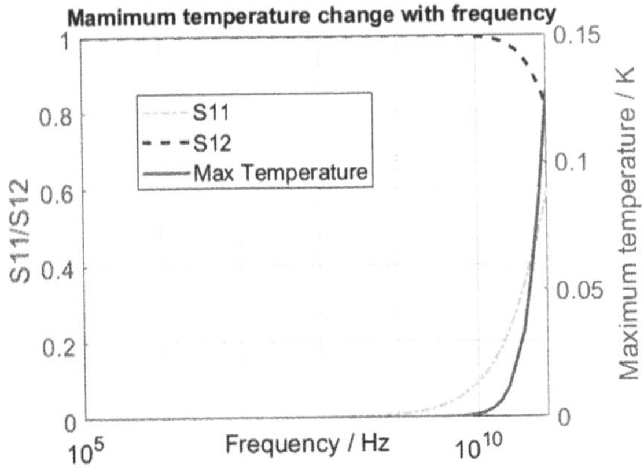

b)

Figure 5.39. *S parameters with a) error curve and b) maximum temperature curve for a via model (valid for a frequency below 20 GHz and quasi-static assumption. For a color version of this figure, see www.iste.co.uk/abouelatta/smartpower.zip*

The key difference between this model and the previous model is the substrate. In the previous model, the substrate is treated directly as a node of ground. But in this model, the substrate is strongly doped. It is considered as a conductor with very low resistivity (1e-4 Ω·m). The substrate is expressed as a parallel RC model. The strength and capacity of the substrate can be expressed by the following equations:

$$C_{Si\_sub} = \frac{\pi \times \varepsilon_0 \varepsilon_{r,si}}{cosh^{-1}(\frac{p_{TSV}}{d_{TSV}})} \times h_{TSV}[F] \qquad [5.66]$$

$$G_{Si\_sub} = \frac{\pi \times \sigma_{si}}{cosh^{-1}(\frac{p_{TSV}}{d_{TSV}})} \times h_{TSV}[S] \qquad [5.67]$$

Figure 5.41 shows the results. We see that by adding a strongly doped substrate, we can reduce the electrothermal effect, especially in the frequency band where the S parameters change strongly. The maximum temperature has been reduced to 5e-4 K, which is about three times less than the TSV model, and the maximum error has been reduced by half, about 1.5e-5. Thus, by adding a highly doped substrate, the ET heating effect can be diminished.

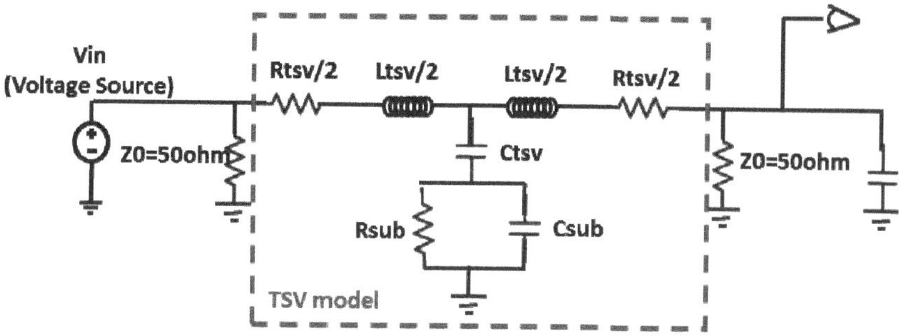

a)

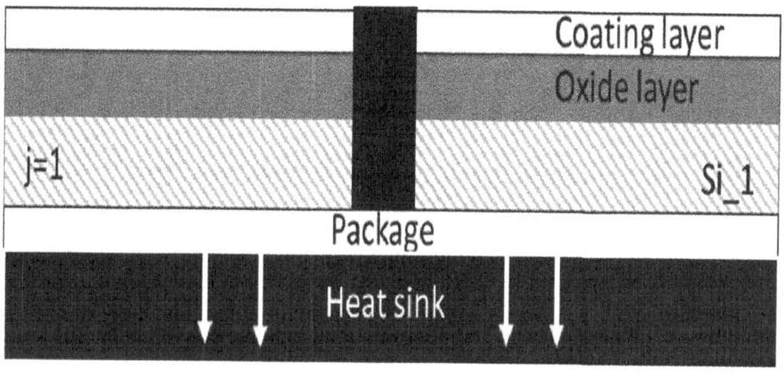

b)

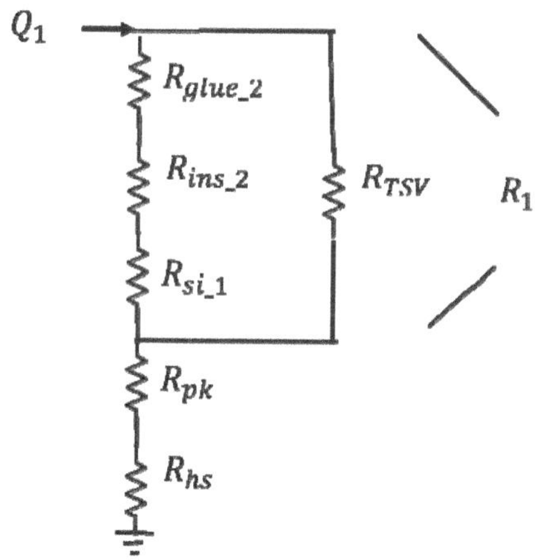

c)

Figure 5.40. *a) Electrical compact mode, b) 2D geometry and c) compact thermal model of a via with substrate model. For a color version of this figure, see www.iste.co.uk/abouelatta/smartpower.zip*

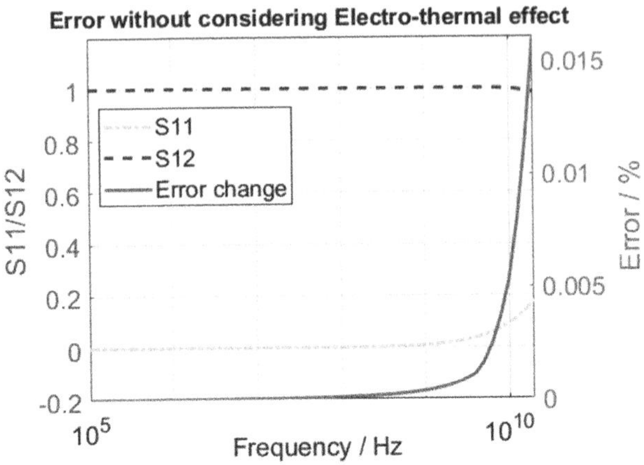

a)

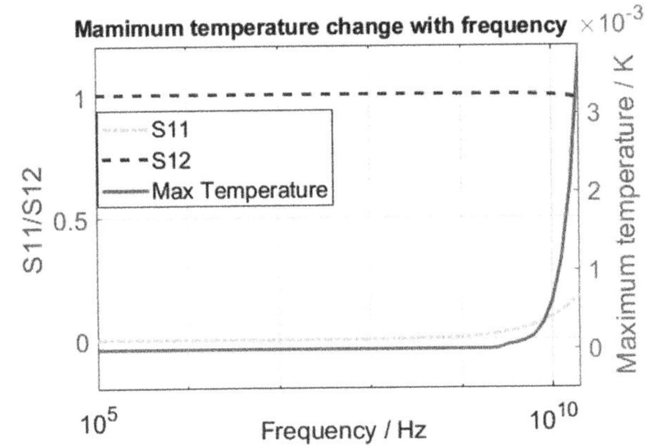

b)

Figure 5.41. *S parameters with a) error curve and b) maximum temperature curve for a via with substrate model (valid for a frequency below 20 GHz and quasi-static hypothesis). For a color version of this figure, see www.iste.co.uk/abouelatta/smartpower.zip*

– Case III: A TSV model with very high thermal resistance

In order to achieve a higher temperature and see a more obvious difference in signal transmission, we set Rhs and Rpk in Figure 5.42(b) to 2,000 K/W.

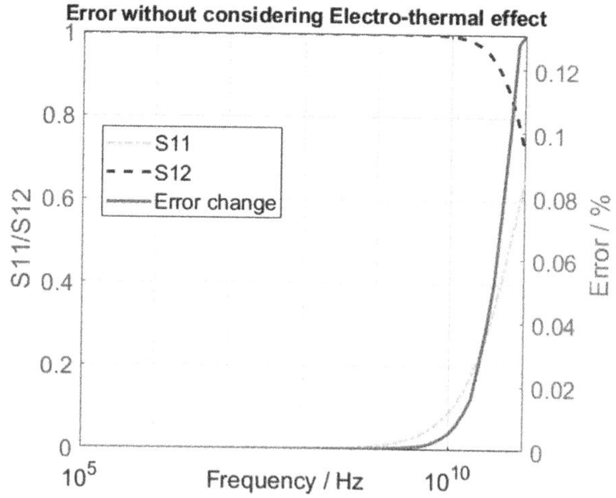

a)

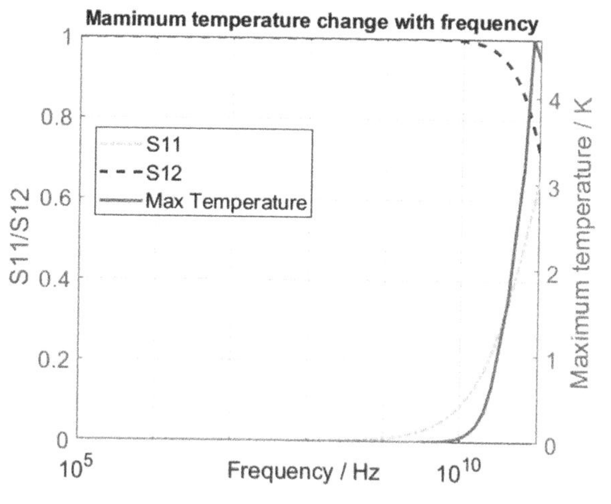

b)

Figure 5.42. *S parameters with a) error curve and b) maximum temperature curve for a via with high thermal resistance model (valid for a frequency below 20 GHz and quasi-static hypothesis). For a color version of this figure, see www.iste.co.uk/ abouelatta/smartpower.zip*

Figure 5.44 shows that the error curve and the maximum temperature curve always have the same trend. They keep stable at low frequency. Once the S parameters change, both curves change drastically. Although we studied electrothermal modeling up to 10 THz, due to the nonlinear system, from 20 GHz to 10 THz, we should use "full wave" models for the electrical study. But it works for frequencies below 15 GHz, enough for our industrial collaborator, for the next generation of mobile sample, for example. Here, we will not talk about quantum temperature interference for nano-devices.

5.4. Conclusion

In this part, 3D IC analytical models and an electrical-thermal analysis technique for the substrate, thermal connections (3D via) and 3D PDN at the system level, are familiarized and used to explore power and thermal integrity issues in 3D integration. 3D IC analytical models also take into account the electrothermal effects of TSV parasites. We delivered an in-depth study of power and heat dissipation issues in 3D integration, taking into account several aspects such as:

– single and multi-layer auto-heating effects on 3D power and thermal dissipation;

– TSVs and substrate impact on power and thermal integrity, and their design implications;

– heat sink's thermal conductivity's influence on the ET effects;

– single-layer and multi-layer self-heating effects on 3D power and heat dissipation;

– the impact of TSVs and substrate on power and thermal integrity and their design implications;

– influence of the thermal conductivity of the heat sink on electrothermal effects.

We combined switching frequency and electrothermal effects. Compared with the S parameters, we have highlighted that the electrothermal effects only appear when these S parameters change, and the reduction of the thermal resistance and the increase in the doping concentration of the substrate will slow down the problem of integrity of the electrothermal signal. In fact, we can use more layers, with lower thicknesses. Here, we have not considered the change in the characteristics of the material depending on the temperature or we can also say that the temperature is fixed relatively low, so the change in the characteristics of the material is small; we simply give a solution for each uniform layer and low temperature design.

In our study, we adopt a linear model; the overall effects of the substrate and connections can be split. But, if the connections become very important, we cannot dissociate the substrate with TSV, RDL, etc. Intercorrelation terms are involved; the system can be considered by nonlinear models. In addition, our extractor (3D-TLE) is now efficient, although some models are now rather simple. But the industry wants simple compact models; we should keep it as simple as possible, considering, physics, in depth.

5.5. Heat pipe

Investigations on flat heat pipes have been conducted to characterize the thermal and hydrodynamic performance of miniature heat pipes, used for the cooling of electronics. This has led to renewed interest in the use of vapor chambers as heat spreaders in applications, ranging from mobile devices to high-performance computing and power electronics systems. While detailed numerical models of vapor chamber operation have been developed, an easy modeling with low computational cost is needed for parametric study and quick assessment of vapor chamber performance in system-level models.

In the last few decades, heat dissipation has become one of the most limiting factors affecting further improvement of integrated device performance. The heat, generated by the Joule effect in the transistor channel, is distributed by conduction through silicon and metal interconnections. As a result, the circuits undergo a temperature rise proportional to the power dissipated; this phenomenon is to be considered from the elementary transistor to the system as a whole. The smaller and more dense the circuits, the higher the level of heat generated; thus, it is worth noting the presence of localized hot spots in areas of higher densities of logical components.

A rise in temperature decreases the mobility of the carriers and the width of the bandgap in the silicon, impacting the I_{on} and I_{off} currents of the transistors. It increases the resistivity of metals and impacts the propagation time in interconnections, proportional to the product RC (Figure 4.46). It exacerbates thermomechanical stresses resulting from differences in thermal expansion coefficient (ETCs) and electromigration phenomena in interconnections.

The presence of localized hot spots, corresponding to areas of high power density (>100 W/cm^2), is particularly important from a performance perspective. The temperature differences associated with these hot spots induce significant variations in device performance between two areas of the same chip. If it is a CMOS image sensor, for example, the non-uniformity of temperature within the pixel matrix can have important consequences on the final quality of the image.

It is not only electronic, but also ET design that is now necessary to fully measure the consequences of temperature in the operation of a chip: thermal models are increasingly integrated into design kits, as well as floorplaning tools. Once the design is complete, the power profile of the chip is frozen and the only levers to modify the temperature profile, especially to standardize it, become technological.

Passive solutions consist of using materials with high conductivity. In modern housings, the use of heat-spreaders is common to distribute the calories generated by a hot spot over the largest possible area to minimize its impact. Our industrial collaborators (joint between STMicroelectronics and CEA-Leti) have significant experience in passive dissipation based on heat distributors based on copper and pyrolytic graphite. The advantage of these splitters is their small footprint, and they are therefore compatible with integration for very compact applications such as smartphones and tablets. The use of splitters makes it possible to extend the power range dissipated in a chip for the same temperature limit of use.

Active solutions are based on the use of natural or forced convections, in the presence of air or more efficient fluids, in particular due to radiators (heat sinks) and fans. In a modern vision, heat dissipation can be optimized by the integration of micro-fluidic solutions as close as possible to the active areas (Figure 5.46). The effectiveness of such a solution is already demonstrated by simulation and implementation tests are ongoing. Due to the need to use external circuitry and a heat exchanger, this solution, although very efficient, cannot be considered for extremely compact applications.

Compact mobile systems such as phones or tablets cannot integrate heat-sinks and fans, which is why it is necessary to develop the most integrated technological solutions possible. The idea is to create an embedded heat distributor at the chip or housing, based on two-phase fluidic operation, but which does not have the disadvantages encountered in fluidic integration based on micro-channels. In this case, an autonomous solution is sought here, without circuitry or an external heat exchanger.

The steam chamber is a closed cavity, partly made up of a porous material and in which a fluid is placed, the vaporization temperature of which is carefully chosen according to the conditions of use. When it is in contact with a hot spot, the liquid vaporizes. A liquid concentration gradient is then created in the porous material, which serves as a wick for the renewal of the liquid to be vaporized by capillarity. The gas condenses on the cold part(s) and is available again to be transported back to the hot spot. The whole creates a self-sustaining movement that lasts as long as the temperature of the hot spot is higher than the vaporization temperature of the fluid. The overall efficiency of this solution is equivalent to an ultimate heat distributor, whose equivalent thermal conductivity would be greater than 10,000 W/m·K (compared to a maximum of 2000 W/m·K for pyrolytic graphite and 390 W/m·K for copper).

Here, we seek an integration of this approach as close as possible to the hot spots, that is to say in silicon and using the technologies and processes of microelectronics and 3D integration. The implementation and sizing of the solution will have to be considered in the context of practical integration on embedded systems. In addition to achieving the targeted performance levels, determined by the application and nature of the circuits, it must be compatible with microelectronics and packaging technologies. The modeling, in particular for two-phase fluidics, will make it possible to analyze the chip system with integrated vapor chamber from the point of view of the choice of materials and the sizing of the cavity as of the porous material. It will aim to establish the performance according to the geometry, the fluid considered and the boundary conditions corresponding to the power operation of the chip. This step will first include a parametric study according to these different elements.

– Thermal resistances of FHP

Before detailing the analytical-numerical model structure, the relation between the thermal resistances is first studied to understand both the model itself and interpret the results. As shown in Figure 5.43, heat goes through different thermal resistances as it flows from the heat source to the cooling fluid and essentially corresponds to Garimella *et al.*'s approach.

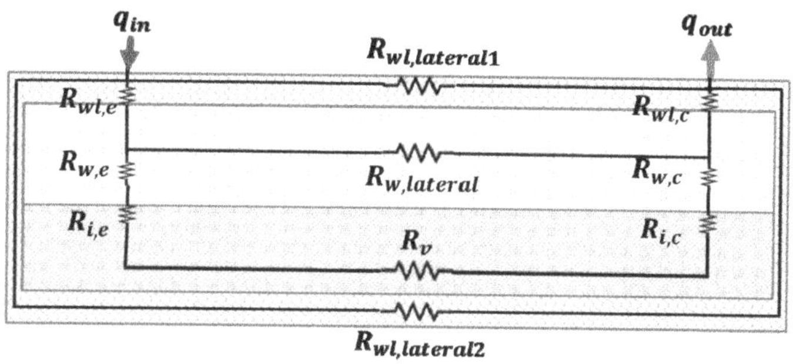

Figure 5.43. *Schematic of approximate network of thermal resistance of FHP. For a color version of this figure, see www.iste.co.uk/abouelatta/smartpower.zip*

The temperature drop due to the primary heat transport mechanisms that occur in each section of the heat pipe can be represented using the simplified effective thermal resistance network, as shown in Figure 5.43. The thermal resistances due to conduction through the thickness of the wall in the evaporator ($R_{wl,e}$) and condenser ($R_{wl,c}$) sections are negligible (because of the thin thickness of wall), which is targeted at ultrathin heat pipes constructed from high-conductivity materials. Lateral conduction along the heat pipe wall is shown as $R_{wl,lateral1}$ and $R_{wl,lateral2}$; these thermal resistances can be calculated using an effective device length that considers the varying heat load along the length of the evaporator and condenser sections, and is given by

$$L_{eff} = L_a + \frac{L_e + L_c}{2} \quad [5.68]$$

Resistances through the wick in the evaporator ($R_{w,e}$), condenser ($R_{w,c}$) and laterally across the wick ($R_{w,lateral}$) are assumed to be due to conduction only. Convective heat transfer in the porous medium is neglected due to the small interstitial liquid velocities; hence, the liquid flow field does not need to be solved. An effective saturated wick thermal conductivity is used to estimate these 1D conduction resistances. Lateral heat flow through the wick is neglected, owing to the high resistance that results from the low thermal conductivity and high aspect ratio wick geometry. To keep the model general, heat loss to the surroundings is neglected.

The interfacial phase-change thermal resistances at the wick–vapor interface in the evaporator and condenser sections, represented by $R_{i,e}$ and $R_{i,c}$, respectively, are typically small and, therefore, neglected. The effective thermal resistance of the vapor core R_v is calculated by representing the vapor flow field as incompressible, laminar, fully developed flow between parallel plates, and governed by continuum physics, to estimate the pressure drop in the vapor core over the effective length. This pressure drop is related to the saturation temperature drop by applying the Clapeyron equation and ideal gas law. By using this approach, an effective thermal conductivity is defined for a 1D lateral resistance along the effective length of the vapor

$$k_{eff,v} = \frac{h_{fg}^2 P_v \rho_v t_v^2}{12 R \mu_v T_v^2} \quad [5.69]$$

where the thermos-physical properties are those of saturated vapor and are evaluated at the local vapor temperature that varies along the effective length of the device. To implement temperature-dependent vapor properties, the 1D vapor thermal resistance is discretized, and each individual resistance is back-calculated from the known condenser-side temperature, and by iterating upon the heat flux passing through the vapor core. This approach does not consider convection effects within the vapor core.

Combining the system resistances yields the effective heat pipe resistance:

$$R_{HP} = \frac{1}{\frac{1}{R_{wl,lateral1}} + \frac{1}{R_{wl,lateral2}} + \frac{1}{(R_v + R_{w,e} + R_{w,c})}} \quad [5.70]$$

– Performance-limiting conditions

By comparing the performance across different input heat fluxes and for a range of geometries, it is possible to identify the limiting conditions that become predominant at small thickness and low input power operation. Figure 5.44. shows a contour map of the ratio between the heat pipe and copper heat spreader thermal resistances, with the adiabatic length and total thickness being varied. The colored region indicates advantageous heat pipe performance, and fades to white as the ratio transitions in favor of the copper heat spreader. The performance thresholds and the capillary limit are also indicated. A pressure of 2250 Pa is set as the maximum capillary pressure.

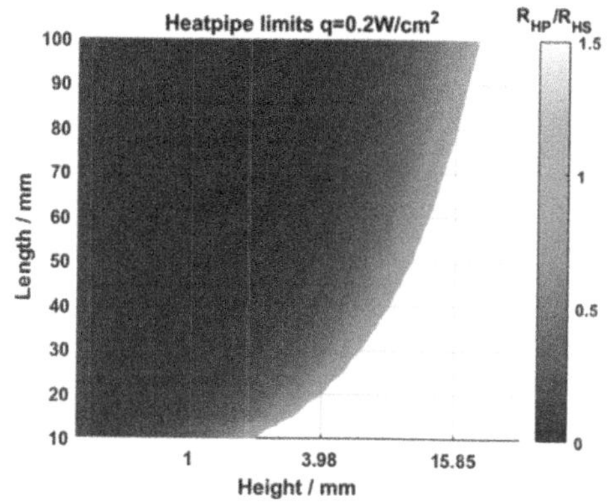

a)

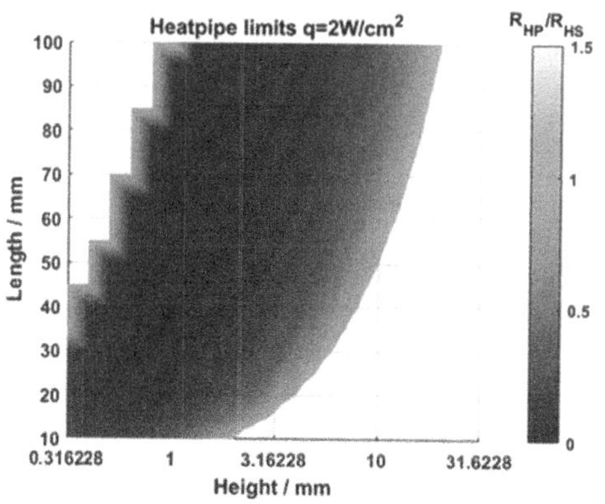

b)

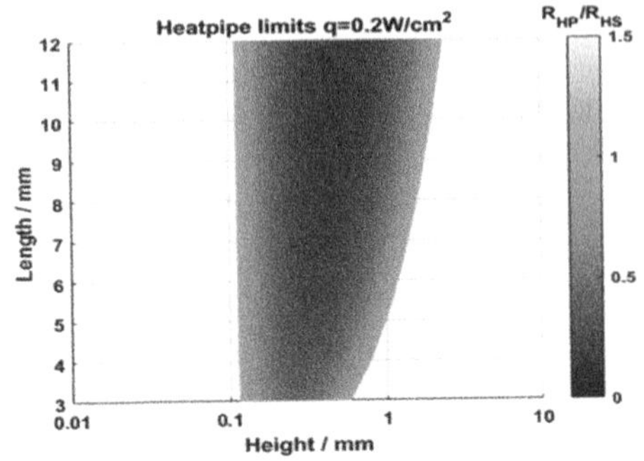

c)

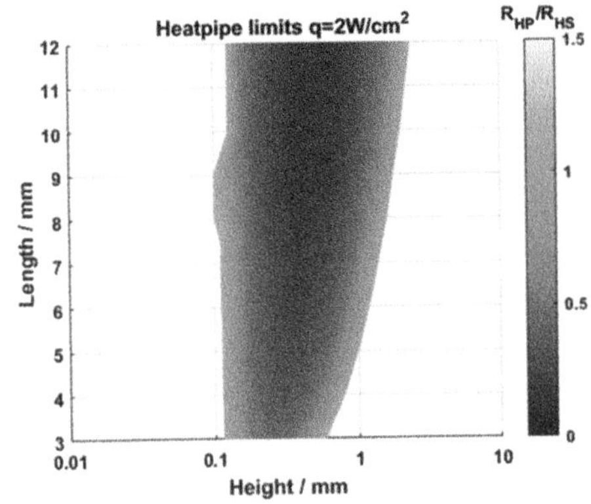

d)

Figure 5.44. *Contour map of the resistance ratio R_{HP}/R_{HS} plotted as a function of adiabatic length and thickness for an input heat flux of a) 0.2 and b) 2 W/cm^2 for length from 10 to 100 mm, and of c) 0.2 and d) 2 W/cm^2 for length from 3 to 12 mm. For a color version of this figure, see www.iste.co.uk/abouelatta/smartpower.zip*

Figures 5.44 (a) and (c) show the limiting conditions for the case of a much lower input heat flux of 0.2 W/cm^2. At low input heat fluxes, there is sufficient capillary pressure due to the reduced fluid velocities, and the capillary limit dramatically shifts to lower thicknesses and higher working lengths.

Figures 5.44 (b) and (d) show the resistance ratio contour plot for a relatively high input heat flux of 2 W/cm^2 to illustrate the limits encountered when designing ultrathin heat pipes for moderate power dissipation. In this case, as the thickness is reduced at a given length, the minimum thickness for which RHP/RHS remains greater than unity is governed by the capillary limit; a sharp transition from white to color indicates this physical transport limit. At small thicknesses and relatively high-heat fluxes, the mass flow is high enough that the pressure drop along the wick and vapor core is higher than the available capillary pressure. The performance advantage of a heat pipe comes from the small resistances incurred due to phase change, and the vapor core between the evaporator and condenser ends when heat is applied to the wick structure. This advantage is lost for short, thick heat pipes, where the effective length is reduced and solid heat spreaders provide a more direct and higher conductivity heat flow path than the heat pipe wick.

Except for the thickness-to-length ratio limit and the capillary limit, there is another limit, which is called vapor-resistance threshold. This threshold is due to the increasing pressure drop (and corresponding saturation temperature drop) along the vapor core. By substituting the individual thermal resistance components and representing the thicknesses of the vapor core, wick, and walls as fractions of the total thickness, that is, $t_v = r_v t$, $t_w = r_w t$ and $t_w l = r_{wl} t$, the threshold conditions may be expressed as:

$$\frac{r_w}{k_w}\left(\frac{1}{L_e}+\frac{1}{L_C}\right)t^4-\frac{L_{eff}}{k_S}\frac{1}{1-r_w}t^2+\frac{L_{eff}}{M_v r_v^3}=0 \qquad [5.71]$$

Solving the equation results in two positive roots representing the thickness at both threshold conditions (vapor-resistance and high thickness-to-length ratio). The limiting minimum thickness governed by the vapor-resistance threshold is

$$t_{limit}=\left(\frac{\frac{L_{eff}}{k_S}a-\sqrt{\left(\frac{L_{eff}}{k_S}\right)^2-4\frac{r_w}{k_w}\left(\frac{1}{L_e}+\frac{1}{L_C}\right)\left(\frac{L_{eff}}{M_v r_v^3}\right)}}{2\frac{r_w}{k_w}\left(\frac{1}{L_e}+\frac{1}{L_C}\right)}\right)^{\frac{1}{2}} \qquad [5.72]$$

where $a = 1/(1 - r_{wl})$ and M_v is a constant representing the vapor properties. This limiting thickness is independent of the heat input; however, the capillary limit should be evaluated in conjunction with this threshold, which would be expected to prevail at any moderate heat inputs. The vapor properties dictating the vapor-resistance threshold are represented as a single factor that can be used as a merit number for fluid selection in the design of the ultrathin heat pipes operating at low heat inputs

$$M_v=\frac{h_{fg}^2\rho_v P_v}{R\mu_v T_v^2} \qquad [5.73]$$

which is independent of geometric parameters.

– 2D modeling approach

The classical configurations of the flat heat pipes, which consist of the wall, the porous wicks saturated with working fluid and the vapor space, are shown; Figure 5.45 presents the configurations for flat heat pipes with a single heat source, multiple heat sources, a heat source at the bottom and a heat sink on the top of the heat pipe, and multiple heat sources with symmetric boundary conditions. The underlying assumptions considered are two-dimensional wall temperature and vapor flow, at steady-state, an incompressible and laminar flow, a saturated wick, constant material properties, constant saturation temperature, and a linear temperature profile across the thin wick structure.

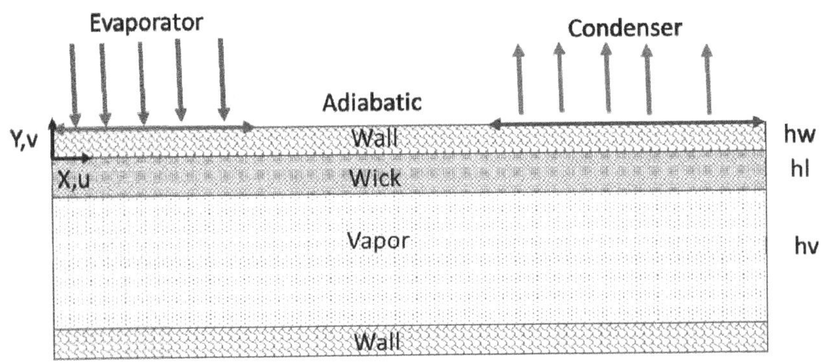

a) Configuration I

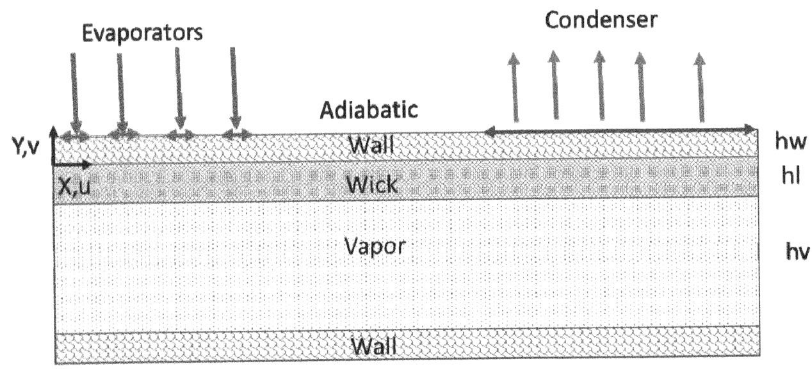

b) Configuration II

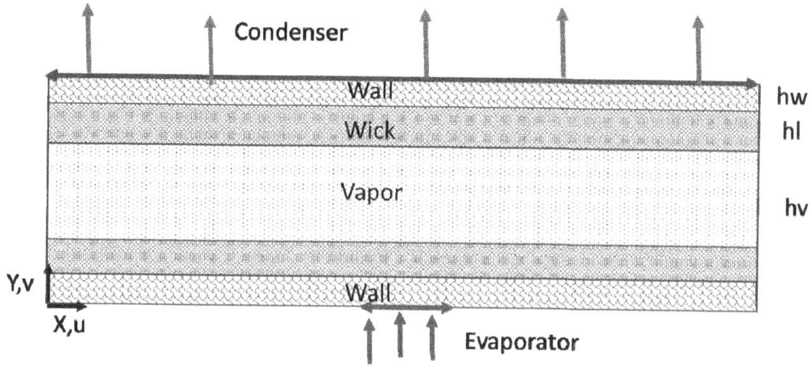

c) Configuration III

Figure 5.45. *Flat heat pipe with different heating/cooling configurations: a) configuration I, single heat source and sink at the top; b) configuration II, multiple heat sources and sink at top; c) configuration III, heat source at the bottom and heat sink on the top. For a color version of this figure, see www.iste.co.uk/abouelatta/smartpower.zip*

– Conduction in wall

The non-dimensional temperature, heat flux and coordinates are defined as:

$$\vartheta = \frac{k_w}{q_e h_w}(T - T_{sat}) \quad [5.74]$$

$$X = \frac{x}{l}; Y = \frac{y}{h_w}; H_w = \frac{h_w}{l}; H_l = \frac{h_l}{l}; L_e = \frac{l_e}{l}; L_c = \frac{l_c}{l}; \gamma = \frac{L_e}{L_c} \quad [5.75]$$

The two-dimensional dimensionless steady-state heat conduction equation in the wall with constant thermal conductivity can be written as:

$$\frac{\partial^2 \theta}{\partial X^2} + \frac{1}{H_W^2}\frac{\partial^2 \theta}{\partial Y^2} = 0 \quad [5.76]$$

The boundary condition at the cap (X = 0 and X = 1) is given as

$$\frac{\partial \theta}{\partial X} = 0 \quad [5.77]$$

At the wick/wall interface (Y = 0), the thermal boundary condition can be determined by assuming a linear temperature profile across the thin wick structure and constant saturation temperature at the liquid–vapor interface. The boundary condition can be written as:

$$\frac{\partial\theta}{\partial Y} = \frac{k_{eff}H_w}{k_w H_l}\theta \qquad [5.78]$$

The boundary conditions at the outer wall (Y = 0) are constant heat fluxes in the evaporators and the condenser, which can be located at different locations on the heat pipe, and a heat flux equal to zero in the adiabatic sections at the rest of the surface. This is shown as:

$$\frac{\partial\theta}{\partial Y} = \begin{cases} 1 \ evaporator \\ 0 \ adiabatic \\ -\gamma \ condenser \end{cases} \qquad [5.79]$$

The adimensional temperature for configurations I and II can be expanded in the form of an infinite Fourier series as:

$$\theta(X,Y) = \sum_{m=1}^{\infty} A_m(Y)\cos{(m\pi X)} \qquad [5.80]$$

The adimensional input heat flux is dependent on the location of heat sources and sinks, and can be written as follows:

$$\frac{\partial\theta}{\partial Y} = \sum_{m=1}^{\infty} B_m \cos{(m\pi X)} \qquad [5.81]$$

where

$$B_m = \frac{2}{m\pi}\{\sin(m\pi L_{2e}) - \sin(m\pi L_{1e}) + \sin(m\pi L_{4e}) - \sin(m\pi L_{3e}) + \sin(m\pi L_{6e}) - \sin(m\pi L_{5e}) + \sin(m\pi L_{8e}) - \sin(m\pi L_{7e}) - \gamma[\sin(m\pi L_{2c}) - \sin{(m\pi L_{1c})}]\} \qquad [5.82]$$

Here, $L_{1e}, L_{2e}, L_{3e}, L_{4e}, L_{5e}, L_{6e}, L_{7e}, L_{8e}, L_{1c}$ *and* L_{2c} express the locations of the heat sources and heat sink.

Substituting [5.80] and [5.81] into [5.76], [5.78] and [5.79] yields the following expression for coefficient A_m:

$$A_m(Y) = B_m \frac{\left(m\pi + \frac{k_{eff}}{k_w H_l}\right)\exp(m\pi H_w Y) + (m\pi - \frac{k_{eff}}{k_w H_l})\exp{(-m\pi H_w Y)}}{m\pi H_w[\left(m\pi + \frac{k_{eff}}{k_w H_l}\right)\exp(m\pi H_w Y) - (m\pi - \frac{k_{eff}}{k_w H_l})\exp{(-m\pi H_w Y)}]} \qquad [5.83]$$

– Vapor flow

A parabolic velocity profile is used for vapor flow within the heat pipe. The velocity distribution is represented by a functional product in the x- and y-directions:

$$u_v(x,y) = -6U_v(x)[\left(\frac{h_l}{h_v}\right)^2 + \frac{h_l}{h_v} + \left(1+2\frac{h_l}{h_v}\right)\left(\frac{y}{h_v}\right) + (\frac{y}{h_v})^2] \quad [5.84]$$

where $U_v(x)$ is the local mean velocity along the x-axis.

The continuity equation for the two-dimensional incompressible vapor flow can be integrated with respect to y to determine $U_v(x)$:

$$\int_{-(h_l+h_v)}^{-h_l} (\frac{\partial u_v}{\partial x} + \frac{\partial v_v}{\partial y}) dy = \int_{-(h_l+h_v)}^{-h_l} \frac{\partial u_v}{\partial x} dy + v_{v,I}(-h_l) - v_v(-(h_l+h_v)) \quad [5.85]$$

where $v_{v,I}$ denotes the vapor interfacial velocity. The heat flux normal to the liquid–vapor interface, q_I, can be calculated from the conduction model with the assumption of a linear temperature profile across the wick:

$$v_{v,I}(-h_l) = \frac{q_I}{\rho_v h_{fg}} = q_e \frac{k_{eff} h_w}{\rho_v h_{fg} k_w h_l} \sum_{m=1}^{\infty} A_m(0) \cos\left(\frac{m\pi x}{l}\right) \quad [5.86]$$

$$v_v(-(h_l+h_v) = 0 \quad [5.87]$$

Integrating equation [5.81] with respect to x and using the boundary condition at the beginning of the evaporator $u_v = 0$ will result in an expression for $U_v(x)$:

$$U_v(x) = q_e \frac{k_{eff} h_w}{\rho_v h_{fg} k_w h_l} \sum_{m=1}^{\infty} (\frac{l}{m\pi}) A_m(0) \sin\left(\frac{m\pi x}{l}\right) \quad [5.88]$$

Therefore, the quasi-two-dimensional velocity profile is:

$$u_v(x,y) = -6q_e \frac{k_{eff} h_w}{\rho_v h_{fg} k_w h_l} \sum_{m=1}^{\infty} \left(\frac{l}{m\pi}\right) A_m(0) \sin\left(\frac{m\pi x}{l}\right) \times [\left(\frac{h_l}{h_v}\right)^2 + \frac{h_l}{h_v} + \left(1+2\frac{h_l}{h_v}\right)\left(\frac{y}{h_v}\right) + \left(\frac{y}{h_v}\right)^2] \quad [5.89]$$

The boundary layer form of the x-momentum equation for steady-state incompressible laminar vapor flow can be integrated to obtain the vapor pressure distribution:

$$\rho_v \int_0^x \int_{-(h_l+h_v)}^{-h_l} \frac{\partial (u_v)^2}{\partial x} dxdy + \rho_v \int_0^x \int_{-(h_l+h_v)}^{-h_l} \frac{\partial (u_v v_v)}{\partial y} dxdy = \int_0^x \int_{-(h_l+h_v)}^{-h_l} -\frac{\partial p_v}{\partial x} dxdy + \mu_v \int_0^x \int_{-(h_l+h_v)}^{-h_l} \frac{\partial^2 (u_v)}{\partial y^2} dxdy \quad [5.90]$$

Assuming the constant vapor pressure at each cross-section, and utilizing the boundary conditions at the beginning of the evaporator, $u_v = 0$ and $p_v = p_{ref}$ yields

$$p_v = p_{ref} + q_e \frac{k_{eff} h_w}{\rho_v h_{fg} k_w h_l h_v^3} \sum_{m=1}^{\infty} A_m(0) \left[\cos\left(\frac{m\pi x}{l}\right) - 1\right] + 18\rho_v U_v^2 \left[\frac{1}{5}\left(\frac{h_l}{h_v}\right)^5 + \frac{1}{30}\right] \quad [5.91]$$

where U_v is obtained from [5.88].

The temperature drop in the vapor can be related to the pressure drop in the vapor region by applying the Clausius–Clapeyron equation and using the ideal gas law:

$$T_v = \frac{1}{\frac{1}{T_{ref}} - \frac{R}{h_{fg}} \log\left(\frac{p_v}{p_{ref}}\right)} \quad [5.92]$$

where R is the ideal gas constant, h_{fg} is the latent heat and p_v is the pressure drop obtained from [5.91].

– Liquid flow

The continuity equation for the incompressible liquid flow can be integrated with respect to y to obtain the liquid axial velocity $u_l(x)$:

$$\int_{-h_l}^{0} \left(\frac{\partial u_v}{\partial x} + \frac{\partial v_v}{\partial y}\right) dy = -v_{l,I}(-h_l) - h_l \frac{du_l}{dx} \quad [5.93]$$

where $v_{l,I}$ denotes the interfacial velocity (normal to the liquid vapor interface) for the liquid and is related to the vapor interfacial velocity by:

$$\rho_l v_{l,I} = \rho_v v_{v,I} \quad [5.94]$$

The axial liquid velocity can be calculated by integrating [5.93] with respect to x using the velocity boundary condition at the beginning of the evaporator (x = 0), $u_l = 0$:

$$u_l(x) = q_e \frac{k_{eff} h_w}{\rho_v h_{fg} k_w h_l^2} \sum_{m=1}^{\infty} \left(\frac{l}{m\pi}\right) A_m(0) \sin\left(\frac{m\pi x}{l}\right) \quad [5.95]$$

The adimensional steady-state conservation of momentum for incompressible liquid flow in the wick can be expressed by Darcy's law, assuming negligible inertial effects in comparison to viscous losses, as:

$$\frac{dp_l}{dx} = -\frac{\mu_l u_l}{K} \quad [5.96]$$

Here, the permeability K is calculated for a mesh screen as:

$$K = \frac{d^2\varphi^3}{122(1-\varphi)^2} \quad [5.97]$$

$$\varphi \approx 1 - \frac{1.05\pi N d}{4} \quad [5.98]$$

The liquid pressure is obtained by integrating [5.97] and utilizing the pressure boundary condition at the end of the condenser (x = l), $p_l = p_v$:

$$p_l = q_e \frac{\mu_l k_{eff} h_w}{\rho_l h_{fg} k_w K h_l^2} \sum_{m=1}^{\infty} \left(\frac{l}{m\pi}\right)^2 A_m(0) \left[\cos\left(\frac{m\pi x}{l}\right) - \cos(m\pi)\right] + p_{ref} + q_e \frac{12\mu_v k_{eff} h_w}{\rho_v h_{fg} k_w h_v^3 h_l} \sum_{m=1}^{\infty} \left(\frac{l}{m\pi}\right)^2 A_m(0)[\cos(m\pi) - 1] \quad [5.99]$$

Capillary pressure is given as

$$p_{cap} = p_v - p_l \quad [5.100]$$

– FEM verification results

The properties of the working fluid, copper and porous wick materials used in the vapor core chamber simulations are shown in Table 5.3, and vapor properties are shown at a temperature of 40°C.

The temperature fields predicted by the modeling approaches are shown in Figure 5.46. Figure 5.46(a) shows the temperature profile comparison along the x direction of 50 W heat source between COMSOL and publication Maziar Agnvami, and the error between them, while Figure 5.46(b) shows the COMSOL simulated temperature profile with 30 W, 40 W and 50 W heat sources. The temperature profile has a maximum temperature at the evaporation side at the surface and a minimum temperature at the condenser side. The plots reveal a high-accuracy match between COMSOL and the theoretical solution. The maximum error is about 6% for configuration I and 0.075% for configuration II. From Figures 5.46(b) and (d), we can see that by increasing 10 W at the heat source, the temperature gains about 3 K for configuration I and 4 K for configuration II.

Property	**Value**
Copper density (ρ_{wall})	8,978 kg/m^3
Water liquid density (ρ_{water})	992.3 kg/m^3
Water vapor density (ρ_v)	0.05122 kg/m^3
Copper thermal conductivity (k_{wall})	400 W/mK
Wick effective thermal conductivity (k_{wick})	1.3 W/mK
Water vapor thermal conductivity (k_v)	0.05 W/mK
Copper specific heat capacity (C_p)	381 J/kgK
Water liquid specific heat capacity (C_p)	4,182 J/kgK
Water vapor specific heat capacity (C_p)	1,889 J/kgK
Water liquid viscosity (μ_l)	0.00065 Pa*s
Water vapor viscosity (μ_v)	0.0000096 Pa*s
Enthalpy of vaporization (h_{fg})	2,473 kJ/kg
Specific gas constant (R_const)	8.3145 J/molK
Water vapor mean molar mass (M_n)	0.018015 kg/mol
Wick porosity (φ)	0.5
Wick permeability (K)	1*10-11 m^2

Table 5.3. *Properties of the working fluid, copper and porous wick materials used in the heat pipe simulations*

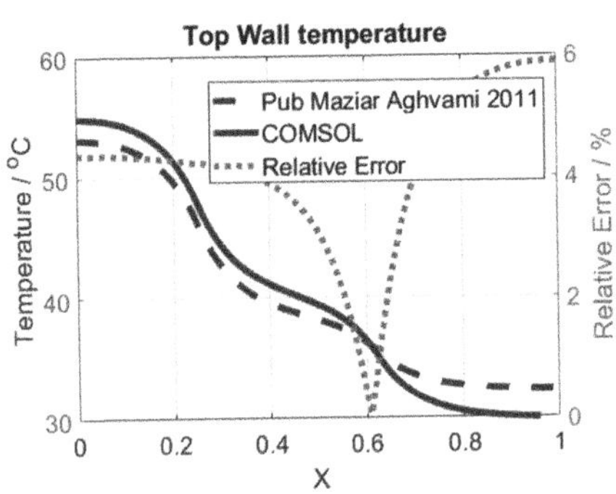

a)

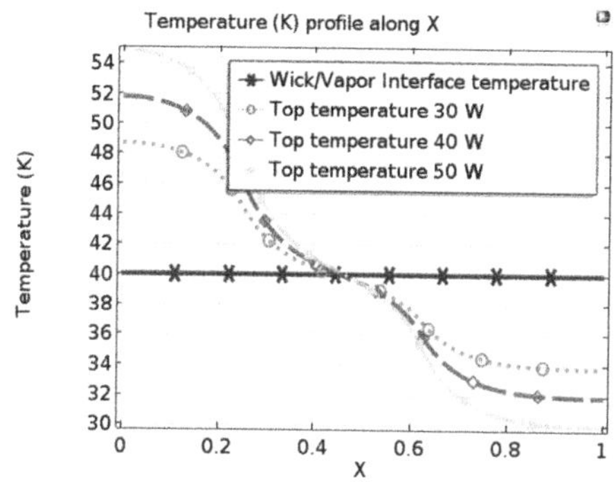

b)

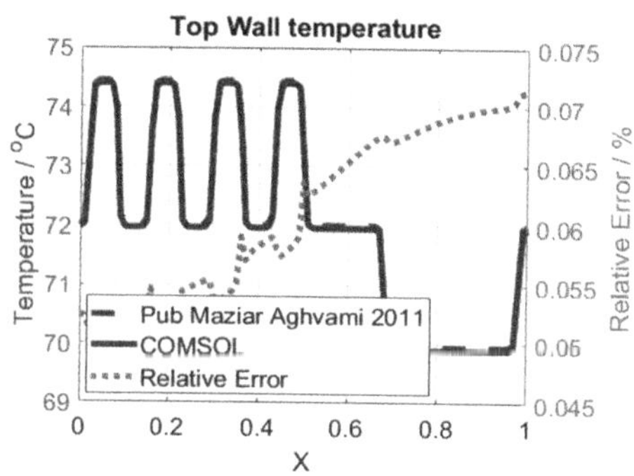

c)

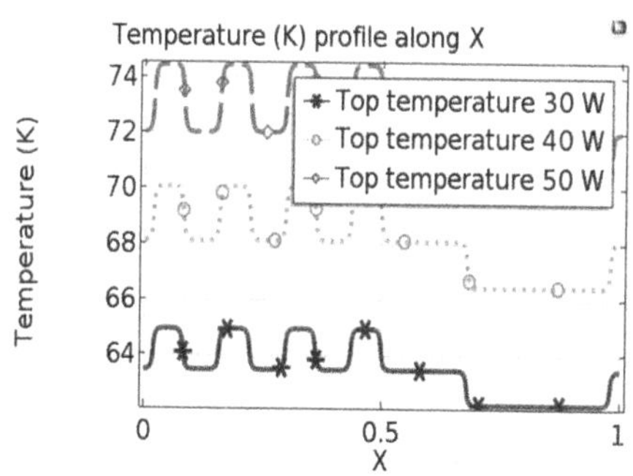

d)

Figure 5.46. *Temperature profile: a) comparison between COMSOL and pub with 50 W heat source in Matlab configuration I; b) parameterization of power in COMSOL configuration I; c) comparison between COMSOL and pub with 50 W heat source in Matlab configuration II; d) parameterization of power in COMSOL configuration II. For a color version of this figure, see www.iste.co.uk/abouelatta/smartpower.zip*

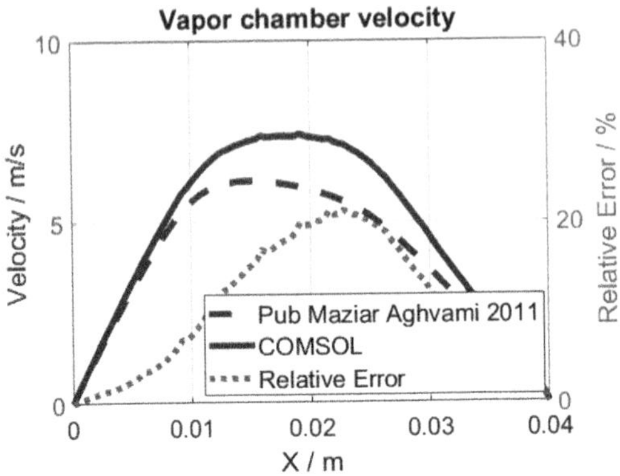
Vapor chamber velocity
Velocity / m/s
Relative Error / %
X / m
Pub Maziar Aghvami 2011
COMSOL
Relative Error
10
5
0
40
20
0
0
0.01
0.02
0.03
0.04

a)

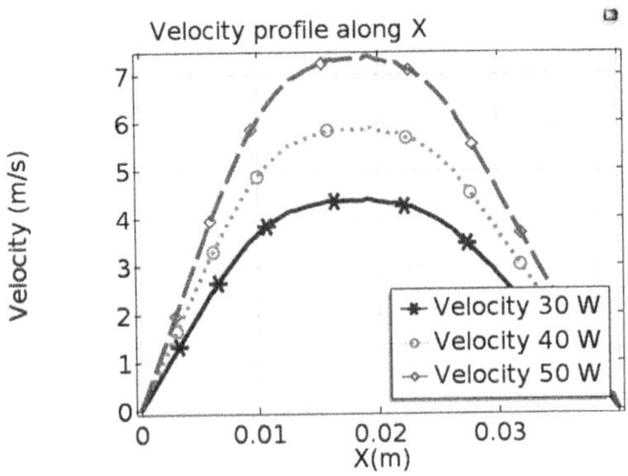
Velocity profile along X
Velocity (m/s)
X(m)
Velocity 30 W
Velocity 40 W
Velocity 50 W
0
1
2
3
4
5
6
7
0
0.01
0.02
0.03

b)

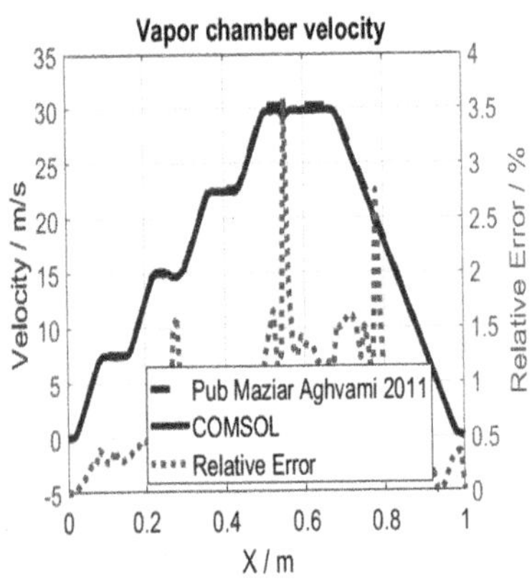
Vapor chamber velocity
Velocity / m/s
Relative Error / %
X / m
Pub Maziar Aghvami 2011
COMSOL
Relative Error
35
30
25
20
15
10
5
0
-5
4
3.5
3
2.5
2
1.5
1
0.5
0
0
0.2
0.4
0.6
0.8
1

c)

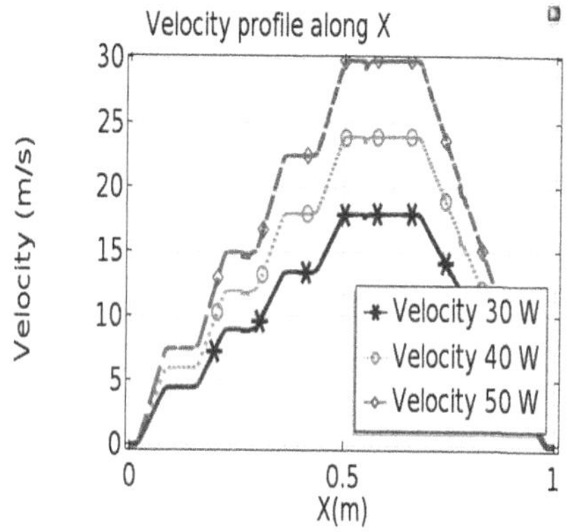

d)

Figure 5.47. *Comparison of the velocities at the center of the vapor chamber: a) comparison between COMSOL and pub with 50 W heat source in Matlab configuration I; b) parameterization of power in COMSOL configuration I; c) comparison between COMSOL and pub with 50 W heat source in Matlab configuration II; d) parameterization of power in COMSOL configuration II. For a color version of this figure, see www.iste.co.uk/abouelatta/smartpower.zip*

As we can see in Figure 5.47, the vapor has a maximum velocity at the vapor core center and it reduces to 0 when it attaches to the wall. For the COMSOL model, the maximum velocity is about 7 m/s, almost the same as the velocities of paper. For configuration I, there is approximately a 20% difference between the analytical solution and COMSOL solution; for the analytical solution in particular, the velocity of configuration I has a non-symmetry form. But for configuration II, the difference in velocity is less than 4%. By increasing 10 W of the heat source, the velocity gains about 1.5 m/s for configuration I and 6 m/s for configuration II.

Finally, the most important part of the heat pipe is the capillary pressure in the vapor/wick interface. Figure 5.48 shows the capillary pressure at the wick/vapor interface. At the evaporation side, the water evaporates and the vapor has a maximum pressure gain at this side, but at the condenser side, the water is condensed and causes the liquid flow in the wick, so we can see a contrary process in the wick flow; the wick has a mass flow at the condensation side, and a mass flow at the evaporation side, so the water flows from the condensation side to the evaporation side, but with a velocity much slower than the vapor core. From Figure 5.48, we can see that the capillary pressures of configuration I of COMSOL simulation and publication have at maximum 20% difference. For configuration II, the capillary pressures are almost the same. The error is less than 0.5%.

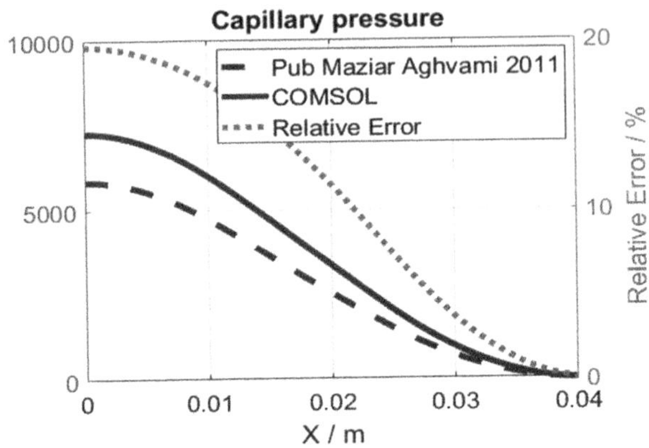
Capillary pressure
Pub Maziar Aghvami 2011
COMSOL
Relative Error
10000
5000
0
20
10
0
Relative Error / %
0
0.01
0.02
0.03
0.04
X / m

a)

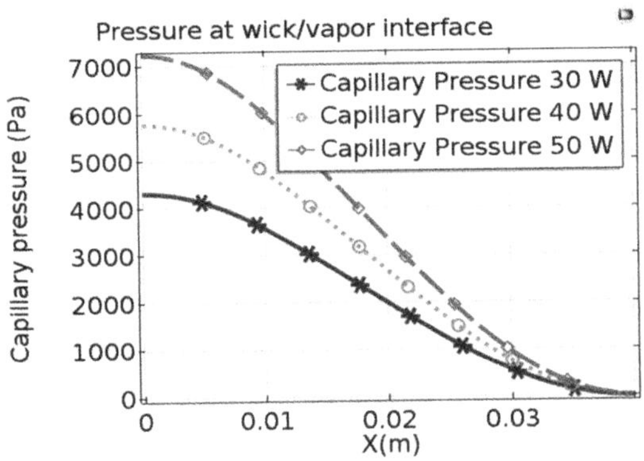
Pressure at wick/vapor interface
Capillary Pressure 30 W
Capillary Pressure 40 W
Capillary Pressure 50 W
Capillary pressure (Pa)
7000
6000
5000
4000
3000
2000
1000
0
0
0.01
0.02
0.03
X(m)

b)

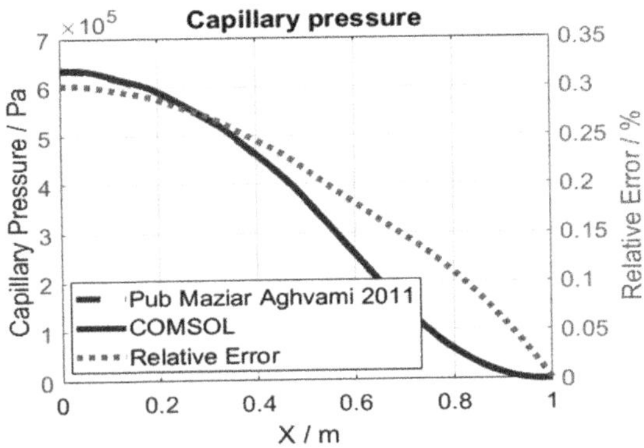
Capillary pressure
×10^5
Capillary Pressure / Pa
Relative Error / %
Pub Maziar Aghvami 2011
COMSOL
Relative Error
7
6
5
4
3
2
1
0
0.35
0.3
0.25
0.2
0.15
0.1
0.05
0
0
0.2
0.4
0.6
0.8
1
X / m

c)

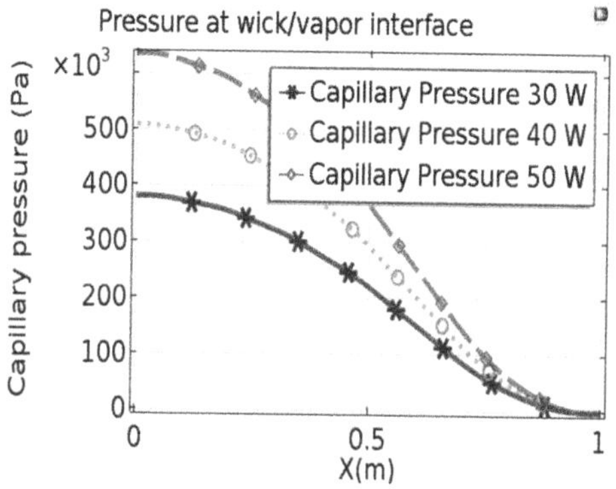

d)

Figure 5.48. *Perssure (Maylab calculatios) at the wick–vapor interface: a) comparison between COMSOL and pub with 50 W heat source in Matlab configuration I; b) parameterization of power in COMSOL configuration I; c) comparison between COMSOL and pub with 50 W heat source configuration II; d) parameterization of power in COMSOL in Matlab configuration II. For a color version of this figure, see www.iste.co.uk/abouelatta/smartpower.zip*

A stationary model for vapor chamber operation was developed that allows for multiple, arbitrarily shaped, heat inputs on the evaporator-side face; the model predicts 2D fields of temperature, pressure and velocity in the vapor chamber, subject to assumptions regarding the vapor chamber size, liquid and vapor flow, and the material properties. The governing mass, momentum and energy equations in the wall, wick and vapor core domains were established. The model is compared with a numerical model.

5.6. Conclusion

Numerical modeling, analysis and experimental simulation of heat pipes have significantly progressed due to a much greater understanding of various physical phenomena in heat pipes, as well as advances in computational and experimental methodologies.

FEM-based models (COMSOL Multiphysics) are very accurate in the 2D and 3D simulations of heat pipes. Precise boundary conditions can be found in the previous part; it considers the pressure field in the calculation, and it can especially extract the capillary pressure directly from the model, which will be very useful in the future study of the wick layer's capillary effects.

As a future work, the simplification for the evaporated mass should be changed to temperature-dependent parameters, and a quantitative analysis of the model error will be performed over a range of operating conditions due to the simplifying assumptions employed. This model will be changed to a 3D model and then used to simulate the behavior of a vapor chamber subjected to multiple heat input boundary conditions to demonstrate the capability of the model to resolve the 3D thermal response to complex boundary conditions expected in real-world applications. Before coming to the real applications, some transfer mechanisms should be included in the model. These include the effects of external boundary conditions on the interfacial mass flux profile, convection in the vapor core and smearing of the interfacial mass flux into the adiabatic section by 2D conduction. However, these additional mechanisms should be accounted for detailed numerical models for the application-oriented design and optimization of ultrathin-form-factor heat pipes.

This study highlighted the relative importance of the three kinds of resistances met in microfluidic chip cooling, and that conventional thinking is challenged when realistic module configurations are considered.

6

Substrate Coupling in Smart Power Integration

6.1. Introduction

In this chapter, two technologies are proposed to implement the smart power integrated circuits (ICs). The first is planar integration using the deep trench isolation (DTI) technique; the second is the new stacked three-dimensional (3D) integration using through-silicon vias (TSVs). Hence, the chapter is divided into two parts; in each part, we study the impact of substrate perturbations due to the HV digital signal on the performance of low-voltage MOS devices.

6.2. Part I: smart power integration using the DTI technique

6.2.1. *DTI technology*

In smart power ICs, there have been substantial improvements in performance and reliability, as well as reductions in cost, compared to the discrete approach. The savings come from the elimination of the many packages needed to house the individual chips. However, cross-talk between power devices, as well as interaction between power and low-voltage devices, have been the most significant hurdles.

The objective of this section is to introduce the DTI scheme for high-voltage isolation. Deep trench-based processes have been integrated previously in CMOS architectures to suppress inter-well parasites and CMOS thyristor latchup. This isolation scheme is used nowadays to reduce the isolation distance between power devices, as well as between power devices and low-voltage CMOS devices, and

For a color version of all the figures in this chapter, see www.iste.co.uk/abouelatta/smartpower.zip.

hence to drastically reduce the total chip area in comparison to technologies with a standard junction isolation scheme.

Although SOI platforms are very effective in electrical isolation and chip area reduction, the complexity of the design, higher wafer cost and lower heat removal capability limit the use of these technologies. The deep trench high-voltage isolation is a low-cost alternative to SOI in realizing significant analog shrink. This type of isolation is suitable for smart power automotive ICs.

In this chapter, the impact of DTI schemes on the performance of both the HV LDMOS and the LV CMOS devices is discussed. The electrical impact of the (0.0/42.0V) dynamic signal of future automotive applications on the CMOS parts is also studied.

6.2.2 *DTI structure*

During the development phase of the deep trench, severe silicon stress situations are encountered, leading to defects within the silicon material. An extensive process is required in order to finally achieve a complete structure that is free of defects. The filler material combination (oxide/polysilicon) is chosen for this purpose.

A high-quality thermal oxide liner is grown along the sidewalls of the deep trench. A polysilicon stress-relief layer is deposited so that it fills the deep trench and is recessed below the silicon surface. The polysilicon has an almost perfect conformal deposition, that is, both step coverage and bottom coverage are 100%.

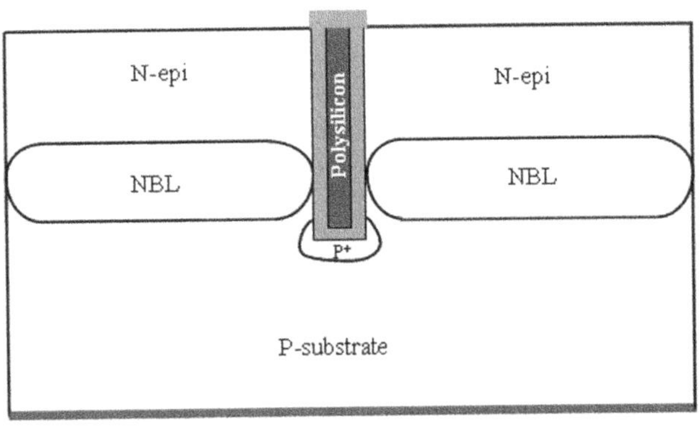

Figure 6.1. *The deep trench isolation structure*

However, when two N-epi regions are placed adjacent to a deep trench separation, as shown in Figure 6.1, punchthrough under the deep trench becomes a concern. This can be avoided in P-substrate with punchthrough retardation implants (P^+ boron implant) or simply by increasing the depth of the trench significantly.

6.2.3. *LDMOSFET performance with DTI*

In Figure 6.2, nLDMOS and pLDMOS devices, isolated with DTI, are shown which construct a complementary LDMOS (cLDMOS). An in-depth investigation was done with the help of technology computer-aided design (TCAD) simulations in order to describe the functionality of the DTI structure fully.

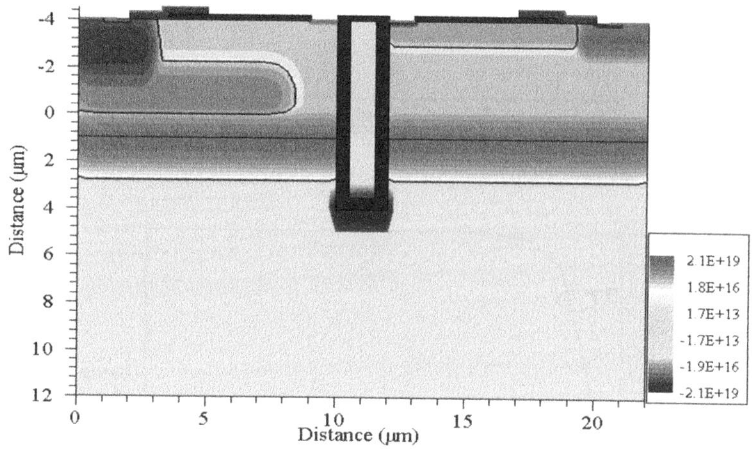

Figure 6.2. *The structure of cLDMOS with deep trench isolation*

6.2.3.1. *OFF-state performance*

Leakage current levels of nLDMOS and pLDMOS are very low with DTI and are no different to junction leakage without deep trench, indicating that there is no additional leakage due to deep trench presence, as illustrated in Figures 6.3a and b. The thickness of the deep trench oxide is sufficient to sustain a breakdown voltage from N-epi to substrate of more than 300 V, which is greater than the breakdown voltage of both devices.

As shown in Figure 6.3, the leakage currents for nLDMOS and pLDMOS are in the order of 1.0E-14 A/μm and 1.0E-19 A/μm, respectively. These values of leakage currents are very small compared to the values of the drain currents. It is clear from the figure that the OFF-state breakdown and the currents of the drain and the bulk are approximately unchanged in the presence of trench isolation.

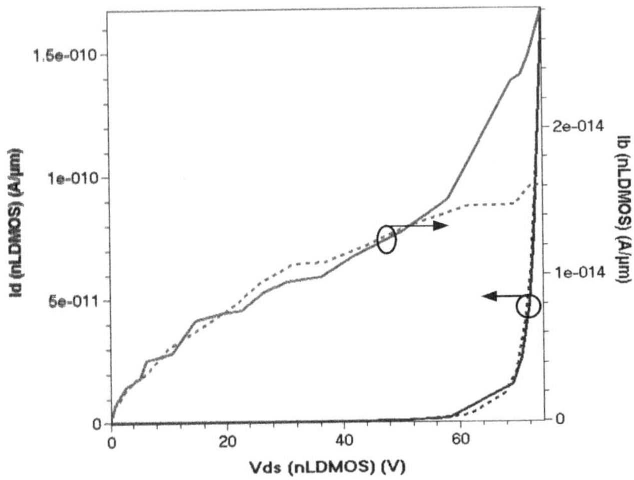

a)

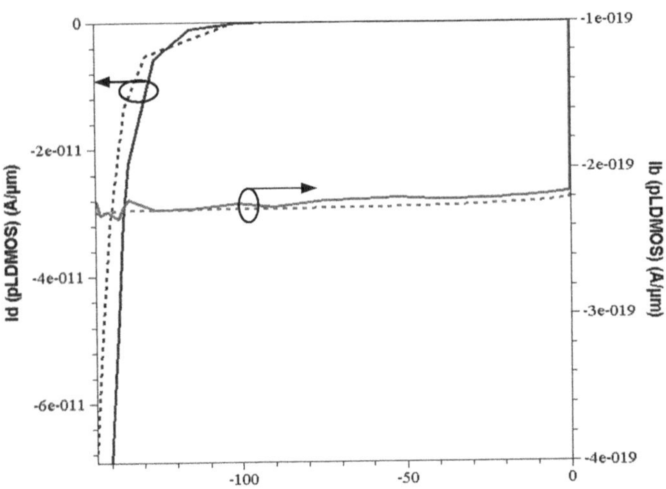

b)

Figure 6.3. *The drain and bulk currents for a) nLDMOS and b) pLDMOS, without trench isolation (dashed curves) and with trench isolation (solid curves)*

The potential and electric field distributions in the OFF state for both nLDMOS and pLDMOS are shown in Figures 6.4 and 6.5. The deep trench does not affect the distribution of the electric field, which is approximately distributed uniformly over the drift region of each device.

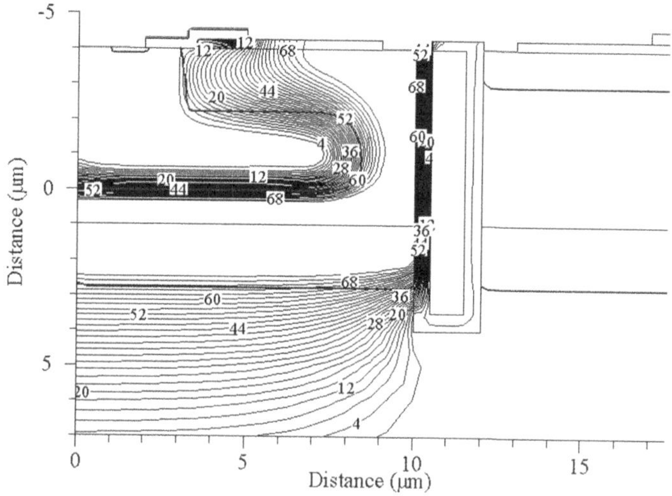

a)

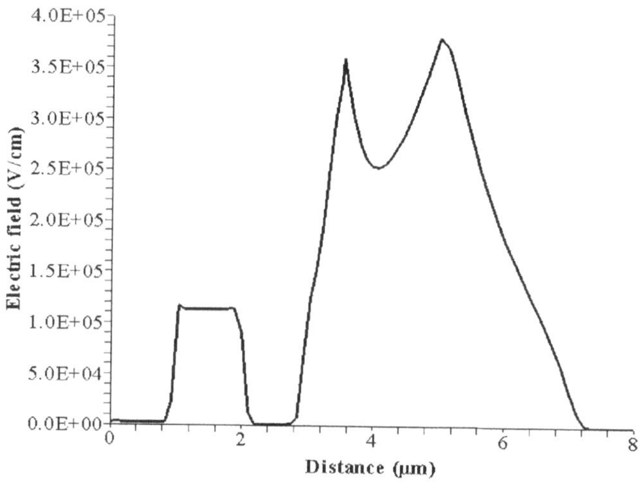

b)

Figure 6.4. *a) Potential distribution and b) electric field distribution in nLDMOS*

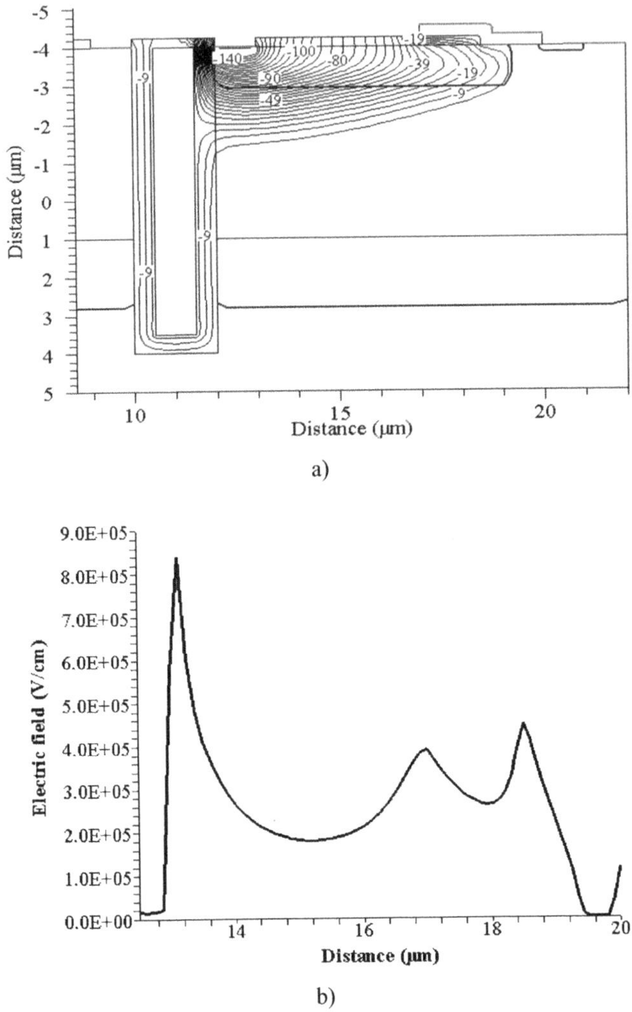

Figure 6.5. *a) Potential distribution and b) electric field distribution in pLDMOS*

6.2.3.2. ON-state performance

The ON-state characteristics are not changed with the use of the DTI. The ON-state characteristics of both nLDMOS and pLDMOS at $|V_{GS}| = 3$ V are shown in Figure 6.6.

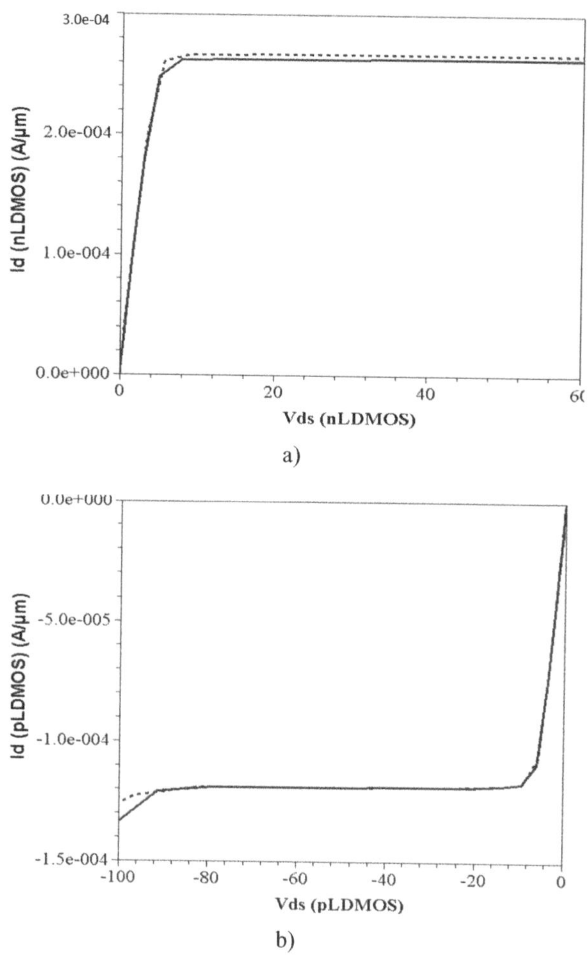

Figure 6.6. *The drain currents of a) nLDMOS and b) pLDMOS, at* $|V_{GS}| = 3$ *V, without trench isolation (dashed curve) and with trench isolation (solid curve)*

6.2.4. *Parasitic suppression in 2D smart power ICs with deep trench*

Minority carrier diffusion is a core problem in smart power ICs where underground conditions on the output driver stage can lead to significant electron current injection into the substrate. Negative voltages (down to –1.5 V) occur in power stages due to inductive load switching during normal operation (motor control in automotive applications), causing the injection of minority carriers into the substrate which leads to their collection by sensitive N-wells in the pLDMOS device or in the CMOS devices; this in turn causes potential functionality failures.

A simplified cross-section of the cLDMOS, CMOS and the parasitic NPN BJT is shown in Figure 6.7. In this structure, the CMOS is simplified to have the source contacts of the MOS devices only.

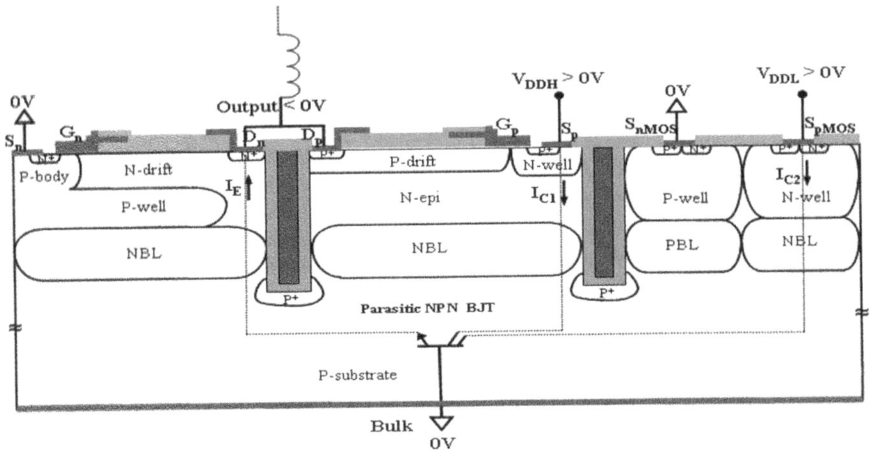

Figure 6.7. *Simplified cross-section, showing the cLDMOS, the CMOS and the parasitic NPN transistor*

Figure 6.7 illustrates that the N^+ source contact of the nMOS and the N^+ source-body contact of the nLDMOS are tied to ground, while the P^+ source contact of the pMOS is connected to V_{DDL} (3.3 V) and the P^+ source-body contact of the pLDMOS is connected to V_{DDH} (42.0 V), which is the new automotive battery voltage.

Due to the negative bias applied to the drain of the cLDMOS, the N-epi/ P-substrate diode becomes forward-biased and induces a huge current. Most of the current is drained toward the source (S_n) through the intrinsic body diode. The remaining part of the injected current (I_E) flows into the substrate and can be collected by any reverse-biased junction.

The N-region of the pLDMOS at V_{DDH} acts as a collector of the lateral parasitic NPN transistor with current I_{C1}, as shown in Figure 6.7. Other N-regions on a positive potential V_{DDL}, for example, those of the controlling circuitry, also collect the minority carriers with current I_{C2}, which can disturb the functionality of its components.

In BiCMOS technology, all devices share the same epitaxial layer, which leads to cross-talk between power devices as well as between power and low-voltage CMOS devices. The injection collected ratio α, defined as the ratio of the collected current I_C to the injected current I_E ($\alpha = I_C / I_E$), reaches about 0.35. However, when using DTI, this ratio is reduced by a factor between 3 and 8.5.

Insight investigations are performed with a TCAD device simulator because a circuit simulator does not take into consideration minority carriers. The simulations are carried out with a voltage of –2 V applied on the drain of the cLDMOS and all the other contacts are grounded; this is because it is difficult to characterize the structure of Figure 6.7 with multi-potential levels (–2 V, 0 V, 3.3 V, 42.0 V) in TCAD because of the convergence problems.

The injection collected ratio (α) of the most sensitive devices as a function of the trench length (L_{DTI}) in two cases, with and without P^+ implant, is shown in Figures 6.8a and b.

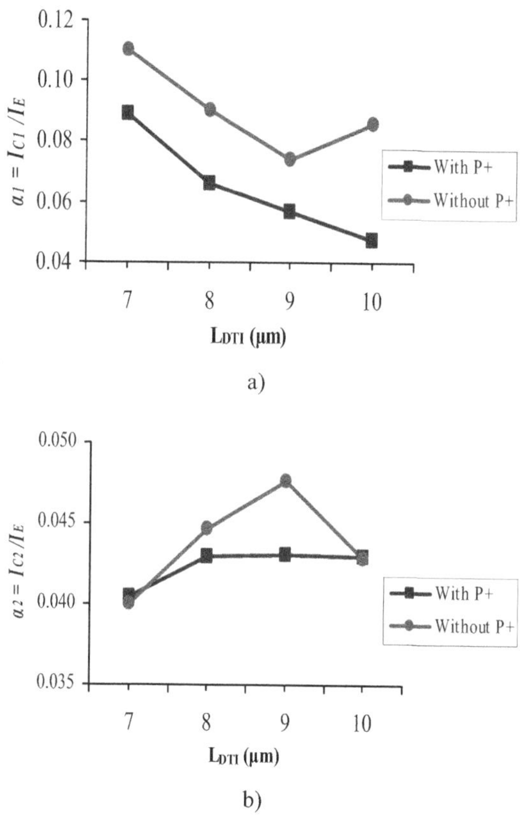

Figure 6.8. *The injection collected ratio of a) pLDMOS and b) pMOS in two cases with and without P^+*

For the first case, trench isolation without P^+ implant, the injection collected ratio α_1 decreases with the increase of L_{DTI} for $L_{DTI} \leq 9$ µm, as shown in Figure 6.8a. For the same range, values of L_{DTI} α_2 are increased, as illustrated in Figure 6.8b. For L_{DTI} = 10 µm, the situation is reversed. This means that the amount of deflected carriers changes as a function of the trench length and the distance between the emitter and the collector regions.

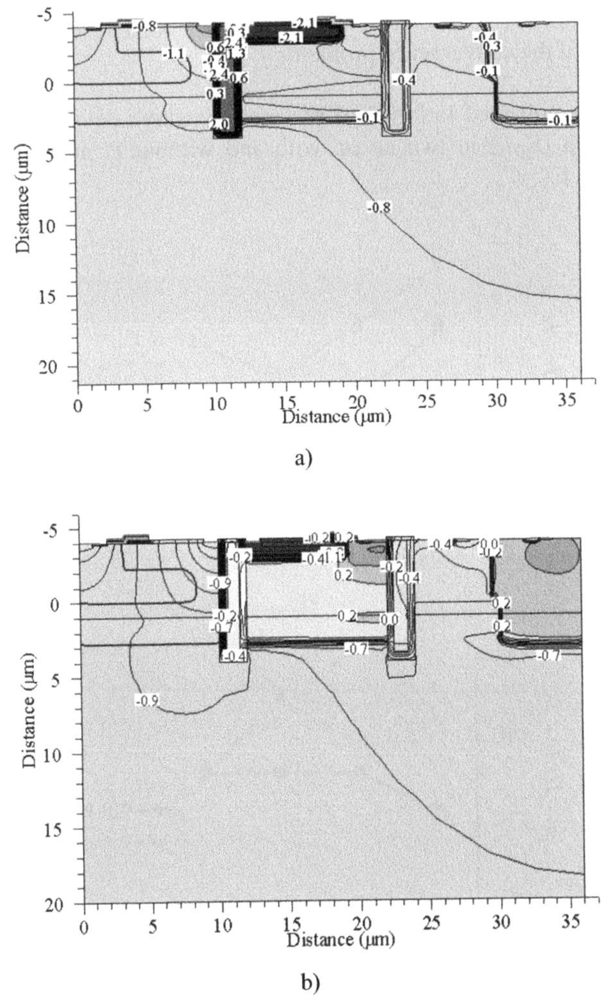

Figure 6.9. *The potential contour distributions in the case of a) without P^+ and b) with P^+ implant*

For the second case, trench isolation with P^+ implant, the injection collected ratio α_l decreases with the increase in L_{DTI} for the whole range and is reduced by more than 23% compared to the case without P^+, as shown in Figure 6.8a; α_2 is nearly constant, as illustrated in Figure 6.8b. As a result, this highly doped implant region reduces the lifetime of the minority carriers and hence reduces the collected carriers of the sensitive devices.

The negative substrate potential shift caused by the injected minority carriers has been simulated for a trench length of 8 μm and for two cases: with and without a P^+ implant. Figure 6.9 illustrates that the potential distribution, and hence the electric field distribution, is a strong function of P^+ implant regions.

Without the high implant region, the potential contours are crowded in the sensitive devices and this increases the electric field in these regions, which in turn increases the collected minority carriers. In the other case, in the presence of the high P^+ implant region, the potential contours are reduced and hence this reduces the collected current.

6.2.5. *HV dynamic signal impact on CMOS devices*

To describe the HV dynamic signal impact on the CMOS parts through the substrate perturbations, we must first describe the different injection mechanisms in a digital circuit. Most CMOS digital gates can be reduced into CMOS inverters. Figure 6.10 shows a CMOS inverter and its different sources of substrate noise.

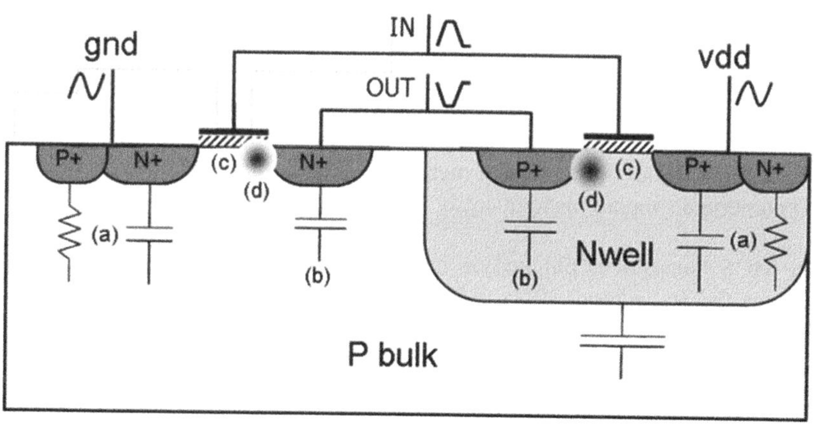

Figure 6.10. *Substrate noise injection mechanisms in a digital inverter*

The substrate coupling is through the parasitic capacitances of the MOSFET. First, the drain depletion (b) capacitance of each MOS transmits voltage variations into the substrate. In the case of an inverter, the output voltage is directly coupled to the substrate through the drain depletion capacitance of the nMOS and the pMOS. Additionally, the gate electrode is coupled to the bulk through the gate oxide (c) and the channel capacitance.

Another noise generator is the well-known impact ionization (d). When a MOS transistor is biased in the saturation regime, a high electric field develops in the depleted region of the channel near the drain. This high field generates hot carriers that can dissipate their energy by creating some electron–hole pairs. For an nMOS, the holes created are swept to the substrate. Other minor noise sources are described in different developments such as gate-induced drain leakage, photon-induced current and diode leakage current.

All carriers created by the injectors are propagated through the substrate, which has to be quantified. Silicon material is characterized by a cutting frequency $f_c = \sigma/_{2\pi\varepsilon}$ where σ is the conductivity and ε the permittivity. With high resistivity substrate (>10 Ω·cm), silicon can be considered as purely ohmic for a signal frequency below 10 GHz. This propagation could vary the body voltage, which in turn varies the MOS threshold voltage according to the following equation:

$$V_T = V_{TO} + \frac{\sqrt{2.q.\varepsilon.N_A}}{C_{ox}}.\left(\sqrt{\left|-2\Phi_f + V_{sb}\right|} - \sqrt{\left|2\Phi_f\right|}\right) \qquad [6.1]$$

Several methods exist to analyze the substrate propagation, commonly considered as propagation within a finite conducting medium. The main methods are as follows: solving the Poisson equation numerically with Green's equations and the boundary element method, or extracting an RC network model of the substrate from a 3D mesh, that is, the finite element method (FEM). In this chapter, we will use the TCAD numerical simulations to quantify the coupling mechanism.

The MOS transistor is plugged on its static mode ($V_{GS} = V_{DS} = 1.8$ V) for the nMOS and the static mode ($V_{GS} = V_{DS} = -0.5$V) for the pMOS, which are the biasing conditions for the convergence simulated structures. A square voltage (f = 200 MHz, 200 ps-long rise and fall ramp time, 42 V peak voltage) is then applied on the drain of cLDMOS, which is represented by the N^+ electrode in the N-epi region. The saturation drain current I_{DSAT} and the local body potential V_{body}, located in between the conduction channel at 0.5 µm under the gate oxide/silicon interface, are extracted during the transient analysis. This position enables us to quantify the smallest electrical disturbances that may modify the electrical behavior of the transistors. We use a 100 µm substrate thickness with its backend connected to ground.

6.2.5.1. *Impact on nMOS*

The 2D cross-section of the simulated structure is shown in Figure 6.11, while Figure 6.12 shows the potential contour distributions. It is observed that all the potential lines are distributed across the deep trench and the N-epi/P-sub junction space charge region, far from the active area of the nMOS, so we expect that the effect of the applied dynamic signal is limited and negligible.

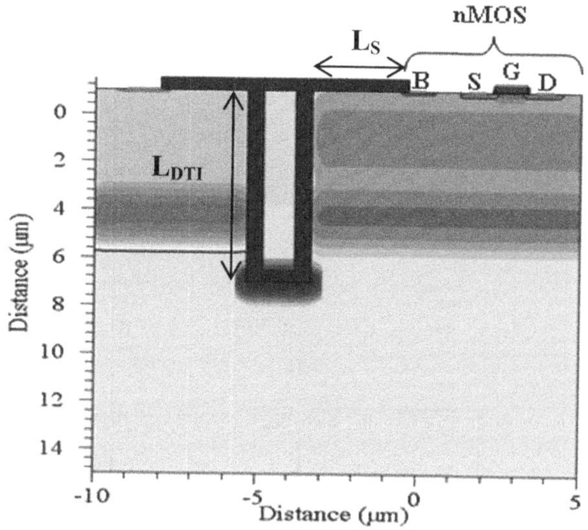

Figure 6.11. *The 2D structure of the nMOS with DTI*

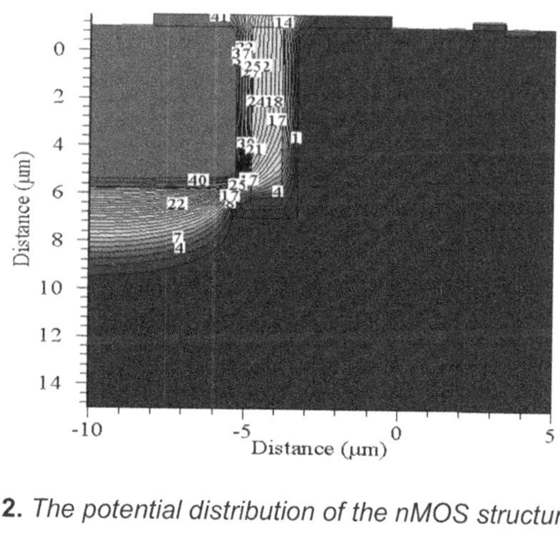

Figure 6.12. *The potential distribution of the nMOS structure with DTI*

The waveforms of the saturation drain and the bulk currents are shown in Figures 6.13a and b. The signal coupling is through the parasitic capacitances of the N-epi/P-sub diode and the DT and the resistive network of the substrate of this system acts as a differentiator for the dynamic signal applied on the drain electrode of the cLDMOS.

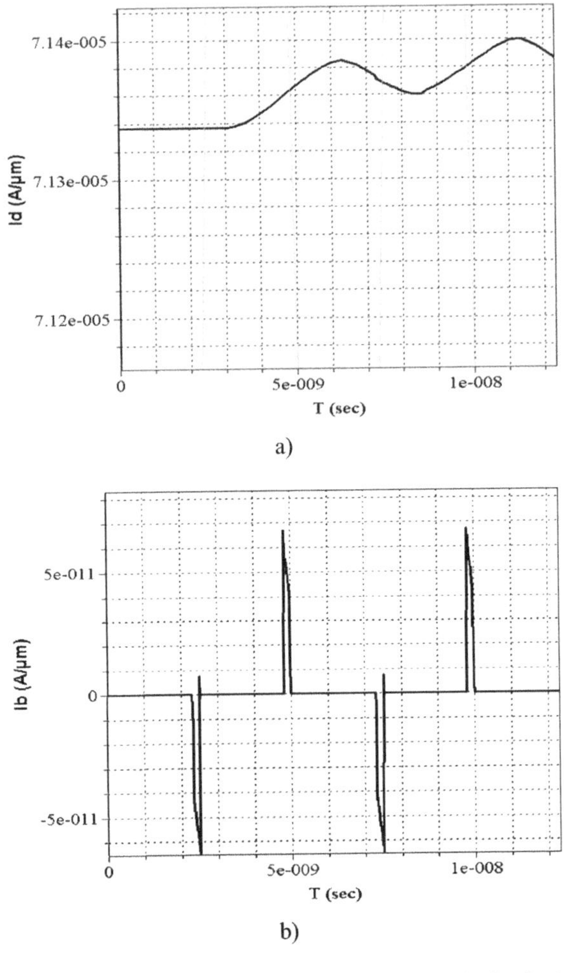

Figure 6.13. *a) The saturation drain current and b) the bulk electrode current due to 0.0/42.0 V square signal applied on the cLDMOS drain for L_{DTI} = 8 µm*

This fact is clearly illustrated through the waveform of the bulk current in Figure 6.13b. The variation in the bulk current is about 65 pA/µm, which can be neglected with respect to the value of the saturation drain current. For the drain current in Figure 6.13a, there is a ringing effect due to the capacitance coupling, but the variation is very small in the range of 0.04%, which can be neglected. Also, the waveform of the body voltage has no variations.

6.2.5.1.1. Effect of DTI length (L_{DTI})

In Figure 6.14, the percentage of the average variation in the drain current and the variation in the bulk current are shown as a function of trench length (L_{DTI}). The figure illustrates that increasing L_{DTI} reduces variation in the saturation current (from more than 4E-2% to less than 1E-2%) because of increasing the base width of the parasitic BJT, consisting of the N-epi (the drift region of nLDMOS) as an emitter region, the (P-sub+P-well + PBL) as a base region and the N^+ source/drain of the nMOS as a collector. By increasing the base width of the parasitic BJT, the minority carriers transit time decreases and hence the collected carriers are decreased.

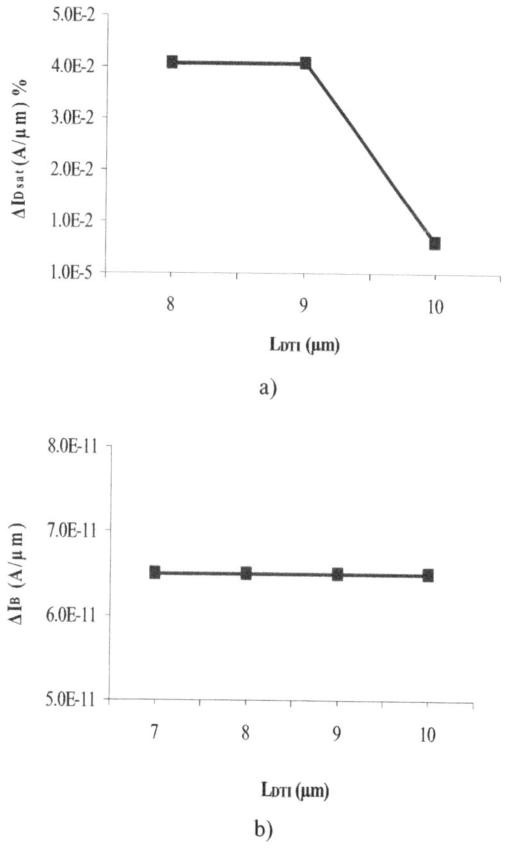

Figure 6.14. *a) The saturation drain current and b) the bulk current, as a function of the deep trench length for L_S = 3 μm*

The variation in the bulk current does not depend on the BJT effect but on the reverse conditions of the (N-epi/P-sub) diode. So, the variation value is independent of the trench length, as shown in Figure 6.14b.

6.2.5.1.2. Effect of DTI spacing (L_S)

The layout parameter L_S (keep-out zone) may impact the coupling, as illustrated in Figure 6.15. It is evident that the keep-out zone seems to have an impact on the coupling mechanism, especially for the bulk current. As L_S increases, the coupling resistance increases and hence the collected currents are decreased.

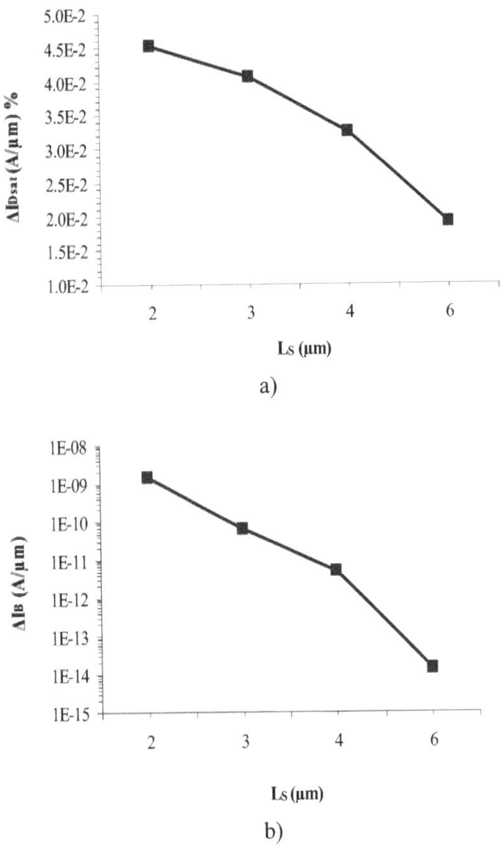

Figure 6.15. *a) The saturation drain current and b) the bulk current, as a function of the keep-out zone length for L_{DTI} = 8 µm*

6.2.5.2. *Impact on pMOS*

The 2D cross-section of the simulated structure is shown in Figure 6.16 and the potential contour distribution in the ON-state is shown in Figure 6.17. It can be seen that all of the potential lines distributed across the deep trench and the N-epi/ P-sub junction space charge region are far from the active area of the pMOS.

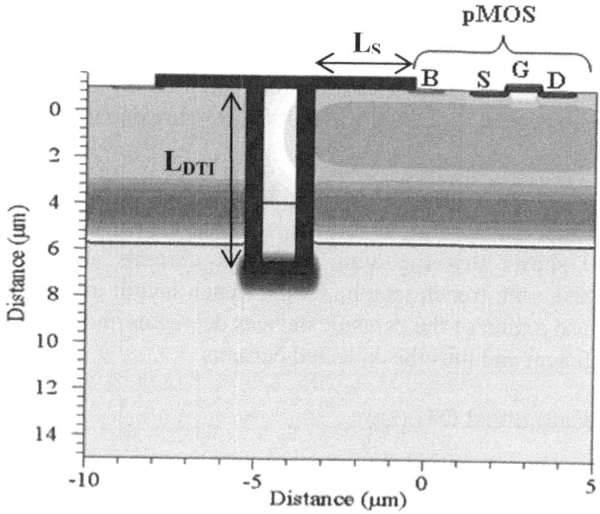

Figure 6.16. *The 2D structure of the pMOS with DTI*

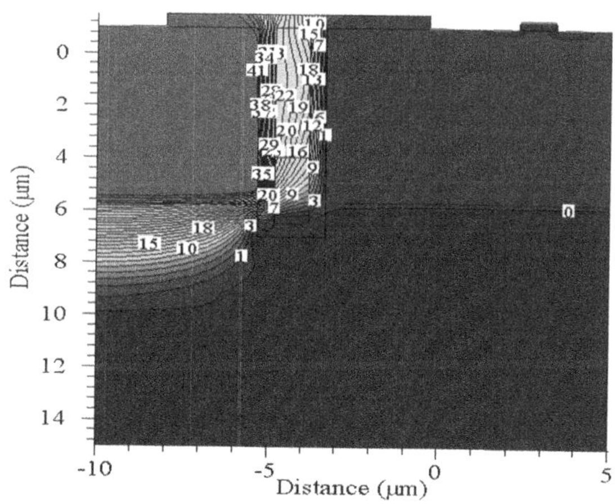

Figure 6.17. *The potential distribution of the pMOS structure with DTI*

The waveforms of the saturation drain current, the bulk current and the bulk voltage are shown in Figures 6.18a-c, respectively. There are two parasitic devices, the first being the parasitic BJT, which consists of N-epi/P-sub/N-well, and the second the thyristor, which consists of the four regions N-epi/P-sub/N-well drain. The parasitic BJT affects both the bulk voltage and the bulk current, and the parasitic thyristor affects the drain current.

The large variation in the bulk current is explained by the vast amount of minority carriers collected by the large collector area. However, in the case of the drain current, the collected area is small, so the variation is small. Also, we observe spikes in the drain current switching waveform; these are due to the thyristor latchup phenomenon.

6.2.5.2.1. Effect of DTI length (L_{DTI})

Figure 6.19 shows that the drain and bulk currents and the bulk voltage variations increase with trench length. As the trench length increases, the effective doping of the base region of the parasitic devices decreases and this in turn enhances the transit coefficient and thus the collected currents.

6.2.5.2.2. Spacing effect DTI (L_S)

From Figure 6.20, we observe that the variations of the electrical parameters of the pMOS have insignificant change with the keep-out zone length compared to the trench length effect.

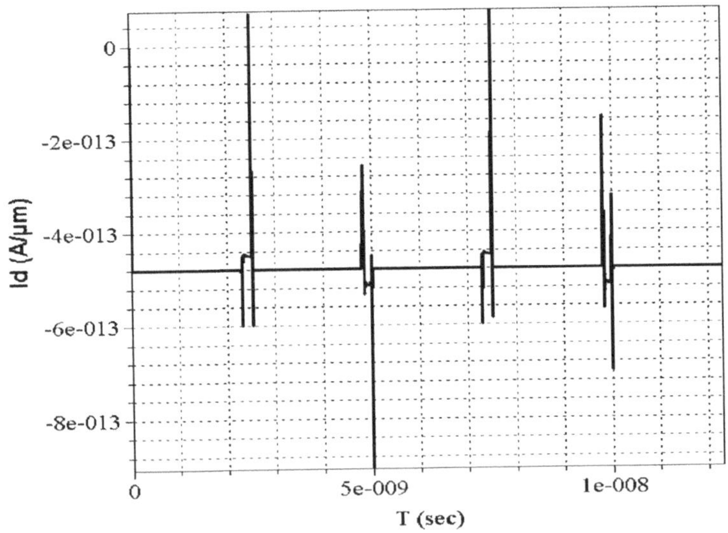

a)

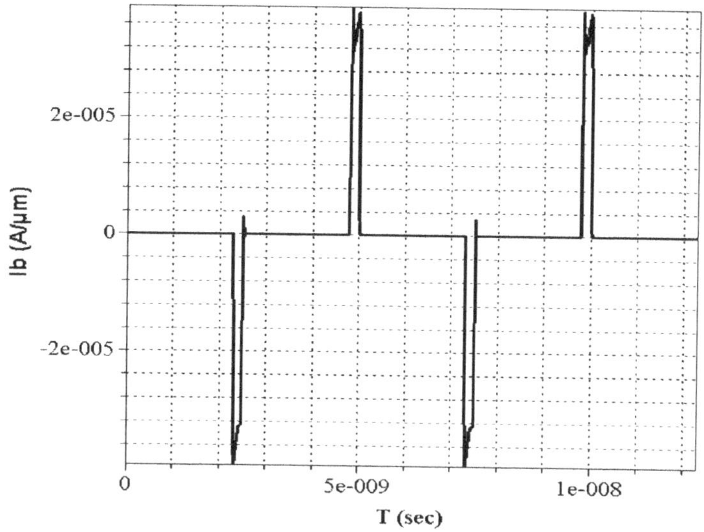

b)

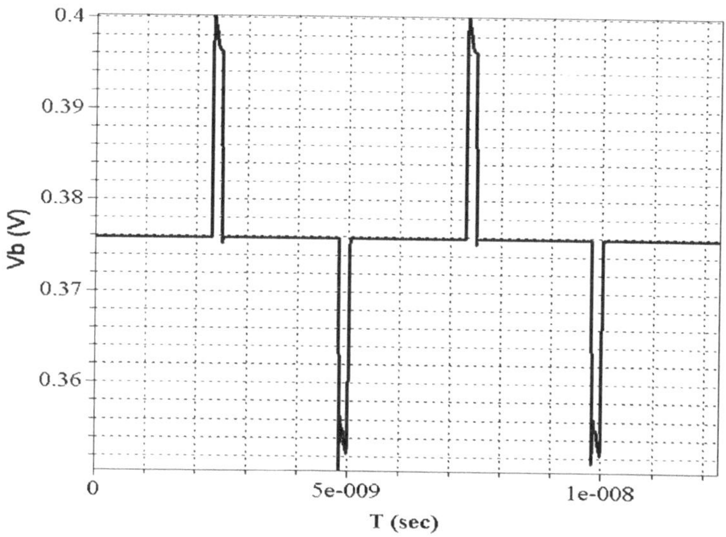

c)

Figure 6.18. *a) The saturation drain current, b) the bulk electrode current and c) the bulk voltage due to 0.0/42.0 V square signal applied on the cLDMOS drain for L_{DTI} = 8 µm*

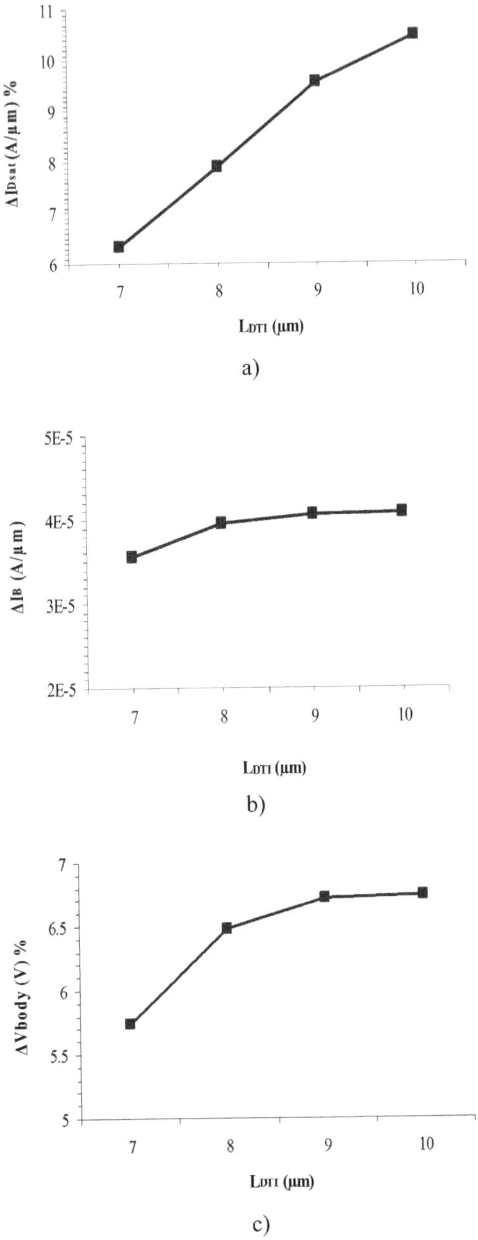

Figure 6.19. *a) The saturation drain current, b) the bulk electrode current and c) the bulk voltage as a function of trench length*

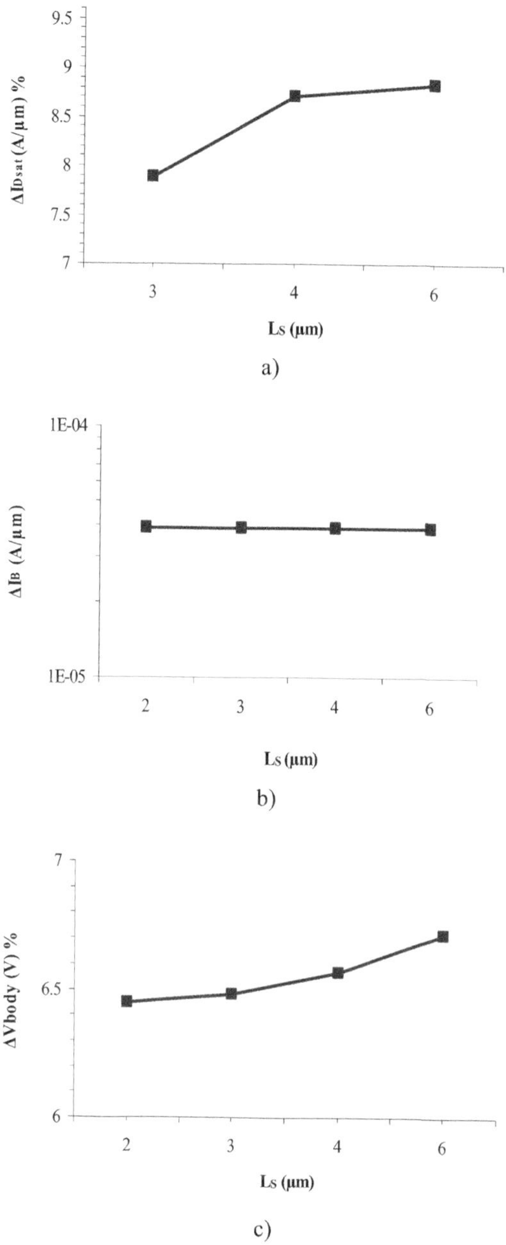

Figure 6.20. *a) The saturation drain current, b) the bulk electrode current and c) the bulk voltage as a function of DTI spacing*

6.2.5.3. *Impact on CMOS sensitive regions*

The 2D cross-section of the simulated structure is shown in Figure 6.21. The potential contours distributions in the ON state are shown in Figure 6.22, and it is observed that all the potential lines are distributed across the deep trench and N-epi/P-sub junction space charge region far from the active area of the CMOS.

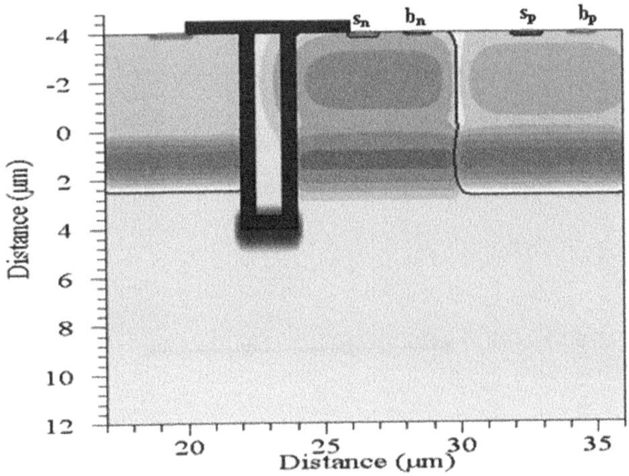

Figure 6.21. *The 2D structure of the CMOS with DTI*

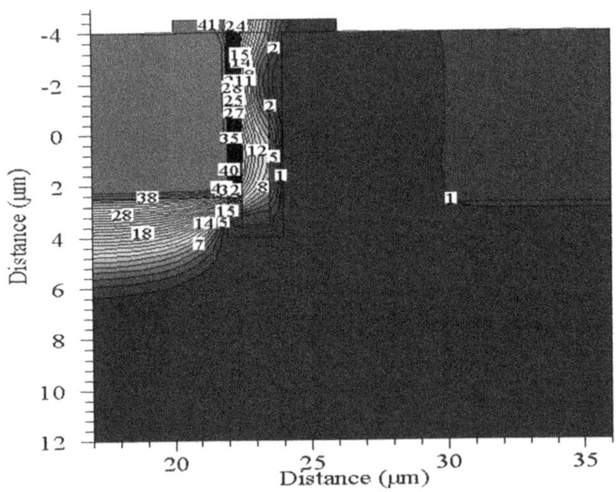

Figure 6.22. *The potential distribution in the CMOS structure with DTI*

The variations of the electrode currents are very small and can be neglected; this is due to the existence of the P-well, which acts as a guard ring which, in turn, decreases the variations. So, the dynamic variations can be tolerated under specific design rules.

6.2.5.3.1. The electrothermal simulation

Using the electrothermal analysis and considering that the bulk electrode is at 300 K (heat-sink at the bulk), the temperature waveform in the CMOS with DTI is shown in Figure 6.23. We observe that the temperature is unchanged throughout the device.

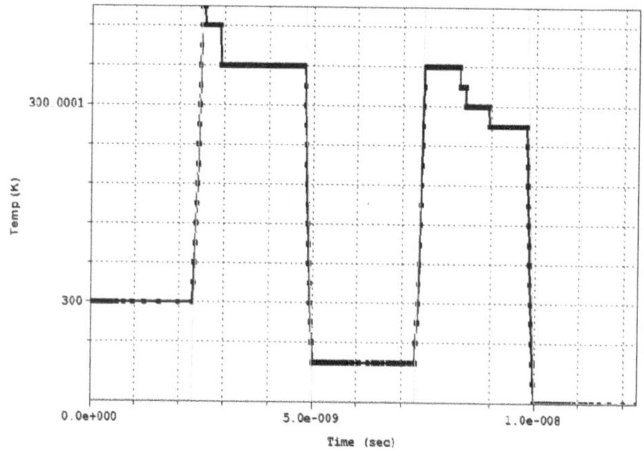

Figure 6.23. *The temperature waveform in the CMOS structure with DTI*

6.2.6. *Mixed-mode CMOS-substrate coupling simulation*

In this section, we evaluate the impact of the (0.0/42.0 V) HV output signal from the cLDMOS drain on the operation of the CMOS inverter. To achieve this, basic simulations are performed using the SPICE models of the CMOS devices, which are previously extracted by means of the ICCAP extraction program and the FEM for the bulk regions of the devices. The CMOS devices used have threshold voltages of 0.65 V for nMOS and –0.6 V for pMOS. A (0.0/42.0 V) square signal is applied to the cLDMOS drain, denoted by v_{DdMOS}, and (0.0/1.2 V) on the inverter input, denoted by v_{in} at 200 MHz frequency for both. A load capacitor of 300 pF is plugged at the inverter output. The CMOS output current is taken as an indicator for the substrate perturbations.

6.2.6.1. *Simulation with a floating nMOS bulk electrode*

The CMOS inverter bulk electrodes are connected to the active region contacts to study the sensitivity of the electrical behavior of the CMOS inverter caused by the fluctuations in the substrate potential as a result of the effect of the (0.0/42.0 V) HV digital signal v_{DdMOS}. In order to simplify the study, the pMOS bulk and source electrodes are shorted and only the impact on the nMOS is considered, as shown in Figure 6.24.

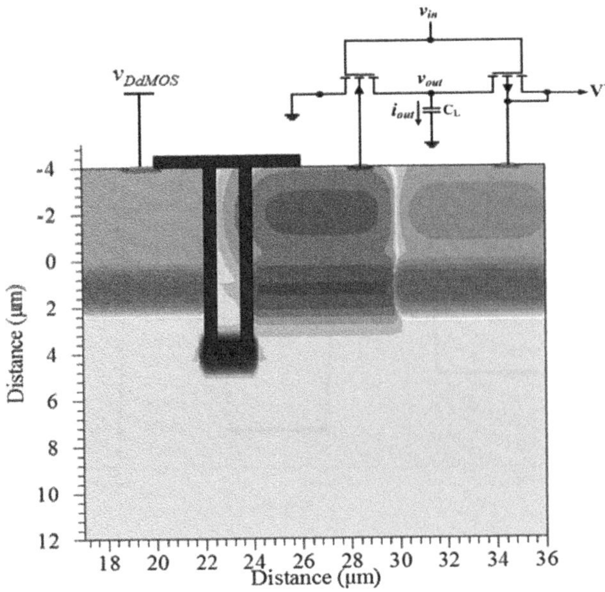

Figure 6.24. *The potential distribution in the CMOS structure with DTI when the nMOS bulk electrode is floating*

The potential distribution in the CMOS structure with DTI for v_{in} = 0.0/1.2V and v_{DdMOS} = 0.0/42.0V is shown in Figure 6.25a. It is observed that all the potential contours are distributed across the deep trench and N-epi/P-sub junction space charge region far from the active area of the CMOS devices. The waveforms of the cLDMOS drain, CMOS input and output voltages, and the output current waveforms (the current through the load capacitor) are shown in Figure 6.25b. It is clear from the figure that the current is switching with respect to the CMOS input waveform and there is no effect of the digital HV output of the cLDMOS.

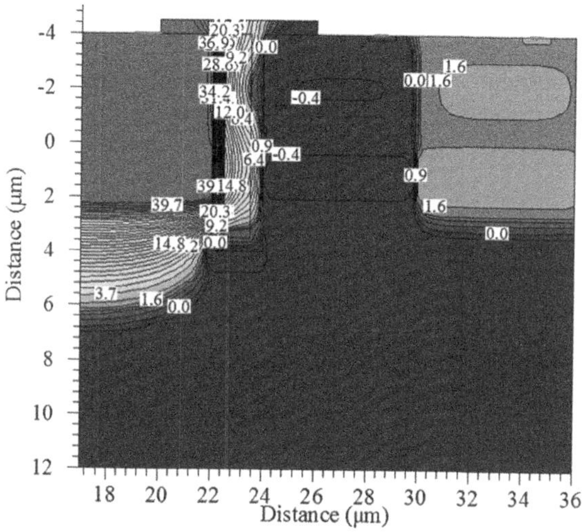

a)

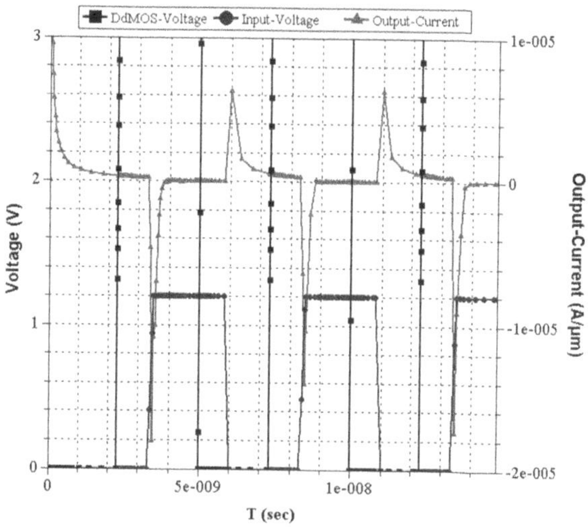

b)

Figure 6.25. *a) The potential distribution and b) the voltages and output-current waveforms for v_{in} = 0.0/1.2 V and v_{DdMOS} = 0.0/42.0 V, with floating nMOS bulk electrode*

6.2.6.2. *Simulation with a floating pMOS bulk electrode*

The simulated structure is shown in Figure 6.26. In this structure, the bulk and source electrodes of the nMOS are shorted and the impact on the pMOS is studied.

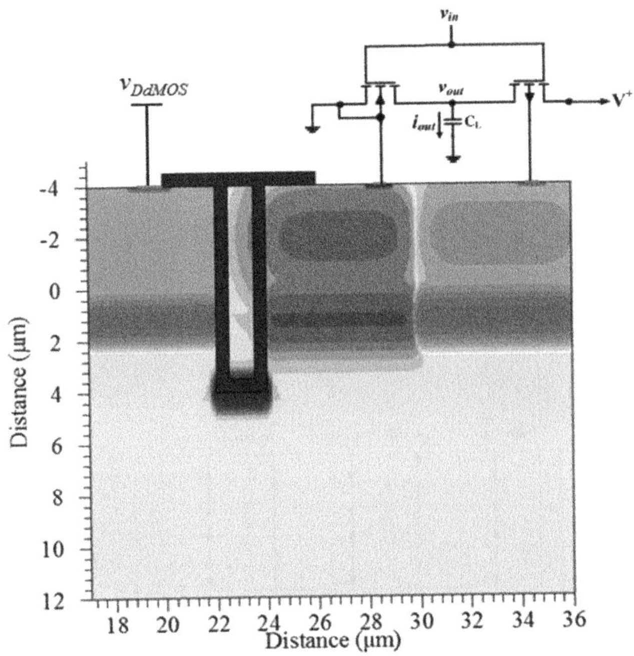

Figure 6.26. *The potential distribution in the CMOS structure with DTI when the pMOS bulk electrode is floating*

In Figure 6.27a, the potential distribution in the CMOS structure with DTI is shown for the same values of v_{in} and v_{DdMOS} in the case of the floating pMOS bulk contact. All of the potential contours are distributed across the deep trench and the N-epi/P-sub junction space charge region far from the active area of the CMOS devices. The waveforms of the cLDMOS drain, CMOS input and output voltages, and the output current waveforms are shown in Figure 6.27b. It is clear from the figure that the current is switching with respect to the CMOS input waveform and there is no effect for the digital HV output of the cLDMOS.

We also observe that the output current switching pulse has vanished because of the fall in the CMOS input signal (when the pMOS in ON); this is due to the increase in the pMOS threshold voltage, which in turn decreases the drain current.

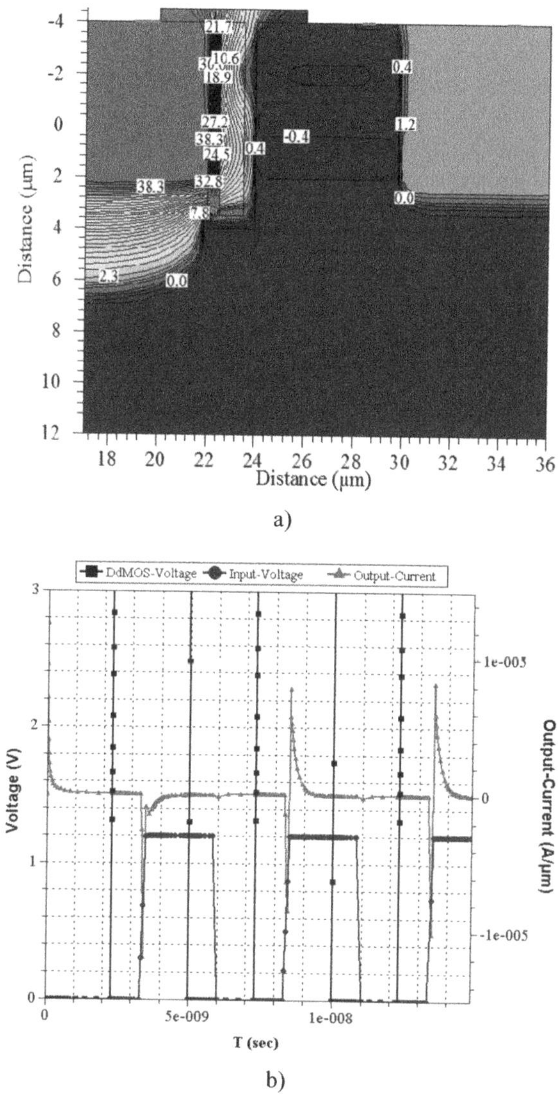

Figure 6.27. *a) The potential distribution and b) the voltages and output-current waveforms for v_{in} = 0.0/1.2 V and v_{DdMOS} = 0.0/42.0 V for the floating pMOS bulk contact*

6.3. Part II: smart power integration using stacked 3D technology

6.3.1. *From 2D planar integration to 3D integration*

In the previous section, we discussed the 2D integration of HV devices with the low-voltage CMOS devices and briefly explained the static and dynamic coupling of this type of 2D planar smart power integration. In this technology, it is difficult to achieve the best performance from all of the HV and LV devices mounted on the chip because of their conflicting requirements, as detailed in Chapter 4. In addition, there is the high cost of implementation and isolation and the delay problem due to the so-called "wiring crisis".

In recent years, 3D integration technology has emerged. This technology uses TSVs and re-distribution layers (RDLs) to interconnect multiple active circuit layers. 3D integration offers significant improvements over two-dimensional (2D) ICs in terms of performance, functionality and integration density. By reducing the length of global interconnects significantly, 3D integration promises a solution to the severe delay problems that are, and will continue to be, encountered as process geometries shrink. Furthermore, 3D integration also provides new architectures for sophisticated ICs and facilitates the integration of heterogeneous materials, devices and signals, enabling the realization of an extended system-on-a-chip (SoC), as shown in Figure 6.28.

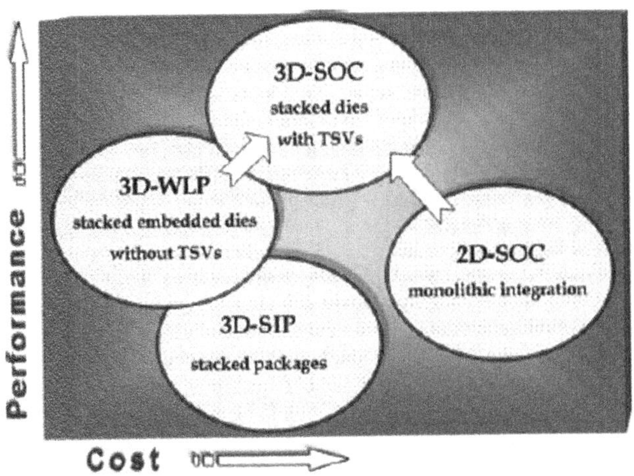

Figure 6.28. *The cost and performance comparison between the various technologies*

Speaking at the 2006 IEEE International Electron Devices Meeting (IEDM), Dr. Chang-Gyu Hwang, President and CEO of Samsung Electronics, stated that "...rapid adoption of 3D integration technology seems to be essential and, thankfully, unavoidable". He believes that the industry paradigm will shift to a new industry-fusing technology era that will offer significant global opportunities for expanded use of 3D silicon-based technologies in highly integrated systems. Indeed, 3D integration is recognized as an enabling technology for future ICs and low-cost micro-/nano-/electro-opto/bio-heterogeneous systems.

The ITRS roadmap shows 3D integration as a key technique to overcome the so-called "wiring crisis". In addition to enabling further improvement of transistor integration densities ("More Moore"), 3D integration is a well-accepted approach for so-called "More than Moore" applications. In general, this technology provides a path for continued improvement in performance, power, cost and size, at the system level, without relying upon conventional CMOS scaling alone.

6.3.1.1. *3D ICs with TSV or interplane vias*

Interplane vias offer the greatest possible reduction in wire length with vertical integration and it is less costly than today's state of the art for mixed technology products, monolithically integrated SoC (2D SoC). In addition, each plane of a 3D system can be processed separately, decreasing the overall manufacturing time. As each IC of the 3D stack is fabricated individually, the yield for each individual IC can be high. A broad spectrum of fabrication techniques for 3D ICs has been developed. Initially, CMOS or SOI wafers are processed separately, producing the physical planes of the 3D stack, while a certain amount of active area is reserved for the through silicon vias. The interplane vias are etched and filled with metal, such as tungsten (W) or copper (Cu), or even low-resistance polysilicon.

The formation of the through silicon or interplane vias is an important issue in the design of high-performance 3D ICs. This fabrication stage includes the opening of deep trenches through the interlayer dielectric (ILD) and the metal and device layers, passivation of the trench sidewalls from the metal layers and conductive substrate (for CMOS circuits), and filling the opening with a conductive material to electrically connect the planes of the 3D system. An example of a heterogeneous 3D system with TSVs is shown in Figure 6.30. The system is a stack of power IC plane, digital IC plane, analog IC plane and RF IC plane.

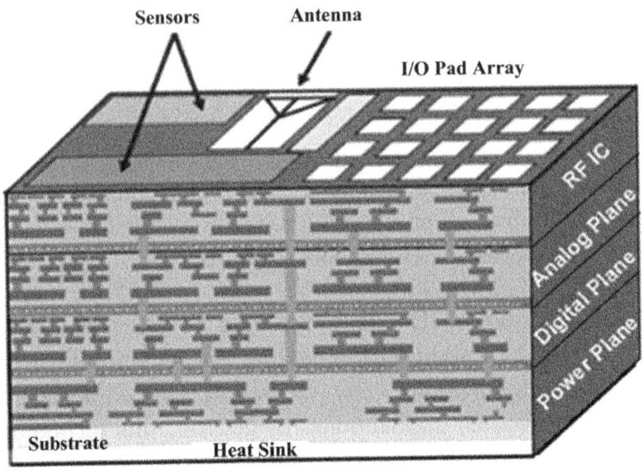

Figure 6.29. *An example of a heterogeneous 3D system-on-chip*

6.3.2. *3D smart power integration*

TSV appears to be one of the greatest technology challenges brought by 3D integration and RDL features: large metal lines implemented on the backside of the thinned active stratum. Both structures lead to electrical parasitic coupling and critical substrate noise on neighboring active devices. This coupling might be restrictive for design capabilities and needs to be quantified as a function of layout and technology parameters and TSV/RDL isolation oxide thicknesses. This will lead to the creation of tuneable 3D-specific design rules that will ensure reliable circuit design according to the application choice.

Two- and three-dimensional TCAD-based simulation of TSV- or RDL-induced coupling is investigated on nMOSFET, pMOSFET and the sensitive regions of the CMOS inverter. The considered region is bounded by a dashed ellipse, as shown in Figure 6.30.

3D integration can be implemented with various device technologies. We focus on 0.35 μm bulk BiCMOS devices integrated with TSV and RDL. Silicon substrate is thinned in the range of 5–20 μm (T_{SUB}). Only electrical coupling phenomena occurring within the thinned stratum are considered, and all other integration process steps do not impact our methodology. This study can be applied to the majority of cases. Simulated MOSFET has a channel length of 1 μm in order to avoid short-channel effects.

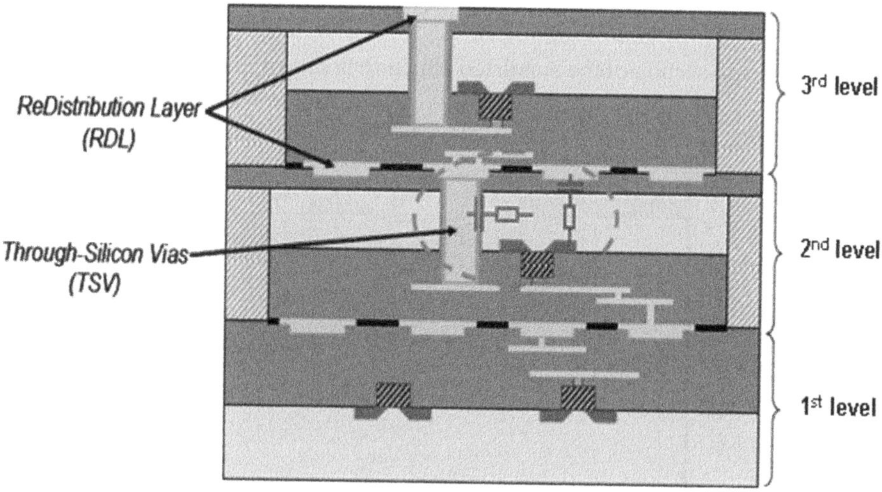

Figure 6.30. *TSV- and RDL-based 3D integration*

The saturation and leakage drain currents demonstrate a rather accurate performance for the channel length in this range. TSV and RDL are, respectively, considered as intra-bulk and backside electrodes, independent from each other and isolated from the thinned substrate with an appropriate oxide layer. TSV and RDL isolation oxide thicknesses are two important technology parameters. They are denoted as T_{OXTSV} and T_{OXRDL}.

In this chapter, we propose to observe how a square signal applied independently on these two electrodes may impact the electrical characteristics of the nMOS and pMOS transistors, respectively. In order to understand how the substrate coupling occurs within the structure, the three technology parameters previously described as T_{SUB}, T_{OXTSV} and T_{OXRDL}, and a layout parameter, L_S, are investigated in the simulation methodology.

The MOS transistor is plugged on its static mode ($V_{GS} = V_{DS} = 1.8$ V) for the nMOS and the static mode ($V_{GS} = V_{DS} = -1.2$ V) for the pMOS, which are the biasing conditions for the convergence simulated structure. A square voltage (f = 200 MHz, 200 ps-long rise and fall ramp time, 42.0 V peak voltage) is then applied on the TSV or on the RDL. The saturation drain current I_{DSAT} and the local body potential V_{body}, located in the conduction channel at 0.5 μm under the gate oxide/silicon interface, are extracted during the transient analysis. This position enables us to quantify the smallest electrical disturbances that may modify the electrical behavior of the transistors.

6.3.2.1. *Impact of TSV and RDL on nMOS*

The 2D cross-section of the simulated structure is shown in Figure 6.31.

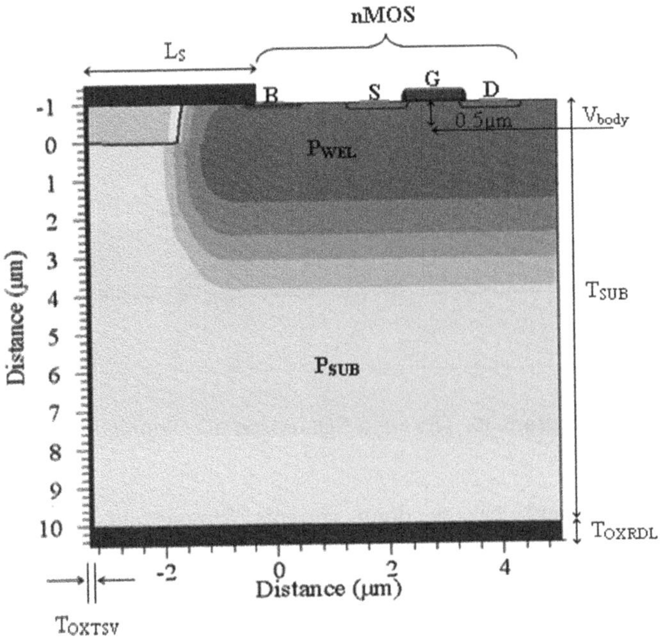

Figure 6.31. *The 2D structure of the thinned stratum with nMOS*

6.3.2.1.1. The impact of TSV (RDL is floating)

The potential contour distribution is shown in Figure 6.32, and it is observed that all of the potential lines distributed across the space charge region are far from the active area of the nMOS, so we expect the effect of the dynamic signal applied on the TSV to be limited and negligible.

The waveforms of the saturation drain current and the bulk current are shown in Figures 6.33a and b. The signal coupling is through the TSV parasitic distributed capacitances and the substrate impedance of this system acts as a differentiator for the dynamic signal applied on the TSV electrode. This fact is clearly illustrated through the waveform of the bulk current in Figure 6.33a. The variation in the bulk current is about 5 nA/µm, which is effective with respect to the value of this current but can be neglected with respect to the value of the saturation drain current. For the drain current in Figure 6.33a, there is a relaxed effect due to the capacitance coupling, but the variation is very small in the range of 0.01%, which can be neglected.

The effect of the technology parameters T_{SUB}, T_{OXTSV} and T_{OXRDL}, and a layout parameter L_S are investigated in the following figures.

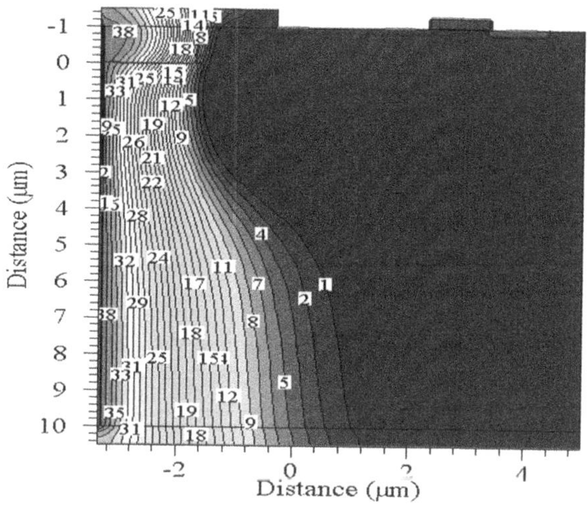

Figure 6.32. *Potential distribution for RDL floating for T_{SUB} = 10 μm and T_{OXTSV} = 0.05 μm*

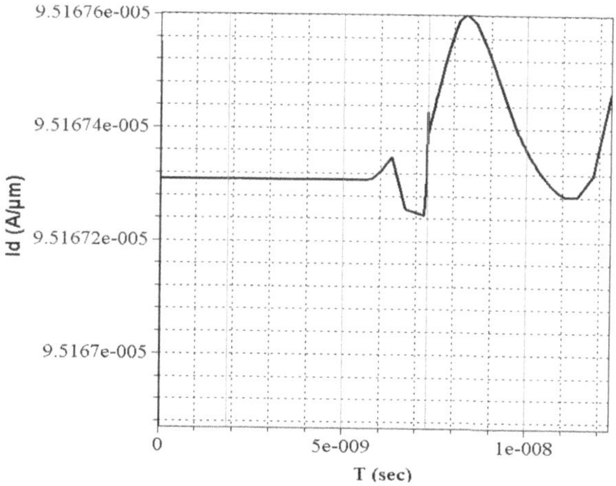

a)

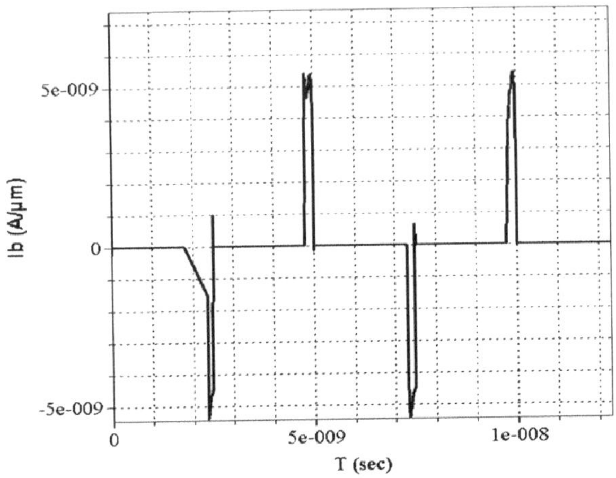

b)

Figure 6.33. *a) The saturation drain current and b) the bulk electrode current due to 0.0/42.0 V square signal on the TSV for T_{SUB} = 10 µm and T_{OXTSV} = 0.05 µm*

Figure 6.34 shows the percentages of the dynamic variation in the saturation drain current as a function of the substrate thickness for two different TSV oxide thicknesses. It illustrates that a thicker oxide for TSV isolation significantly reduces variations on saturation current (5.0e-3% to 2.5e-4%) because of the decrease in TSV oxide capacitance.

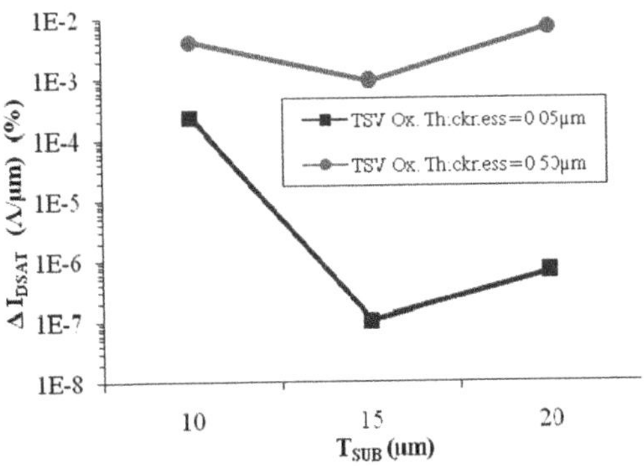

Figure 6.34. *The saturation drain current as a function of the substrate thickness for different TSV oxide thicknesses for L_S = 3.0 µm and T_{OXRDL} = 0.5 µm*

The dynamic variation in the case of the 5 μm substrate is not included because no variation is observed in this case. This is due to an increase in the coupling resistance and there is thus a subsequent decrease in the TSV coupling. As previously stated, the coupling mechanism depends mainly on the bulk conductance, the TSV distributed capacitances and the MOSFET parasitic capacitances.

The bulk current I_B is an important tool to measure the coupling; this is shown in Figure 6.35. It has been reported that substrate thickness has a relatively low impact on dynamic variations of the bulk current for thicknesses in the range of 5–20 μm. The current is decreased from 5e-9 to 2.5e-9 A/μm when increasing the T_{OXTSV} from 0.05 to 0.5 μm.

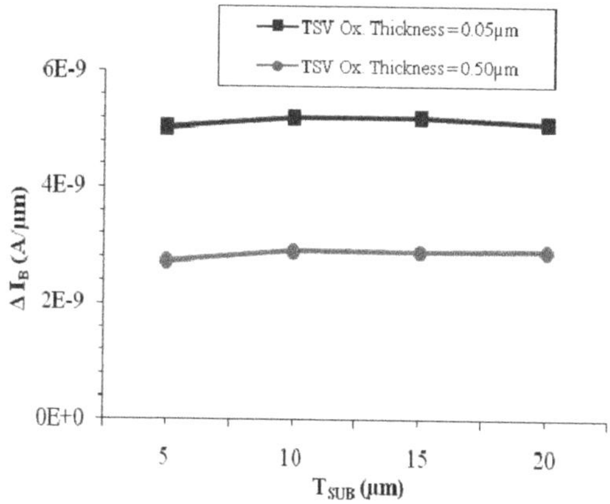

Figure 6.35. *The bulk current as a function of the substrate thickness for different TSV oxide thicknesses for L_S = 3.0 μm and T_{OXRDL} = 0.5 μm*

The layout parameter L_S (so called keep-out zone) may impact the coupling, as illustrated in Figure 6.36. It seems obvious that the keep-out zone does not have enough impact to decrease the coupling, compared to the influence of the technology parameters. The fact is that TSV-induced coupling occurs all along the TSV, so that substrate noise propagates at the same time through the low-resistivity doped well, and through the high-resistivity bulk silicon.

6.3.2.1.2. The impact of RDL (floating TSV)

The distribution of the potential lines is shown in Figure 6.37. The potential contours are distributed across the RDL oxide and the depletion region produced from the repulsion of the substrate majority carriers, and the potential in the active area of the nMOS is small enough to impact the characteristics of the device. The impact of RDL-induced substrate coupling is reported in Figure 6.37. There is no effect of RDL isolation thickness on variations of saturation drain current. Compared to the previous TSV study, RDL-induced coupling seems to be less substantial as the bulk resistance increases with thicker silicon substrate.

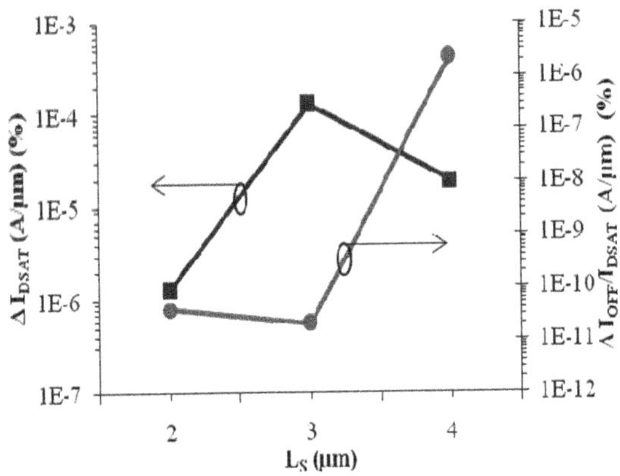

a)

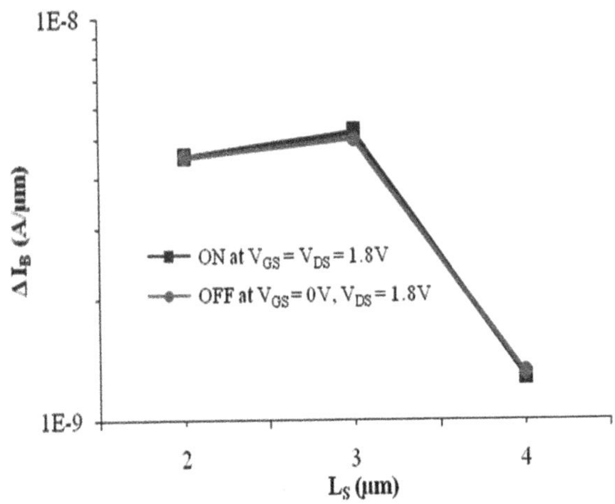

b)

Figure 6.36. *a) The drain current and b) the bulk current as a function of the TSV spacing in the ON and OFF states for T_{SUB} = 10 μm, T_{OXTSV} = 0.05 μm and T_{OXRDL} = 0.5 μm*

In conclusion, in nMOS, the coupling is through a reversed biased P-N junction (N-epi/P-well), so the current variations are very small.

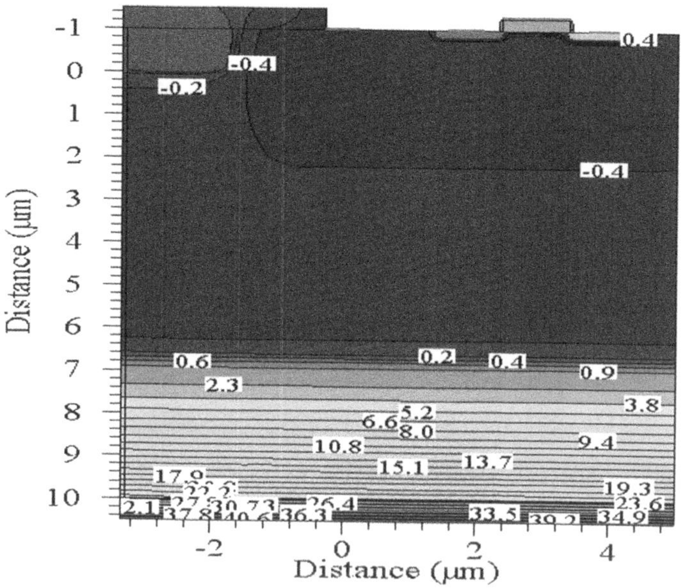

Figure 6.37. *Potential distribution for TSV floating for T_{SUB} = 10 µm, T_{OXTSV} = 0.05 µm, and T_{OXRDL} = 0.5 µm*

6.3.2.1.3. Impact of TSV and RDL

As mentioned, TSV and RDL structures produce substrate noise independently and we have outlined the typical response on active devices. Considering the TSV is connected to the RDL line, that is, the same signal is applied on both structures, their respective coupling noise can be added to obtain the global substrate noise produced by 3D architecture. The global response on the nMOS saturation drain current and bulk current is depicted in Figures 6.38a and b for equivalent oxide thicknesses of 0.05 µm and 0.5 µm for TSV and RDL, respectively.

By comparing the variations in the electrical quantities due to the TSV with RDL floating in Figure 6.33, and the variations in Figure 6.39 due to both TSV and RDL, one can conclude that the variations are due to TSV impact.

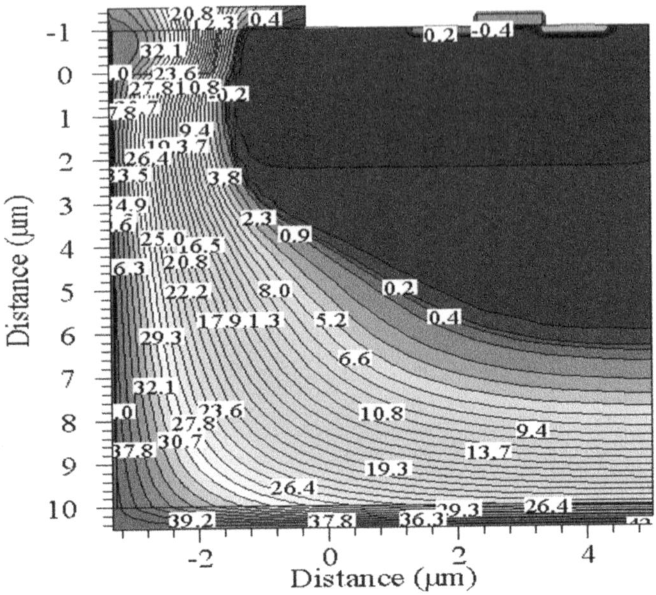

Figure 6.38. *Potential distribution for T_{SUB} = 10 µm and T_{OXTSV} = 0.05 µm*

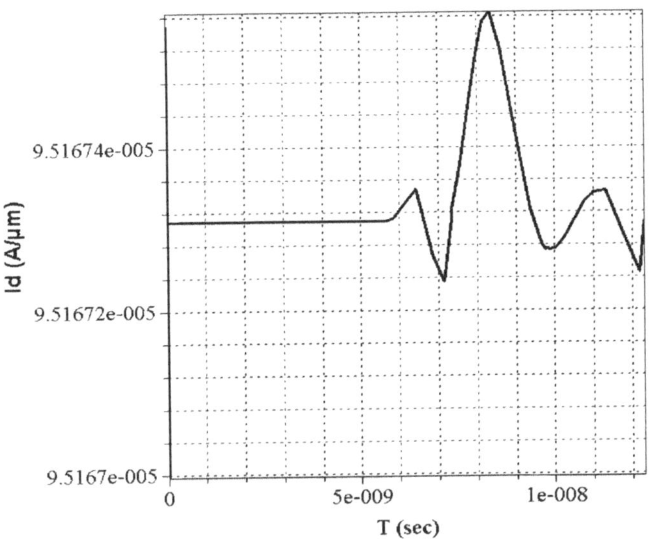

a)

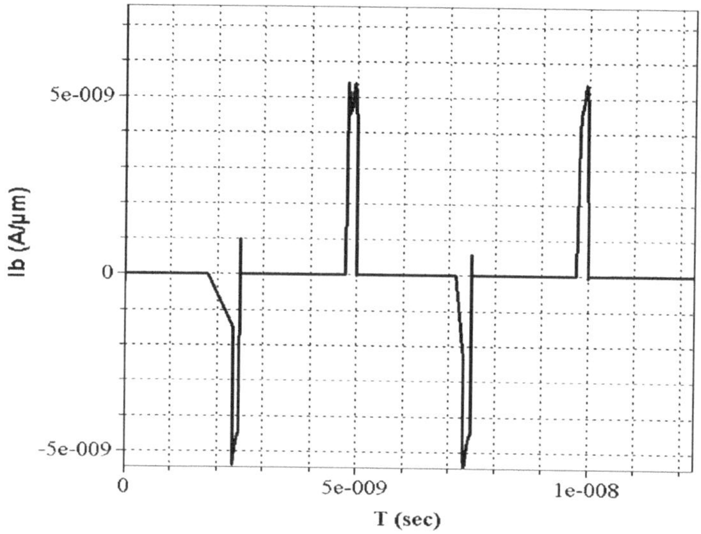

b)

Figure 6.39. *a) The saturation drain current and b) the bulk electrode current due to 0.0/42.0 V square signal on the TSV and RDL for T_{SUB} = 10 µm and T_{OXTSV} = 0.05 µm*

6.3.2.2. *Impact of TSV and RDL on pMOS*

The 2D cross-section of the simulated structure is shown in Figure 6.40.

6.3.2.2.1. The impact of TSV (RDL is floating)

The potential contour distribution is shown in Figure 6.41; it is observed that all of the potential lines distributed across the space charge region in the substrate are produced from the repulsion of the substrate majority carriers. This potential is negative and the potential lines of the 0.0/42.0 V TSV signal are applied across the TSV oxide. So, the TSV oxide thickness must be chosen to support the maximum applied voltage.

The waveforms of the saturation drain current, the bulk current and the bulk voltage are shown in Figures 6.42a–c. Due to the direct coupling between the active area of the pMOS and the epitaxial layer, variations in the electrical quantities are significant, as shown in the figure.

The effect of the technology parameters T_{SUB}, T_{OXTSV} and T_{OXRDL}, and a layout parameter L_S are investigated in the following figures. From Figures 6.43–6.48, the variations in the saturation drain current I_{DSAT}, the bulk current and the body potential V_{body} are extracted during transient analysis as a function of L_S. It is observed that potential and current variations follow the same behavior and the impact of TSV decreases with increasing the keep-out zone length.

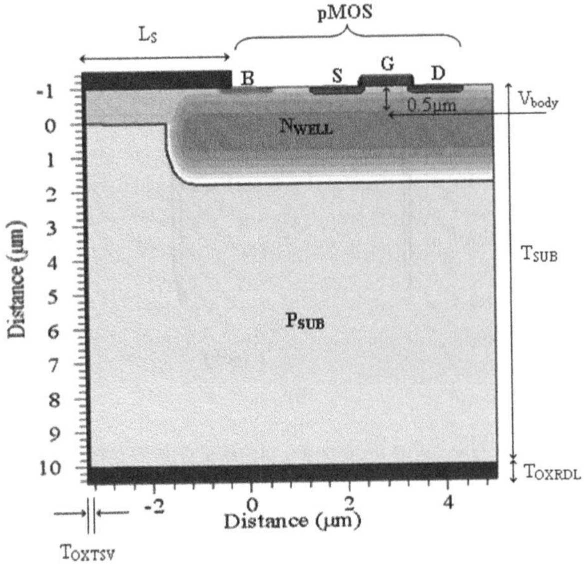

Figure 6.40. *The 2D structure of the thinned stratum with pMOS*

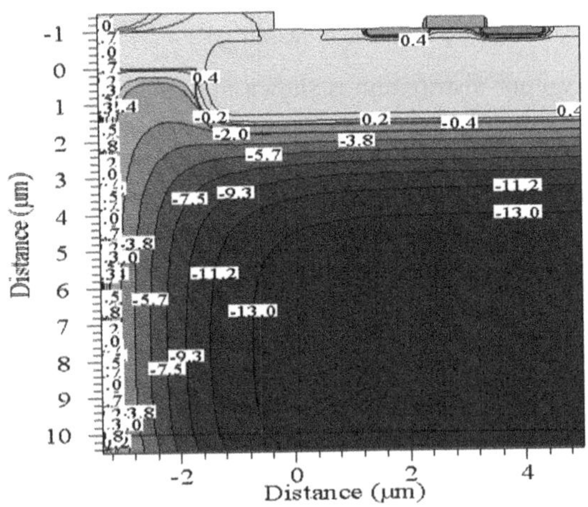

Figure 6.41. *Potential distribution for RDL floating for T_{SUB} = 10 µm and T_{OXTSV} = 0.05 µm*

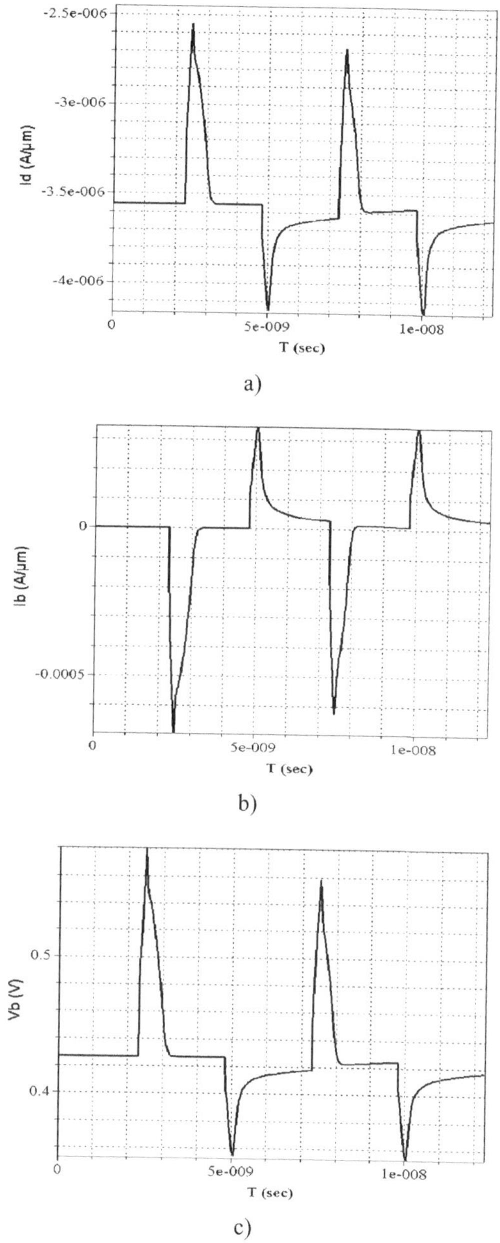

Figure 6.42. *a) The saturation drain current, b) the bulk electrode current and c) the body voltage due to 0.0/42.0 V transient signal on the TSV for T_{SUB} = 10 μm and T_{OXTSV} = 0.05 μm and L_S = 3.0 μm*

The substrate has a significant effect on the body voltage and the bulk and the drain currents as illustrated in the figures. The coupling is a function of the bulk conductance and the TSV distributed capacitances.

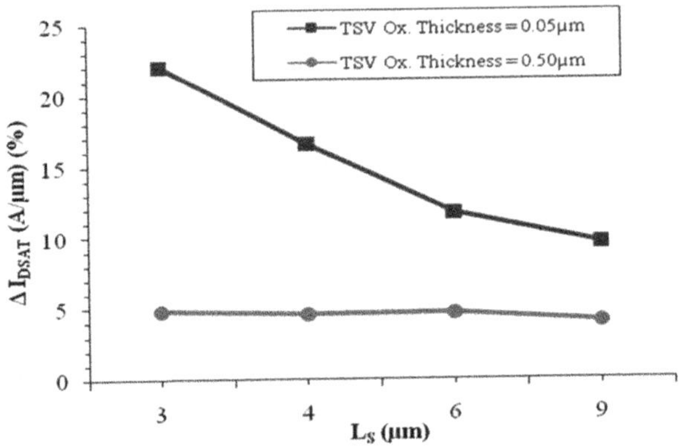

Figure 6.43. *The saturation drain current as a function of the TSV spacing at various TSV oxide thicknesses for T_{SUB} = 10 μm, T_{OXTSV} = 0.05 μm and T_{OXRDL} = 0.5 μm*

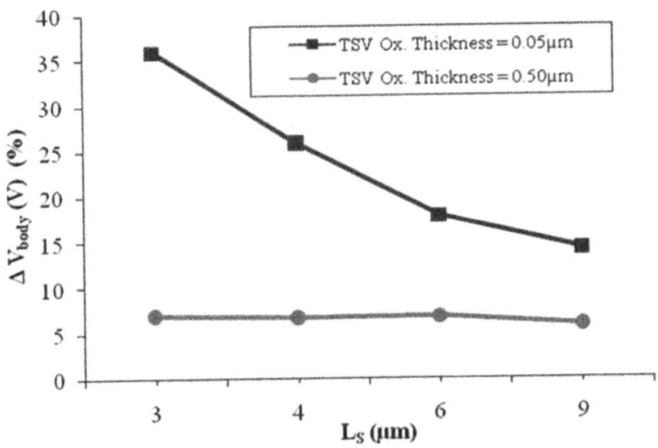

Figure 6.44. *The bulk voltage as a function of the TSV spacing for different TSV oxide thicknesses for T_{SUB} = 10 μm, T_{OXTSV} = 0.05 μm and T_{OXRDL} = 0.5 μm*

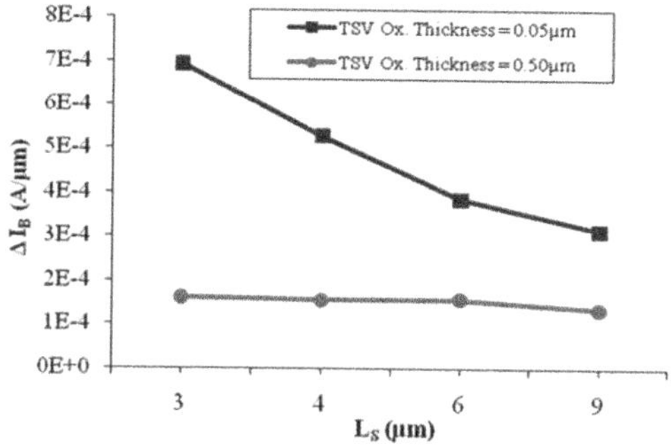

Figure 6.45. *The bulk current as a function of the TSV spacing for different TSV oxide thicknesses for T_{SUB} = 10 μm, T_{OXTSV} = 0.05 μm and T_{OXRDL} = 0.5 μm*

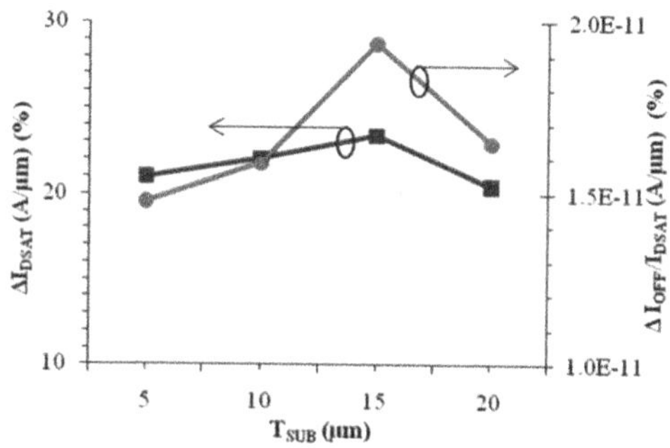

Figure 6.46. *The saturation drain current as a function of the substrate thickness at various TSV oxide thicknesses for L_S = 3.0 μm, T_{OXTSV} = 0.05 μm and T_{OXRDL} = 0.5 μm*

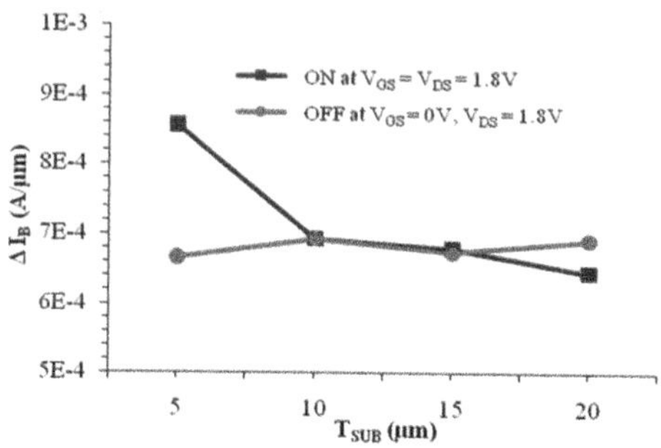

Figure 6.47. *The bulk current as a function of the substrate thickness at various TSV oxide thicknesses for L_S = 3.0 μm, T_{OXTSV} = 0.05 μm and T_{OXRDL} = 0.5 μm*

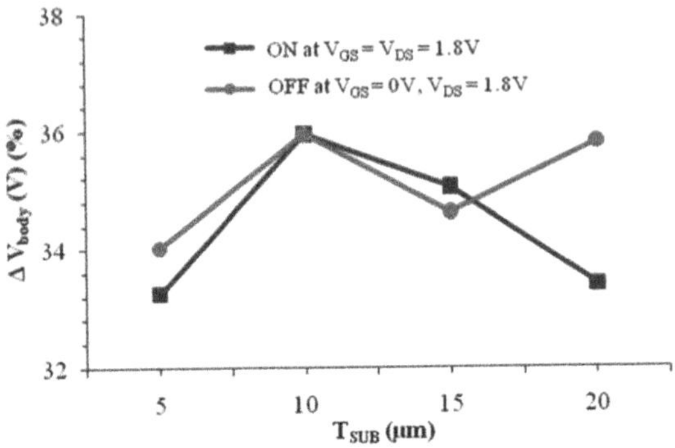

Figure 6.48. *The bulk voltage as a function of the substrate thickness at various TSV oxide thicknesses for L_S = 3.0 μm, T_{OXTSV} = 0.05 μm and T_{OXRDL} = 0.5 μm*

6.3.2.2.2. The impact of RDL

The distribution of the potential lines is shown in Figure 6.49. The potential contours are distributed across the RDL oxide and the depletion region produced from the repulsion of the substrate majority carriers, and the potential in the active area of the pMOS is small enough to impact the characteristics of the device.

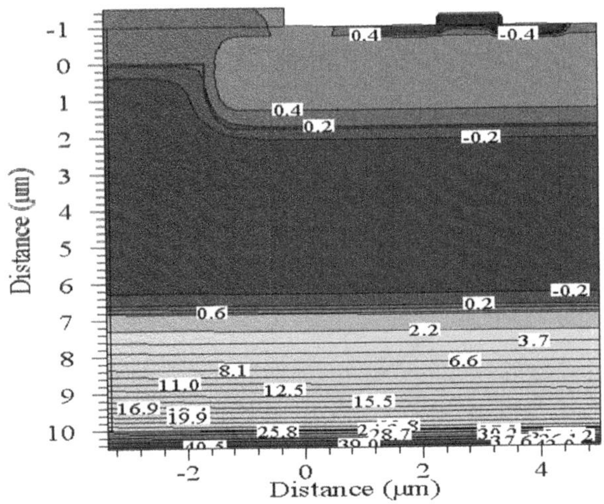

Figure 6.49. *Potential distribution for TSV floating for T_{SUB} = 10 μm, T_{OXTSV} = 0.05 μm and T_{OXRDL} = 0.5 μm*

RDL isolation thickness has no effect on variations in saturation drain current for $T_{SUB} > 5$ µm. The waveform of the drain current is shown in Figure 6.50. Compared to previous TSV studies, RDL-induced coupling seems to be less substantial as the bulk resistance increases with thicker silicon substrate.

In conclusion, in pMOS, the coupling is direct through N-epi and N-well, so there are significant variations in the ON and OFF biasing conditions. For $T_{SUB} > 5$ µm, the effect of RDL is neglected.

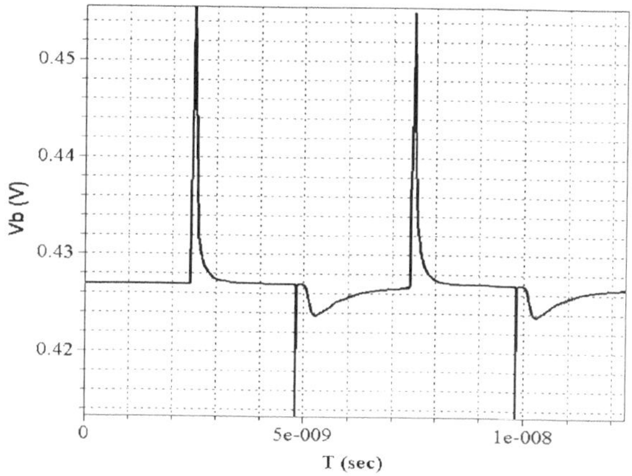

Figure 6.50. *Body-voltage for RDL 0.5 µm ON case $T_{SUB} = 5$ µm*

6.3.2.2.3. Impact of TSV and RDL

We have explained how TSV and RDL structures produce substrate noise independently and what the typical response is on active devices. Considering the TSV is connected to the RDL line – meaning the same signal is applied on both structures – their respective coupling noise can be added to obtain the global substrate noise produced by the 3D architecture. The potential distribution is shown in Figure 6.51 and the global response on the pMOS saturation drain current and bulk current is depicted in Figures 6.52a-c for equivalent oxide thicknesses of 0.05 µm and 0.5 µm for TSV and RDL, respectively.

By comparing the variations in the electrical quantities due to the TSV with floating RDL in Figure 6.42 and the variations in Figure 6.52 due to both TSV and RDL, we can conclude that the variations are due to TSV for large enough substrate thickness ($T_{SUB} > 5$ µm).

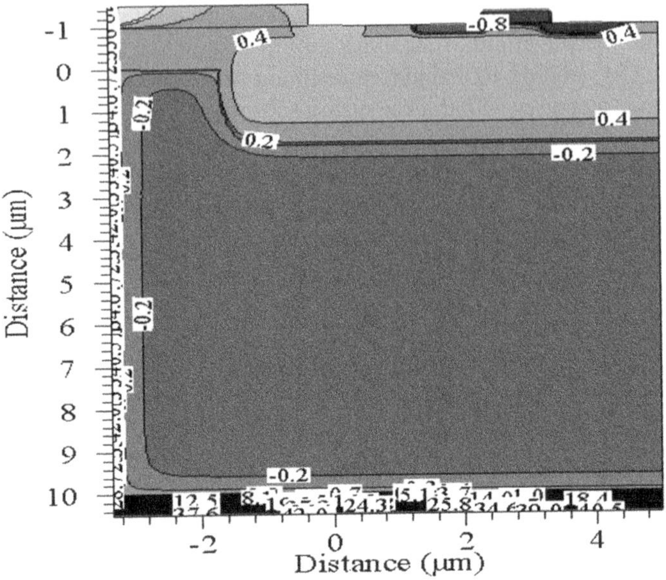

Figure 6.51. *Potential distribution for T_{SUB} = 10 µm and T_{OXTSV} = 0.05 µm*

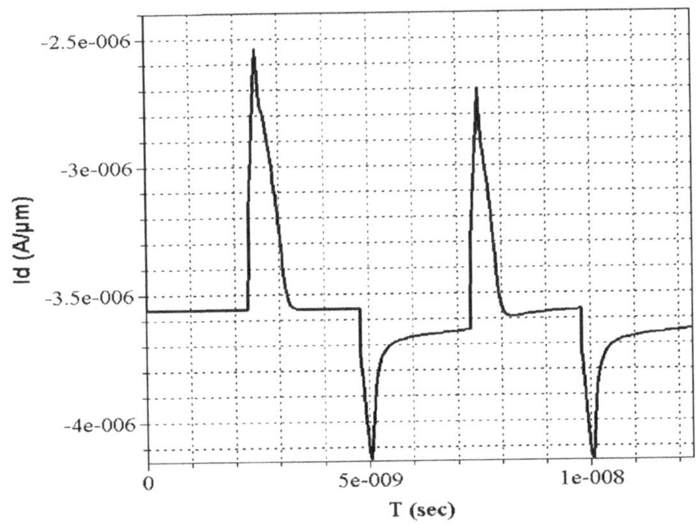

a)

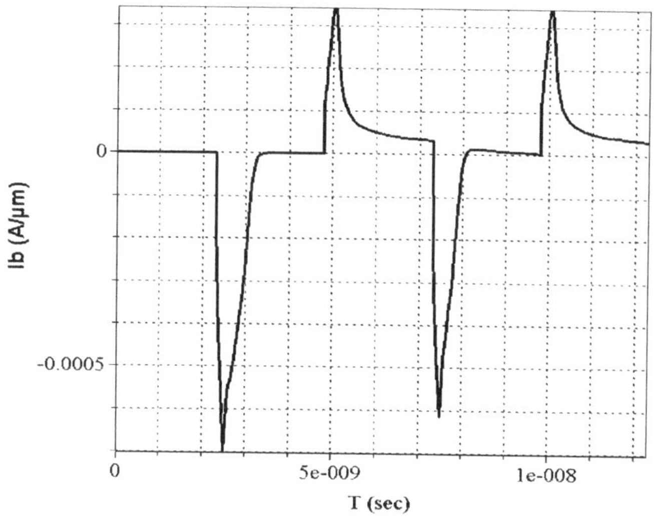

b)

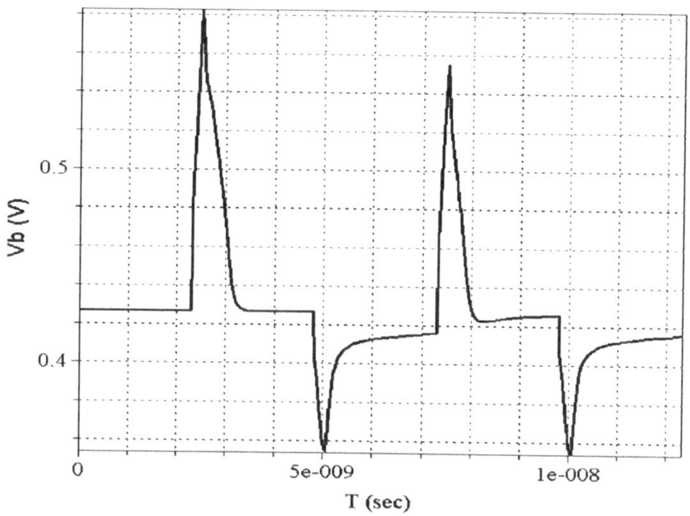

c)

Figure 6.52. *a) The saturation drain current, b) the bulk electrode current and c) the body voltage due to 0.0/42.0 V square signal on the TSV and RDL for* T_{SUB} *= 10 µm and* T_{OXTSV} *= 0.05µm*

6.3.2.3. *Impact of TSV and RDL on CMOS*

The simulated structure is shown in Figure 6.53, and the potential distributions in the CMOS active regions for floating RDL and floating TSV are shown in Figures 6.54a and b.

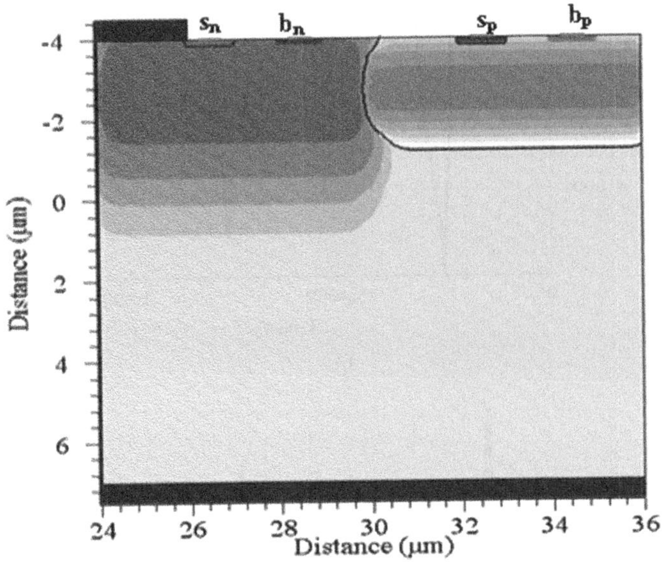

Figure 6.53. *The CMOS sensitive regions*

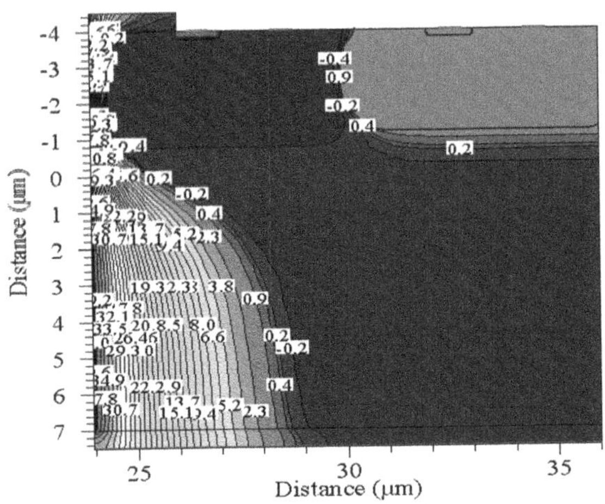

a)

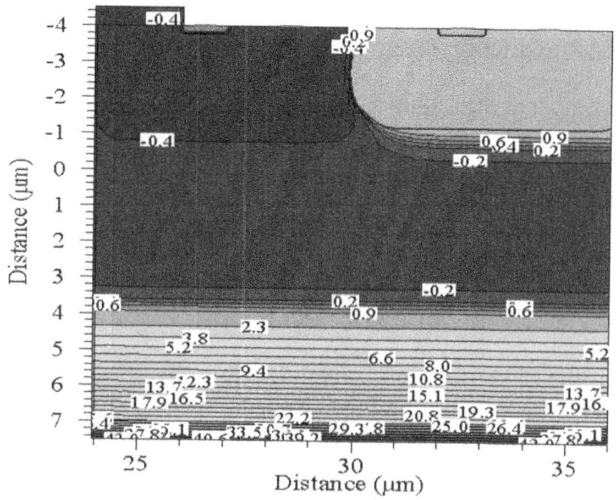

b)

Figure 6.54. *The CMOS potential distribution for a) floating RDL and b) floating TSV*

In case of CMOS, the floor planning is very important. The Pwell region, which is the active area of the nMOS, must be implanted beside the TSV to act as a guard ring for the Nwell, which is the active area of the pMOS.

6.3.3. *TSV-CMOS mixed-mode coupling*

The concept outlined in this section derives from the fact that when the MOS bulk electrode is floating, the bulk potential becomes a function of the substrate perturbations and this affects the value of the threshold voltage, which in turn varies the MOS drain current. Also, the switching of the MOS drain current due to an applied square signal is a strong function of the MOS and the bulk capacitances. Therefore, the aim of this section is to use the CMOS output current as an indicator to determine the source of the bulk variation: the CMOS input voltage or the TSV HV signal and hence evaluate the impact of the TSV on the CMOS inverter. The CMOS devices used are previously implemented using the standard layers of the 0.35 μm BiCMOS technology delivered by ST Microelectronics and with threshold voltages of 0.65 V for nMOS and –0.6 V for pMOS.

To achieve this aim, basic simulations are performed using the SPICE models for the CMOS devices and the FEM for the bulk regions of the devices and this mixed-mode simulation is performed using TCAD program packages. In the mixed-mode simulation, we use the SPICE model parameters that are extracted by means of the ICCAP extraction program.

In the simulations, the CMOS inverter bulk electrodes are connected to the contacts of the active regions with a load capacitor of 300 pF, plugged at the inverter output and a frequency of 200 MHz is used for the applied square signals. A supply voltage (V^+) of 1.2 V and a CMOS input voltage (v_{in}) of 0.0/1.2 V are chosen for the simulation conversion problems.

6.3.3.1. *2D TSV-CMOS mixed-mode coupling*

In this section, the 2D structure of the active regions of the CMOS is considered to study the effect of the position of TSV (near the nMOS or the pMOS) on the body of low-voltage MOS devices.

6.3.3.1.1. The TSV in the nMOS side

The 2D structure is shown in Figure 6.55, with the TSV placed near the nMOS device with T_{OXTSV} = 0.5 µm and V^+ = 1.2 V. In this structure, the pMOS bulk and source electrodes are shorted and only the impact on the nMOS is considered as shown in the figure.

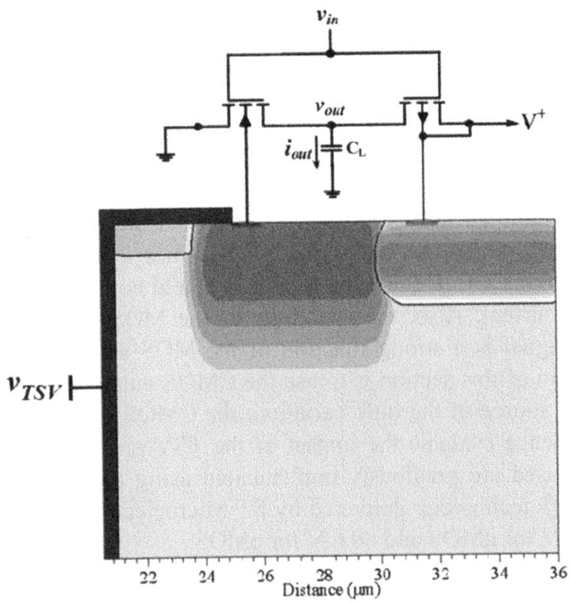

Figure 6.55. *The 2D structure of TSV-CMOS with floating nMOS bulk contact*

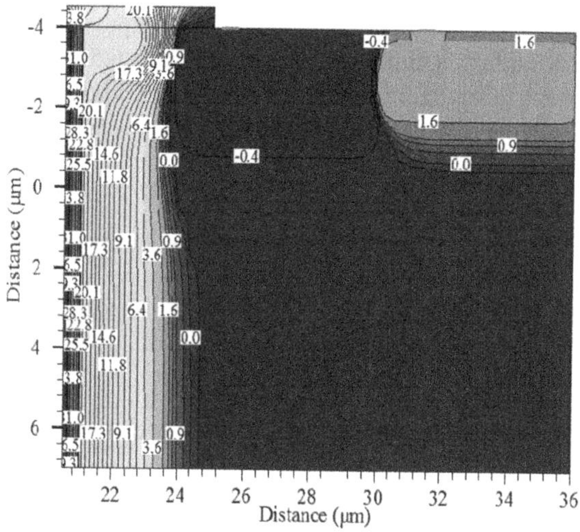

a)

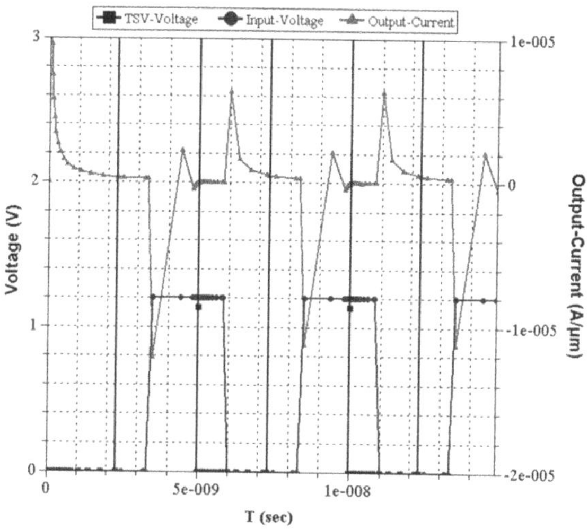

b)

Figure 6.56. *a) The potential distribution and b) the voltages and output current waveforms for V_{in} = 1.2 V and v_{TSV} = 42.0 V, with TSV located in the nMOS side*

The potential distribution in the TSV-CMOS structure is shown in Figure 6.56a. It is evident that nearly half the TSV voltage is dropped across the TSV oxide and the other half across the depleted region in the substrate, and a few potential contours are observed in the active regions of the MOS devices. The waveforms of the TSV, CMOS input and output voltages and the output current waveform (the current through the load capacitor) are shown in Figure 6.56b. It is clear from the figure that the current is switching with respect to the CMOS input waveform and there is no effect from the digital HV signal of the TSV.

6.3.3.1.2. The TSV in the pMOS side

In this case, the TSV is placed near the pMOS device. In this structure, the TSV coupling is studied using two cases: one when the bulk contact of the nMOS is floating and the other case when the bulk contact of the pMOS is floating.

The bulk contact of the nMOS is floating

The 2D structure is shown in Figure 6.57 for the case of floating nMOS bulk contact.

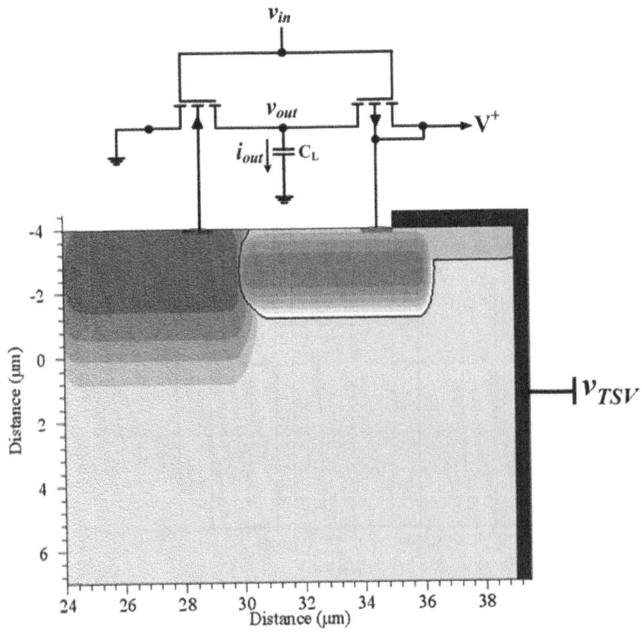

Figure 6.57. *The structure of TSV-CMOS with floating nMOS bulk contact*

The potential distribution in the TSV-CMOS structure is shown in Figure 6.58a. In this case, approximately all the TSV voltage is applied on the TSV oxide and nearly no potential contours in the substrate and the active regions of the devices; so, it is very important to implement the TSV with oxide thickness sufficient to support the applied voltage.

The waveforms of the TSV, CMOS input and output voltages and the output current waveforms are shown in Figure 6.58b. It is cleared from the figure that the current is switching with respect to the CMOS input waveform and there is no effect from the digital HV signal of the TSV.

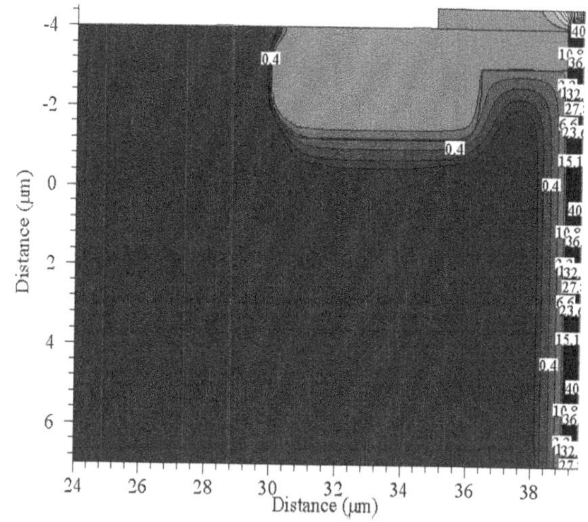

a)

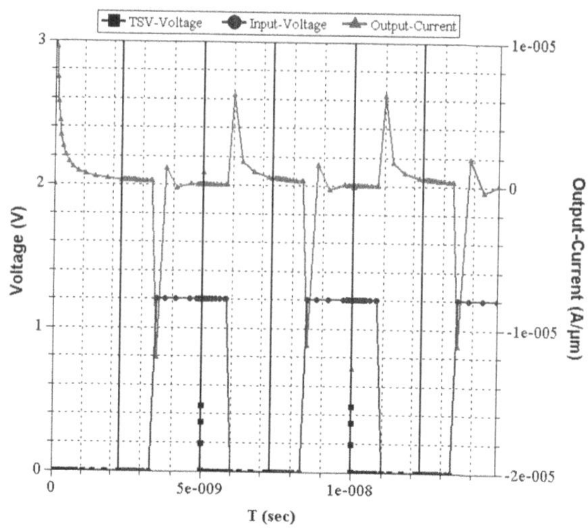

b)

Figure 6.58. *a) The potential distribution and b) the voltages and output-current waveforms for v_{in} = 0.0/1.2 V and v_{TSV} = 0.0/42.0 V, with TSV located in the pMOS side and with floating nMOS bulk contact*

The bulk contact of the pMOS is floating

The 2D structure of this case is shown in Figure 6.59. The potential distribution is applied on the TSV oxide as shown in Figure 6.60a. In this case, approximately all the TSV voltage is applied on the TSV oxide.

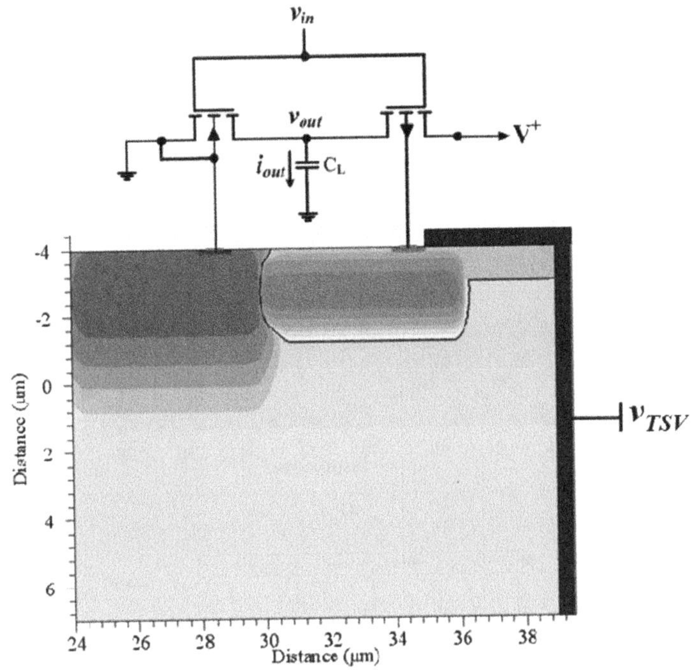

Figure 6.59. *The structure of TSV-CMOS with floating pMOS bulk contact*

The waveforms of the TSV, CMOS input and output voltages and the output current waveform are shown in Figure 6.60b. It is clearly observed that the CMOS output voltage and current are strong functions of the v_{TSV} signal (as illustrated from the output current spikes at switching instants of the TSV square signal). This means that there is a strong impact for the TSV on the bulk region of the pMOS and this impact increases with increasing the peak value of the TSV square signal. So, in smart power ICs, if the TSV is placed near the pMOS transistor, the TSV signal has a strong impact on the performance of the low-voltage control circuits.

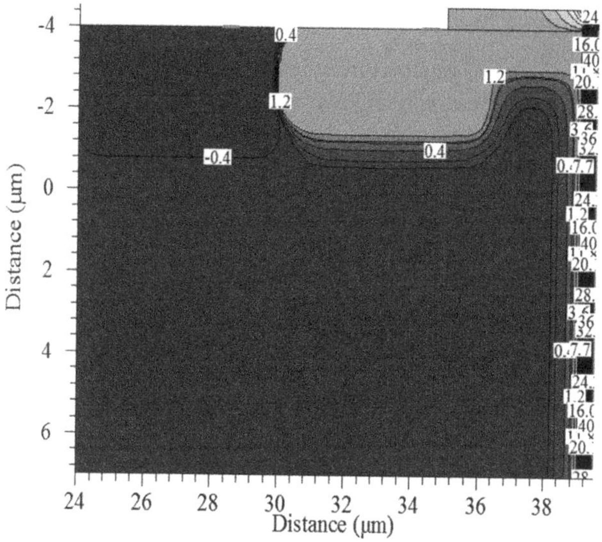

a)

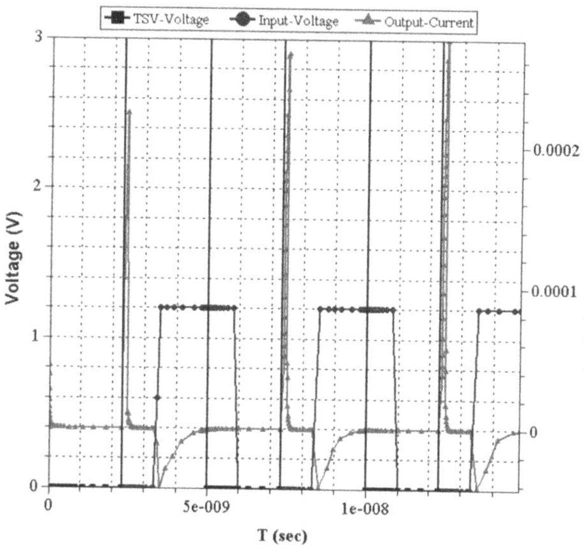

b)

Figure 6.60. *a) The potential distribution and b) the voltages and output current waveforms for v_{in} = 0.0/1.2 V and v_{TSV} = 0.0/42.0 V, with TSV located in the pMOS side and with floating pMOS bulk contact*

Effect of P⁺ guard ring

To protect the low voltage structure from impact of the TSV HV signal, we use a P^+ guard ring. This region is implemented using the same mask of the nMOS Pwell active area, with the same doping level. So, there is no extra cost for the implementation of the guard regions. The structure of TSV-CMOS with floating pMOS bulk contact and with the P^+ guard ring is shown in Figure 6.61.

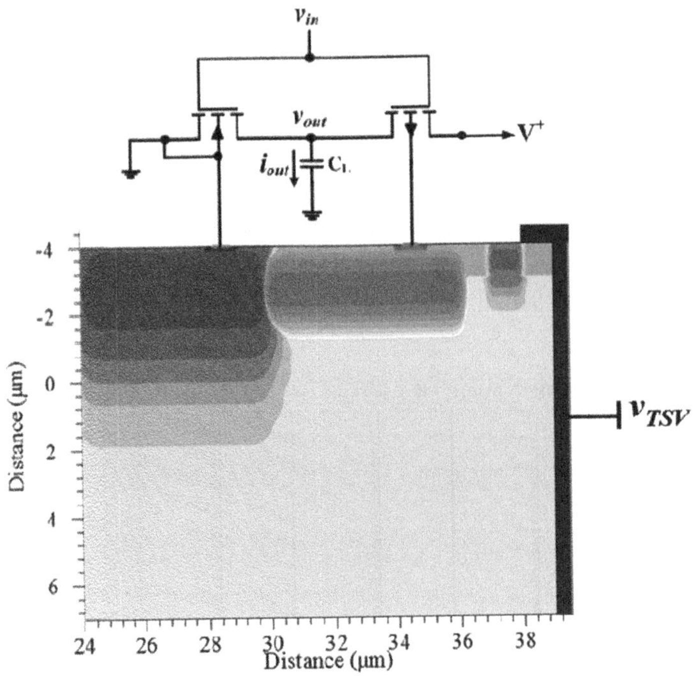

Figure 6.61. *The structure of TSV-CMOS with floating pMOS bulk contact and with a P^+ guard ring*

The potential distribution in the TSV-CMOS structure is shown in Figure 6.62a. The waveforms of the TSV, CMOS input and output voltages and the output current waveforms are shown in Figure 6.62b. The figure shows that in steady state the current is switching with respect to the CMOS input waveform and there is no effect of the TSV applied square signal. In conclusion, only the CMOS input signal has an effect on the CMOS performance.

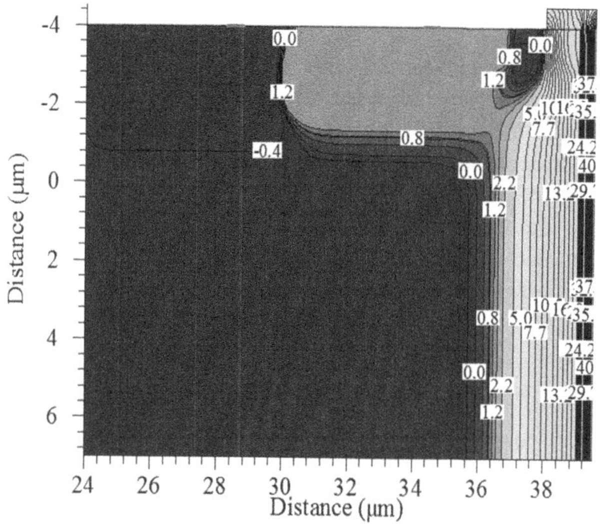

a)

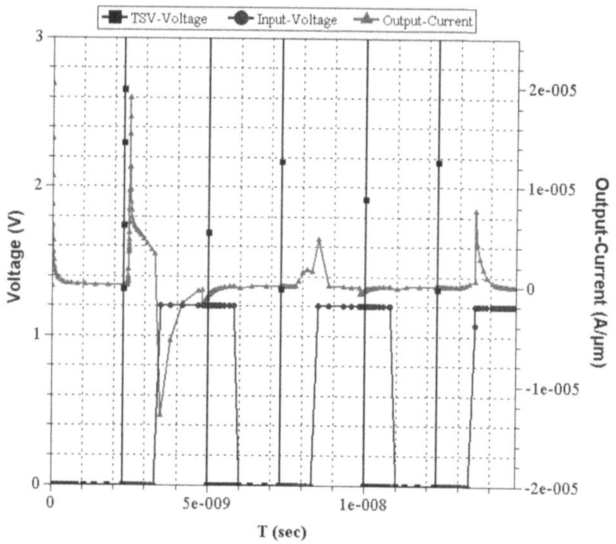

b)

Figure 6.62. *a) The potential distribution and b) the voltages and output current waveforms for v_{in} = 1.2 V and v_{TSV} = 42.0 V*

6.3.3.2. *3D TSV-CMOS mixed-mode coupling*

The 3D structure is shown in Figure 6.63. We use the Sentaurus simulator: Sentaurus is a software package based on the FEM for the device study, and the SPICE-based model for the CMOS devices is connected to the bulk regions. The TSV has a cylindrical shape filled with copper and with oxide thickness T_{OXTSV} = 0.5 μm, as shown in Figure 6.63. In order to simplify the study, only the coupling between the TSV and the nMOS is considered. The nMOS transistor is placed near the TSV as illustrated in the figure. A Y-cut cross-section potential distribution is shown in Figure 6.64(a), for v_{in} = 0.0/1.2 V and v_{TSV} = 0.0/42.0 V.

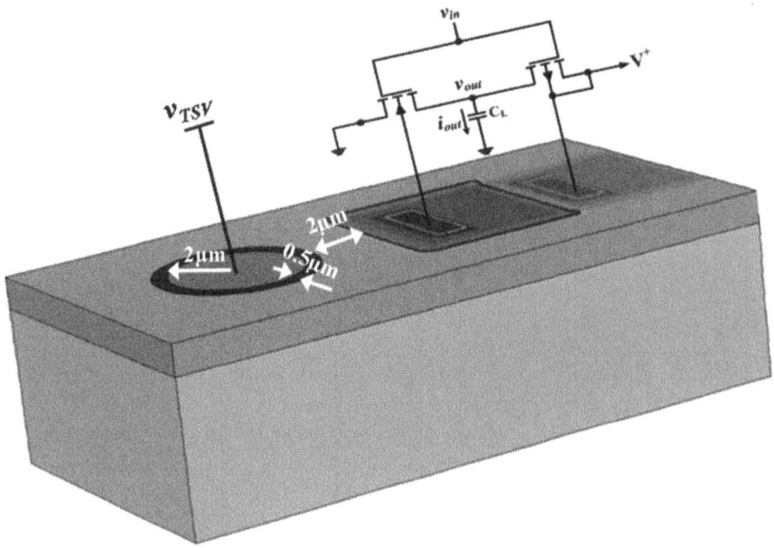

Figure 6.63. *3D cross-section of TSV-CMOS mixed-mode coupling*

The waveforms of the TSV, CMOS input and output voltages, and the output current waveform (the current through the load capacitor) are shown in Figure 6.64b. It is clear from the figure that the current is switching with respect to the CMOS input waveform and there is no effect of the digital HV of the TSV.

Comparing the 2D TSV-CMOS results in Figure 6.56 and the 3D TSV-CMOS results in Figure 6.64, we observe that the 3D output current has a larger peak value and smaller rise and fall times. This is because the area of coupling in the case of the 3D structure is large and hence the current increases and the coupling capacitance decreases; this in turn decreases the time constant.

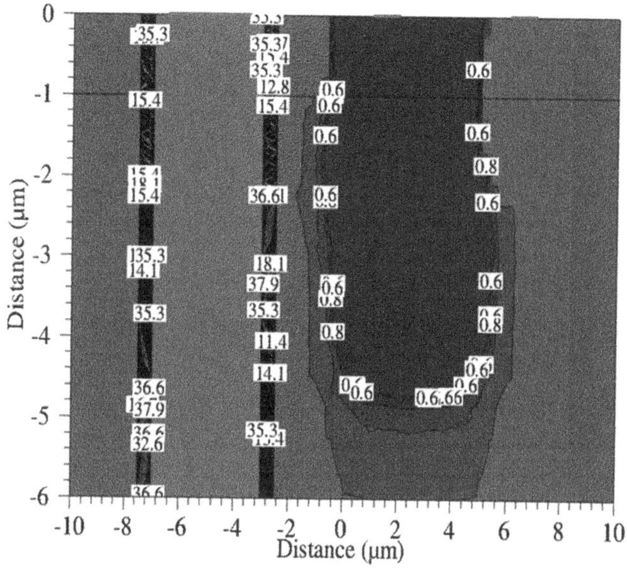

a)

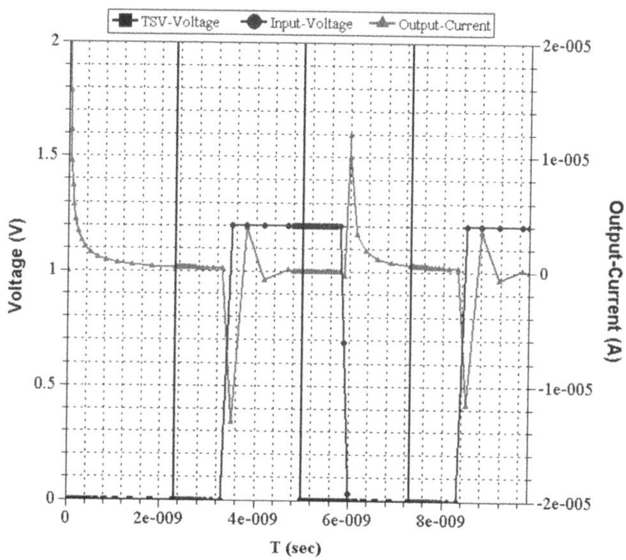

b)

Figure 6.64. *a) Y-cut potential distribution and b) the voltages and output current waveforms at v_{in} = 0.0/1.2 V and v_{TSV} = 0.0/42.0 V*

6.3.4. *Electromagnetic impact of TSV in RF range*

The track of this section is to use a finite difference time domain (FDTD) method that resolves the two Maxwell equations relative to the field curls – the electric field E and the magnetic excitation H; on the basis of a pioneering paper by Yee, we conduct a relatively in-depth study of a via with two striplines. For that, we use the simulator Sentaurus and FDTD for the electromagnetic field.

6.3.4.1. *The FDTD: a brief reminder*

Mawxell's equations relative to the vertical fields, considering media as perfect (homogeneous, linear and isotropic), are given as:

$$\nabla\times\vec{E}=-\mu\cdot\frac{\partial\vec{H}}{\partial t} \qquad [6.2]$$

$$\nabla\times\vec{H}=\varepsilon\cdot\frac{\partial\vec{E}}{\partial t}+\ \sigma\vec{E} \qquad [6.3]$$

where μ denotes the permeability, ε is the permittivity and σ is the electric conductivity.

Although there are different ways to solve Maxwell's equations numerically, the FDTD is one of the most suitable schemes for the purpose of analyzing 3D structures. Yee's pioneering approach uses centered differences to approximate space and time derivatives. The region being modeled is represented by two interleaved grids, one containing the edges of the electric field and the other containing the edges of the magnetic one. Time derivatives are not computed at the same time-frame, but at different interlaced intervals. Each unknown field component depends only on previously calculated components of the dual kind: two sets of equations are thus recovered, each set being composed of independent equations, the so-called "leapfrog" algorithm. This algorithm makes the set of linear equations easy to handle and solve sequentially. For instance, the curl relative to the H-field can be approximated as:

$$\frac{H_{y:i+\frac{1}{2}}-H_{y:i-\frac{1}{2}}}{dx}-\frac{H_{x:i+\frac{1}{2}}-H_{x:i-\frac{1}{2}}}{dy}=\varepsilon\cdot\frac{E_z^{n+1}-E_z^{n}}{\Delta t}+\vec{J}_{le,z}\left(\vec{E}_z\right) \qquad [6.4]$$

$\vec{J}_{le,z}(E_z)$ can be evaluated at the instant (n+½). Δt using a semi-implicit approximation is given as:

$$\vec{J}_{le,z}^{n+1/2} = \vec{J}_{le,z}^{n+1/2}\left(\frac{1}{2}\left[E_z^{n+1} + E_z^{n}\right]\right) \quad [6.5]$$

A stability criterion rewired is:

$$\Delta t \leq \frac{1}{c\sqrt{\frac{1}{\Delta x^2} + \frac{1}{\Delta y^2} + \frac{1}{\Delta z^2}}} \quad [6.6]$$

where c is the propagation speed in the medium and Δx, Δy and Δz are the cell minimum spaces in the FDTD grid. Finally, the chosen computation domain is large enough to potentially have two ports on a stripline(s) sufficiently far away from the via discontinuity. The space grid size must be such that, over one increment, the electromagnetic field does not change too much; this means that in order to achieve meaningful results, the linear dimension of the grid must be only a fraction of the wavelength.

6.3.4.2. *Via structure*

For the purpose of this study, FDTD simulations have been carried out on a realistic pattern (one via filled with tungsten, flanked with two copper strip lines). From the EM study, parameters such as dispersion can be computed. First, parallelepiped TSV have been considered (Figure 6.65). It is completely filled with tungsten and embedded in silicon substrate. Two-port S-parameter simulation is performed and electrical parameters such as the self inductance value and the electrical resistance can be extracted from the Z-matrix considering Z_{12} and Z_{21} terms, for instance.

The geometrical characteristics of the structure are as follows:

– Via:

- width: 4 µm;
- height: 30 µm.

– For each stripline:

- width: 4 µm;
- thickness: 1 µm;
- length: 100 µm.

– Silicon substrate:

- thickness: 60 µm;
- long: 200 µm;
- length: 100 µm.

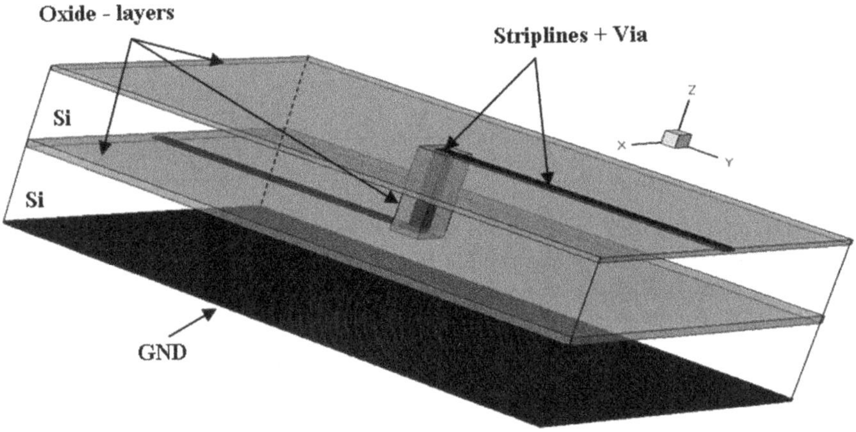

a)

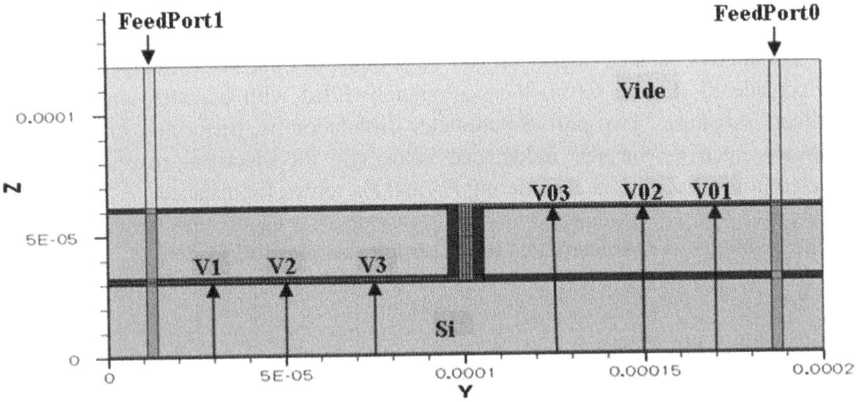

b)

Figure 6.65. *System via and striplines:*
a) 3D-structure and b) cut along the X-axis

We perform transient simulations for each stripline, without being connected to the via, for extraction of incident pulses and, for instance, characteristic impedance, propagation constants. Other simulation procedures for the striplines connected to the via, with each port excited, are needed for extraction of scattered pulses. We can follow the propagation of the fields, particularly across the via itself; the traveling electric wave is represented in Figure 6.66, and voltages are observed at the two ports for the transient regime in Figure 6.67. The pulse is quite well transmitted to the stripline; we also observed important oscillations at very high frequency indicating RLC cavity behavior. And the so-called S-parameters are shown in Figure 6.68, indicating a quasi-transmission of the signals. The simulator can handle such a complex structure.

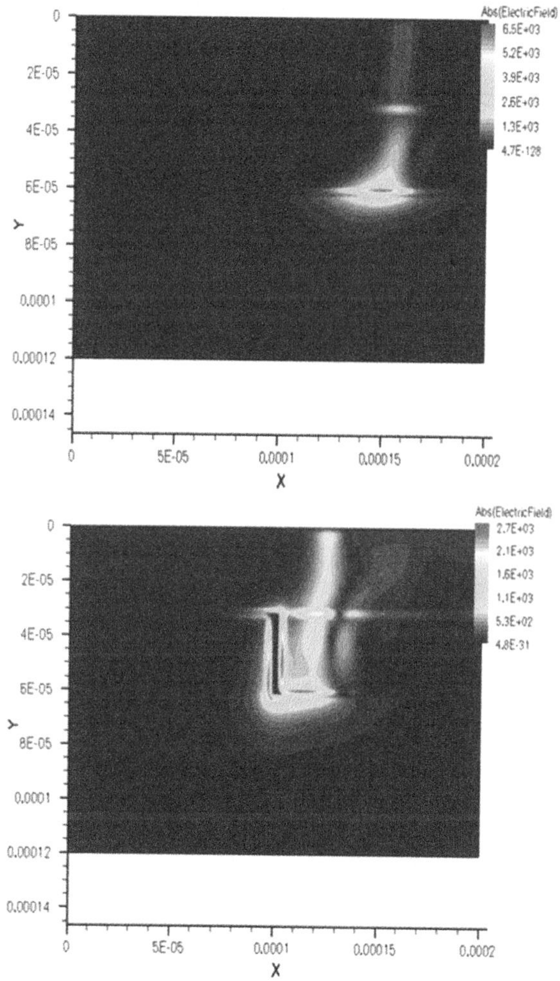

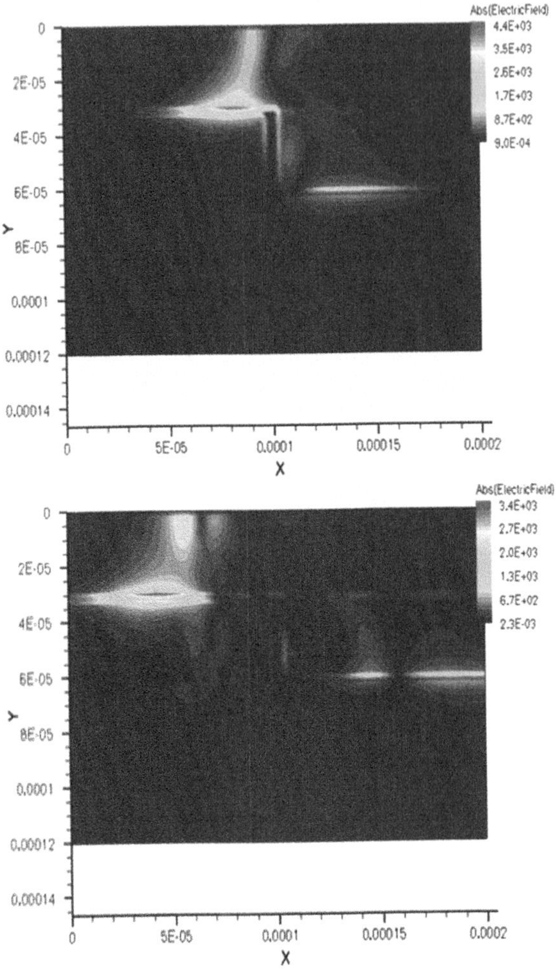

Figure 6.66. *Electric field map, transient regime*

Another point of interest is to study ground bounces induced by parasitic electromagnetic waves. In Figure 6.69, another configuration is presented: the same via but with orthogonal striplines. The difference appears essentially at the output port (see Figure 6.70), compared to the striplines aligned (Figure 6.67); in this latter case, the system acts as a single short strip line: at the output, a negative reflected wave appears. In the former case, striplines bend at 90°, and the negative reflected line does not appear. It does not seem necessary to bend the structure any more (>90°, up to 180°), as the crosstalk will be enhanced.

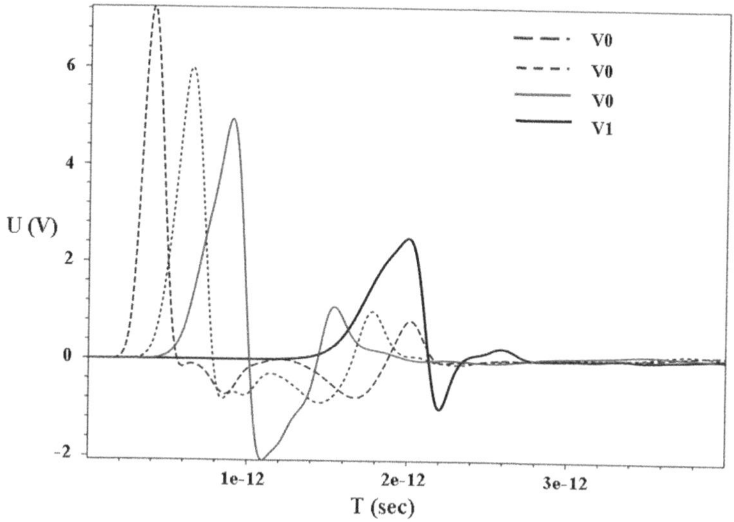

Figure 6.67. *Voltage at different probe points: excitation: pulse on Port0 (see Figure 6.65b)*

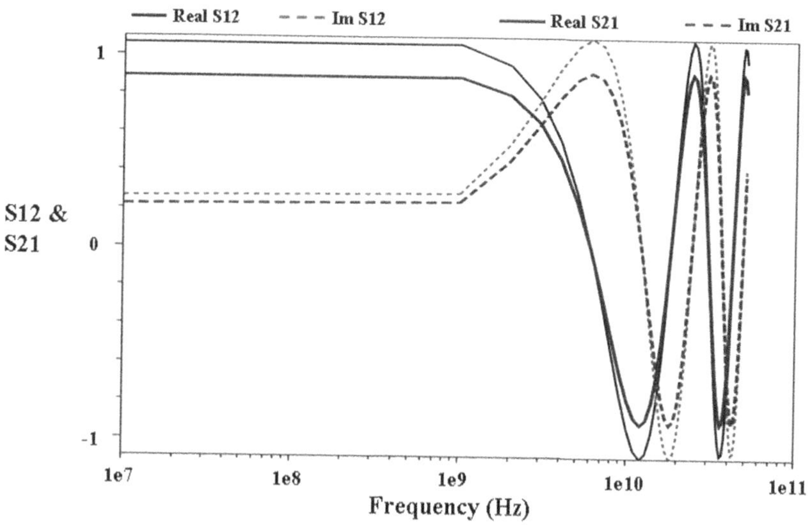

Figure 6.68. *Transmission parameters: S_{12} and S_{21}*

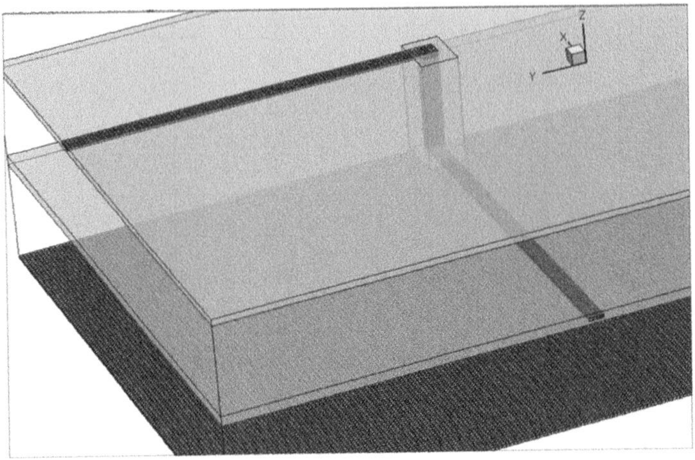

Figure 6.69. *System via and striplines bended at 90°*

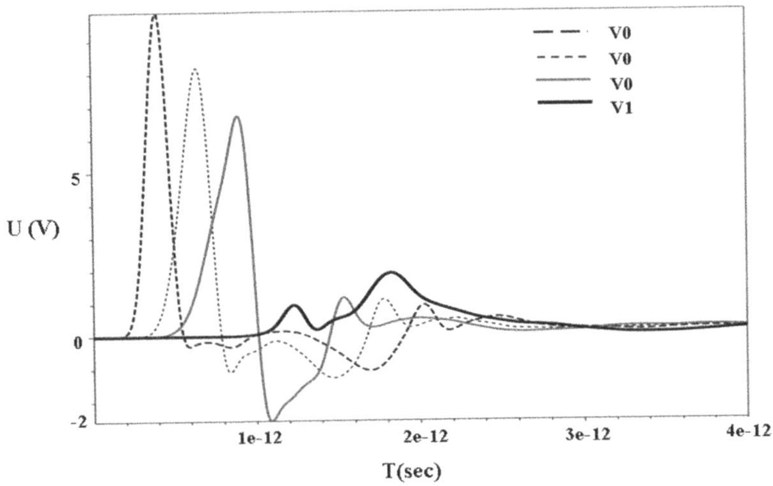

Figure 6.70. *Voltage at different probe points: excitation: pulse on Port0 (see Figure 6.65b)*

Conclusion

C.1. Conclusions

Smart power integrated circuits (SPICs) that monolithically integrate low-loss power devices and control circuitry have attracted much attention across a wide variety of applications. These ICs improve the reliability of, reduce the volume and weight of and increase the efficiency of a system. Considerable effort has been put into the development of smart power devices for automotive electronics, peripheral computer appliances and portable equipment.

Lateral double-diffused MOS transistors (LDMOSFETs) are smart power devices that are commonly used, implemented in bulk silicon or silicon on insulator (SOI). The main issues in the development of these devices are obtaining the best trade-off between specific ON-resistance $R_{ON,SP}$ (R_{ON} × area), and breakdown voltage BV, and shrinking the feature size without degrading the device characteristics.

To conclude, we highlight some of the contributions this book offers to the field of smart power IC design:

– New device structures with recently developed novel device concepts have been designed. The proposed devices were implemented in a 0.35 μm BiCMOS technology provided by ST Microelectronics.

– A comprehensive study of the physical effects that take place in HV devices has been conducted. The effect of the junction curvature on the breakdown voltage has also been explained and the factors that affect the safe operating area (SOA) of the device were discussed. Moreover, an analytical treatment of the REduced SURface Field (RESURF) principle was carried out.

– The intrinsic drain voltage concept (K-point voltage, V_K) has been proposed and its variation explained and related to the charge variation and physical effects inside the device. The K-point voltage was used as a tool to investigate the saturation mechanisms of LDMOS. Once again, the correlation between the capacitances variation and the intrinsic drain voltage V_K was demonstrated.

– The proposed LDMOSFETS were developed by slight modifications of the base 0.35 µm BiCMOS technology. Two extra masks were used for the body and the drift regions formation with a slightly increased thermal budget and without resorting to high-tilt implants.

– The breakdown voltage of the proposed nLDMOS did not depend on the epitaxial layer thickness as in the conventional RESURF LDMOS, so the proposed device was suitable for integration with low-voltage CMOS with shallow epitaxial layers. A pLDMOS has been designed in the same optimized epitaxial layer of the nLDMOS.

– The proposed devices were simulated with advanced TCAD tools in order to investigate the physical phenomena and electrical characteristics. We utilized the commercial device simulators ISE and SENTAURUS-TCAD (GENESISe). The simulators provided the ability to perform simulation in both 2D and 3D. The software provided a convenient framework for designing, organizing and automatically running complete TCAD simulation projects. It provided a graphical user interface (GUI) to drive a variety of simulation and visualization tools, and to automate the execution of fully parameterized projects.

– The modeling of the lateral structures came into play as these devices were integrated in standard technologies. Remarkable progress in the modeling field conducted to geometrical approximations for the drift zone (regional approach) seemed to be the most promising approach.

– Impurity concentration in the channel was considered. The analytical model was implemented using MATLAB and MAPLE programs. The electrical approximations that have to be taken into account to build the DC model were also explained.

– With the help of the proposed analytical model, a SPICE model was developed. The model parameters were extracted using a system that links the ICCAP extraction tool with the ISE-TCAD tools. An interface circuit was proposed to convert low-level logic voltage to high level voltage, which would be suitable for future automotive applications.

– An insulation mechanism, between high-voltage LDMOS devices as well as between LDMOS devices and the low-voltage CMOS, was suggested, using a deep trench filled with silicon dioxide and undoped polysilicon. Good electrical insulation has been obtained using this kind of insulation technique. This mechanism would be suitable in the harsh automotive environment.

– Stacked 3D integration technique offered significant improvements over two-dimensional (2D) ICs on performance, functionality and integration density. 3D integration promised a solution to the so-called "wiring crisis" problem. Furthermore, 3D integration also provided new architectures for sophisticated ICs and facilitated the integration of heterogeneous materials and devices. This new 3D technology was used to stack the HV LDMOS devices and the low-voltage digital and analog CMOS parts.

– Two- and three-dimensional TCAD-based simulation of TSV- or RDL-induced coupling was investigated on nMOSFET, pMOSFET and the sensitive regions of the CMOS inverter. The impact of the TSV on the CMOS inverter operation has been performed using mixed-mode simulation (SPICE models for the CMOS devices and the finite element method (FEM) for the bulk regions of the devices).This impact was dependent on the layout and the position of the TSV.

– Three-dimensional simulation was preferred for studying the substrate coupling in the smart power integrated circuits, especially when using the new 3D-stacked integration technology as it demonstrated the real case but this simulation wasted time as there were too many mesh points.

– Stating that a whole electro-magnetic approach is a key goal in the very high frequency domains, we applied a finite-difference time-domain (FDTD) method to two striplines separated by a via. Results of transient EM simulations have been analyzed based on transmission/reflection coefficients via FFTs which can be taken as post processing.

C.2. Future work

The work suggests continuations of the research and developments in some critical directions:

– This study may help the designers of HV devices to produce high-quality lateral MOS devices. Complete understanding of the effects specific to a HV device is the key to obtain an optimized design.

– It improves the analytical model by including the special physical effects that take place in the LDMOS device, such as accumulation layer injection.

– It studies the substrate coupling in the 3D-stacked technology due to all types of signals: analog or digital, low voltage and high voltage.

– It develops a model of the substrate coupling and compares the model results with those obtained from the TCAD tools in various integration schemes.

– A further study is planned to modify either TSV shape (square, hexagonal, tapered, etc.) or configuration in order to improve the overall performances of TSVs. As the TSV will not be used in an application alone, the global chain (several levels of TSV, interconnect, bump, etc.) will be also simulated in order to provide guidelines in terms of connection.

Appendix

Semiconductor Physical Models

The design of semiconductor devices and an understanding of their operating characteristics are based on knowledge of the fundamental properties of the semiconductor material used in fabrication. Here, we first address the basic properties of semiconductor materials at equilibrium, which includes a discussion on the density of electrons and holes. A brief review of intrinsic and extrinsic semiconductors is presented, followed by an overview of incomplete ionization. We then consider the mobility models that describe carrier behavior inside the semiconductor material, with Si as a default material. Third, carrier lifetimes are discussed, which can generally be divided into recombination and generation lifetimes. Finally, the band-gap narrowing that results from high doping is introduced. In this brief discussion, the implementation for each physical parameter in TCAD simulators is presented.

A.1. Electron and hole densities

General expressions for the equilibrium densities of electrons and holes in a semiconductor can be derived using the Fermi–Dirac (FD) distribution function and the density-of-states function. For undoped and lightly doped semiconductors, the Maxwell–Boltzmann (MB) distribution function is used instead of the FD distribution function. If we assume that the constant-energy surfaces near the bottom of the conduction band and at the top of the valence band are spherical, then the equilibrium distribution functions for electrons in the conduction band and holes in the valence band may be described in terms of the FD distribution function. The default in SILVACO TCAD is Boltzmann statistics. Using FD statistics in ATLAS is done by specifying the `FERMIDIRAC` parameter in the `MODEL` statement.

The FD distribution function for electrons in the conduction band is given by

$$f_n(E) = \frac{1}{1+e^{(E-E_F)/kT}} \qquad \text{[A.1]}$$

and the FD distribution function for holes in the valence band can be expressed by

$$f_p(E) = \frac{1}{1+e^{(E_F-E)/kT}} \qquad \text{[A.2]}$$

Assuming parabolic bands for both the conduction and valence bands, and using the conduction and valence band edge as a reference level, the density of states (DOS) function per unit volume in the conduction band can be expressed by

$$g_n(E-E_c) = \left(\frac{4\pi}{h^3}\right)\left(2m_n^*\right)^{3/2}\left(E-E_c\right)^{1/2} \qquad \text{[A.3]}$$

Also, the DOS in the valence band is given by

$$g_p(E_v-E) = \left(\frac{4\pi}{h^3}\right)\left(2m_p^*\right)^{3/2}\left(E_v-E\right)^{1/2} \qquad \text{[A.4]}$$

The equilibrium electron density n_o in the conduction band can be obtained by integrating the product $dn = f_n(E)g_n(E)dE$ (i.e. the electron density per unit energy interval) with respect to energy over the entire conduction band using equations [A.1] and [A.3], which yields,

$$n_o = N_c F_{1/2}(\eta) \qquad \text{[A.5]}$$

where,

$$N_c = 2\left(\frac{2\pi m_n^* kT}{h^2}\right)^{3/2} \qquad \text{[A.6]}$$

is the effective density of the conduction band states, and

$$F_{1/2}(\eta) = \frac{2}{\sqrt{\pi}}\int_0^\infty \frac{\varepsilon^{1/2}d\varepsilon}{1+e^{\varepsilon-\eta}} \qquad \text{[A.7]}$$

is the Fermi integral of order one-half, $\varepsilon = (E - E_c)/kT$ is the reduced energy, m_n^* is the density-of-states effective mass of electrons, and $\eta = -(E_c - E_F)/kT$ is the reduced

Fermi energy. Equation [A.5] is the general expression for the equilibrium electron density in the conduction band applicable over the entire doping density range.

Similarly, the hole density in the valence band can be derived in a similar way from equation [A.2] and [A.4], and the result is given by

$$p_o = N_v F_{1/2}\left(-\eta - \varepsilon_g\right) \quad \text{[A.8]}$$

where

$$N_v = 2\left(\frac{2\pi m_p^* kT}{h^2}\right)^{3/2} \quad \text{[A.9]}$$

is the effective density of the valence band states, m_p^* is the density-of-states effective mass of holes in the valence band and $\varepsilon_g = (E_c - E_v)/kT$ is the reduced band gap energy.

A.2. Intrinsic semiconductors

A semiconductor may be considered as an intrinsic semiconductor if its thermally generated carrier density (i.e. n_i) is much larger than the background doping or residual impurity densities. The intrinsic carrier density can be expressed by

$$n_i = \left(N_c N_v\right)^{1/2} e^{-E_g/2kT} \quad \text{[A.10]}$$

Since the energy band gap for most semiconductors decreases with increasing temperature, a correction of E_g with temperature is necessary when the intrinsic carrier density is calculated from [A.10]. In general, the variation of intrinsic energy band gap with temperature can be calculated using an empirical formula given by

$$E_g(T) = E_g(0) - \frac{\alpha T^2}{(T + \beta)} \quad \text{[A.11]}$$

where $E_g(0)$ is the band gap energy at 0 K. For Si, $\alpha = 4.73 \times 10^{-4}$ eV/K and $\beta = 636$ K.

A.3. Extrinsic semiconductors

The most important and unique feature of a semiconductor material lies in the fact that its electrical conductivity can be readily changed by many orders of magnitude by simply doping the semiconductor with shallow-donor or shallow-acceptor impurities. By incorporating the doping impurities into a semiconductor, the electron or hole density will increase with increasing shallow-donor or shallow-acceptor impurity concentrations. For example, electron or hole densities can increase from 10^{13} cm^{-3} to more than 10^{20} cm^{-3} if a shallow-donor or shallow-acceptor impurity with an equal amount of impurity densities were added into a silicon crystal.

The temperature behavior of equilibrium carrier density in an extrinsic semiconductor can be determined by solving the charge neutrality equation, using the expressions for the electron and hole densities in the conduction and valence band states. For an extrinsic semiconductor, if both the donor and acceptor shallow impurities are present in a host semiconductor, then the charge neutrality condition in thermal equilibrium is given by

$$q\left[(p_o - N_A^-) + (N_d^+ - n_o)\right] = 0 \qquad \text{[A.12]}$$

where the ionized donor and acceptor concentrations are given by:

$$N_D^+ = \frac{N_D}{1 + g_D e^{(E_F - E_D)/kT}} \qquad \text{[A.13]}$$

$$N_A^- = \frac{N_A}{1 + g_A e^{(E_A - E_F)/kT}} \qquad \text{[A.14]}$$

The degeneracy factors account for the possibility of electrons with different spin, occupying the same energy level (i.e. a true statement of the Pauli exclusion principle is that no electron with the same quantum numbers [energy and spin] can occupy the same state). In most semiconductors, $g_D = 2$ and g_A is 4, due to the above reason combined with the fact that there are actually 2 valence bands in most semiconductors. Thus, 2 spins × 2 valance bands makes $g_A = 4$.

In general, the temperature dependence of the carrier density and the Fermi level for an extrinsic semiconductor can be predicted using equation [A.12], substituting ionized concentrations from equations [A.13] and [A.14], and substituting carrier concentrations from equations [A.5] and [A.8].

A.4. Incomplete ionization

The concentration of free majority carriers is one of the most important quantities that determine the electrical properties of semiconductor devices. Both theoretical calculations and measurements (Kuźmicz 1986) indicate that in silicon, in the doping range from about 10^{17} to about 10^{19} cm^{-3} at room temperature and below, the dopant atoms are not completely ionized and the majority of the carrier concentrations can be significantly lower than the doping concentrations.

Sheng, Li and Thurber (1977) have calculated the percentage of ionized dopant atoms in boron-doped Si and phosphorus-doped Si for doping concentrations up to about 3×10^{18} cm^{-3}. Their model is based on a simple single-impurity-level theory, with the additional assumption that the activation energy of a donor or acceptor level decreases with doping and eventually vanishes as the doping concentration approaches the level of about 3×10^{18} cm^{-3}.

Kuzmicz (1986) proposed a model for calculation of the percentage of ionized dopant atoms. The model includes the effects of formation of the impurity band, change of the mean activation energy with doping concentration and merging of the impurity band with the majority carrier band. The assumptions of Kuzmicz for n-type semiconductors can be summarized as follows:

– the DOS in the conduction band $g_n(E)$ is parabolic regardless of the doping concentration as given in equation [A.3];

– the density of quantum states in the impurity band $g_d(E)$ is a Gaussian distribution of the energy around the donor level E_d;

– the energy level E_d is the donor level in a semiconductor with low donor concentration. The donor activation energy is given by:

$$\Delta E_d = E_c - E_d \qquad [A.15]$$

and it is assumed that this energy decreases with doping according to $N^{1/3}$ law:

$$\Delta E_d = \Delta E_{do} - b_d \left(N_D^+\right)^{1/3} \qquad [A.16]$$

The results of the ionization fraction (the ratio of ionized donors or acceptors to the total concentration of donors or acceptors, respectively) based on this model are shown in Figure A.1. The results for phosphorus (Figure A.1a) are calculated assuming that the activation energy of phosphorus atoms ΔE_d at low phosphorus concentrations is 44 meV. The results for boron (Figure A.1b) are calculated with the activation energy of 43.8 meV. The value of b_d used in equation [A.16] is

3.48×10^{-8} eV cm. When the temperature is decreased to below about 250 K, the results obtained by this method fail in describing the behavior of the incomplete ionization (actually, for temperatures about 150 K and less, negative ionization fractions are obtained).

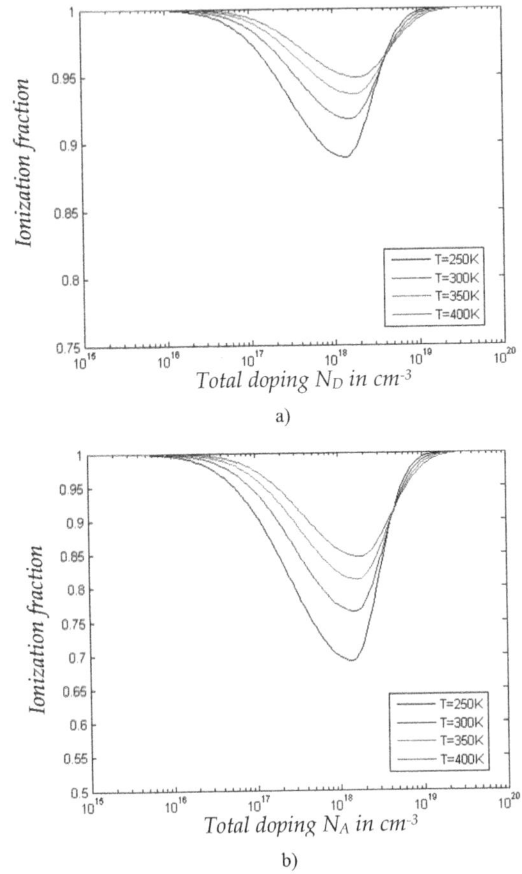

Figure A.1. *Dependence of ionization fraction on total doping for a) phosphorus in Si and b) boron in Si*

Recently, Altermatt *et al.* (2006a and b) proposed another approach for the incomplete ionization in silicon. A parameterization of the DOS near the band edge of doped silicon is derived from photoluminescence and conductance measurements. It is shown that the dopant band only touches the conduction band at the Mott (metal-insulator) transition and that it merges with the conduction band at considerably higher dopant densities. With the proposed DOS, incomplete ionization of donors and acceptors is calculated in the temperature range from 300 to 30 K, which is given by:

$$\frac{N_D^+}{N_D} = 1 - \frac{bn_o}{n_o + gn_1} \qquad [A.17]$$

$$\frac{N_A^-}{N_A} = 1 - \frac{bp_o}{p_o + gp_1} \qquad [A.18]$$

where

$$n_1 = N_c e^{-E_{dop}/kT} \qquad [A.19]$$

$$p_1 = N_v e^{-E_{dop}/kT} \qquad [A.20]$$

The energy of the dopant level E_{dop} is defined as the energy where the dopant DOS peaks. E_{dop} could be expressed as follows:

$$E_{dop} = \frac{E_{dop,o}}{1 + \left(N_{dop} / N_{ref}\right)^c} \qquad [A.21]$$

where N_{dop} could be donor or acceptor concentration. The parameters $E_{dop,o}$, N_{ref}, c, b and g depend on the type of dopants. Values of these parameters are given in Table A.1 for different types of dopants in silicon (phosphorus and arsenic as donors, and boron as acceptors).

Parameter	P in Si	As in Si	B in Si
$E_{dop,o}$ (meV)	45.5	53.7	44.39
N_{ref} (cm^{-3})	2.2×10^{18}	3×10^{18}	1.3×10^{18}
C	2	1.5	1.4
D	2.3	1.8	2.4
G	1/2	1/2	1/4

Table A.1. *Constant parameters used to calculate the amount of incomplete ionization*

To compare the results of Sheng, Li and Thurber (1977), Kuźmicz (1986), and Altermatt *et al.* (2006a and b), we first demonstrate the effect of doping on the activation energy (E_{act}) given by equation [A.16] and the energy of the dopant band peak (E_{dop}) given by equation [A.21]. Figure A.2 shows the difference between the approaches. As shown, E_{act} starts to change at N_{dop} about 10^{17} cm^{-3} due to band broadening and E_{dop} stays rather constant up to N_{dop} about 10^{18} cm^{-3}. At higher N_{dop}, both parameters change rather abruptly. E_{act} becomes zero at the Mott transition, while E_{dop} approaches zero (i.e. E_c) gradually above N_{dop} about 10^{19} cm^{-3}. E_{dop} cannot move into the conduction band because the principal reason for its variation is screening. Furthermore, we compare the ionization fraction of the three models at T = 300 K. Figure A.3 shows the differences between them. We note that when using Sheng, Li and Thurber's model, the ionization fraction is forced to be unity just after E_{act} touches the conduction band edge.

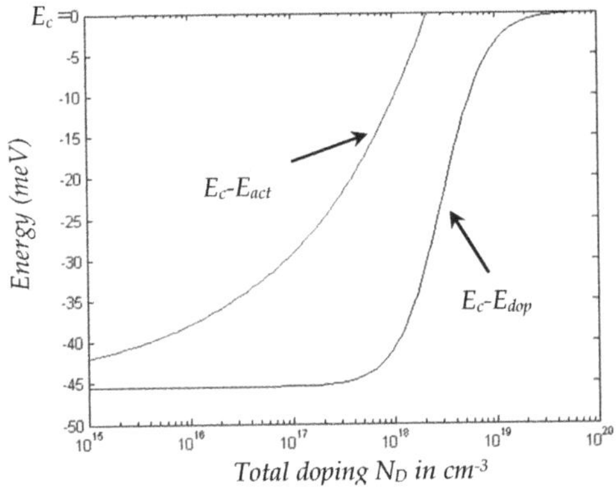

Figure A.2. *The energy of the dopant band peak E_{dop} and the activation energy of the dopant band E_{act}*

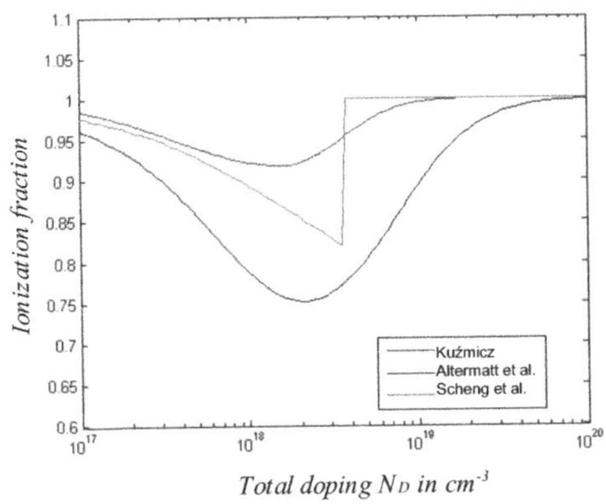

Figure A.3. *Comparison of ionization fractions for three models at T = 300 K*

Moreover, Figure A.4 demonstrates some experimental results from different authors. In these measurements, the amount of incomplete ionization was extracted from the ratio between conductivity mobility μ_{cond} and Hall mobility μ_H. The figure is evidence that Altermatt *et al.* (2006a and b) is the most reliable model; it predicts about 25% incomplete ionization at room temperature as shown in the figure.

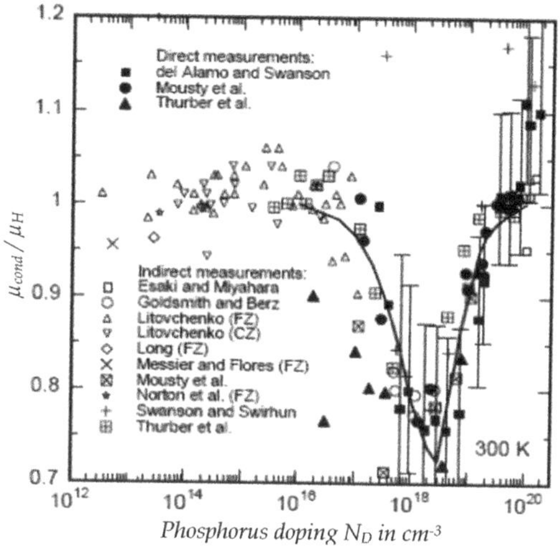

Figure A.4. *The fraction between measured conductivity mobility and Hall mobility (symbols) for P in Si, reflecting incomplete ionization, which is compared with Altermatt* et al.*'s calculations (solid line)*

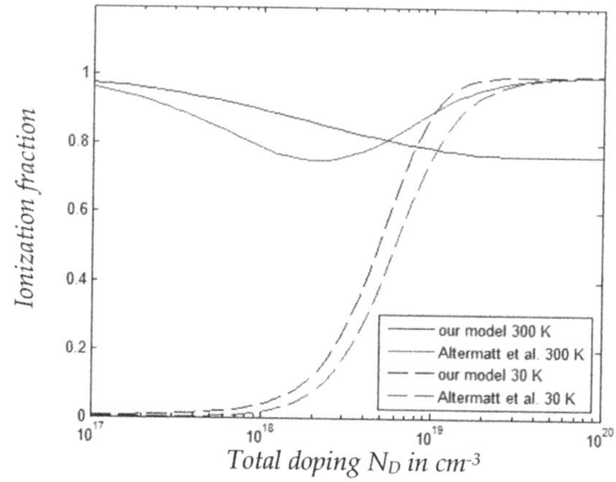

Figure A.5. *Comparison between our modified model and Altermatt* et al.*'s model for T = 300 K and T = 30 K*

The disadvantage of Altermatt *et al.*'s model is that it depends on fitting parameters, which should be adjusted for any material to characterize the incomplete ionization. Here, we present a modified model that could be used to a good level of accuracy despite its simplicity. We use equation [A.16] to represent the ionization energy and assume that the activation energy could enter into the conduction band. That is to say $(E_c - E_{act})$ could be negative. Figure A.5 shows a comparison between the proposed model and Altermatt *et al.*'s model for T = 300 K and T = 30 K for P in Si. The figure indicates a good match between our modified model and Altermatt *et al.*'s model.

A.5. Mobility

By definition, mobility is a measure of the average velocity of free carriers in the presence of an applied electric field. In the presence of an electric field in a semiconductor, the free electrons and holes are accelerated in opposite directions. As the free carriers are transported along the direction of the electric field, their velocity increases until they undergo scattering. In the bulk, the scattering can occur either by interaction with the lattice or at ionized donor and acceptor atoms. As a result, the mobility is dependent on the lattice temperature and the ionized impurity concentration.

In analyzing the influence of the preceding parameters on the mobility, one assumes the electric field strength to be small. The mobility is then defined as the proportionality constant relating the average carrier velocity to the electric field. At high electric fields, the velocity is no longer found to increase in proportion to the electric field and in fact attains a saturation value.

In the following sections, we will discuss the dependence of the electron and hole mobility on the preceding parameters.

A.5.1. *Temperature dependence*

At low doping concentrations, the scattering of free carriers in the bulk occurs predominantly by interaction with lattice vibrations. The constant mobility model assumes that the carrier mobility is only affected by phonon scattering and therefore dependent only on the lattice temperature.

$$\mu_{const} = \mu_L \left(\frac{T}{T_o}\right)^{-\zeta} \quad [A.22]$$

where μ_L is the mobility due to bulk-phonon scattering, T is the lattice temperature and T_o = 300 K. The default values of μ_L and the exponent ζ are given in Table A.2. Figure A.6 shows the variation of the mobility when the temperature is increased. In ATLAS, the choice of mobility model is specified on the MODELS statement. The parameters associated with mobility models are specified on a separate MOBILITY statement. The default is to use constant low-field mobilities within each region of a device. This default model is independent of doping concentration, carrier densities and electric field.

Symbol	Electrons	Holes	Units
μ_L	1,417	470.5	$cm^2/(Vs)$
ζ	2.5	2.2	1

Table A.2. *Constant mobility model: default coefficients for silicon*

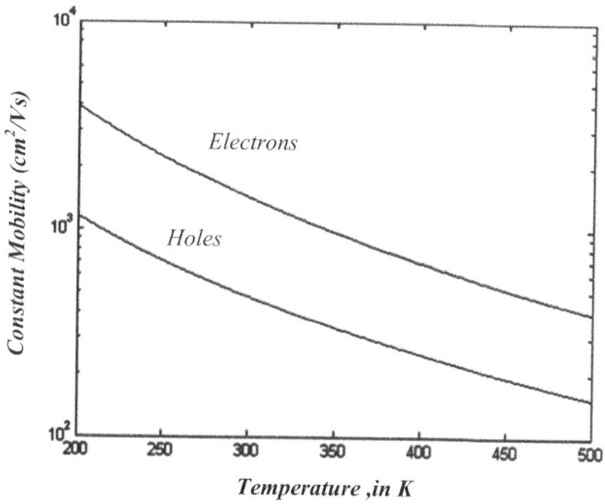

Figure A.6. *Dependence of constant mobility on temperature*

A.5.2. Concentration-dependent low-field mobility

ATLAS provides empirical data for the doping-dependent low-field motilities of electrons and holes in silicon at T = 300 K only. These data are used if the CONMOB parameter is specified in the `MODELS` statement. Figure A.7 illustrates the doping-dependent mobility degradation in silicon for electrons and holes.

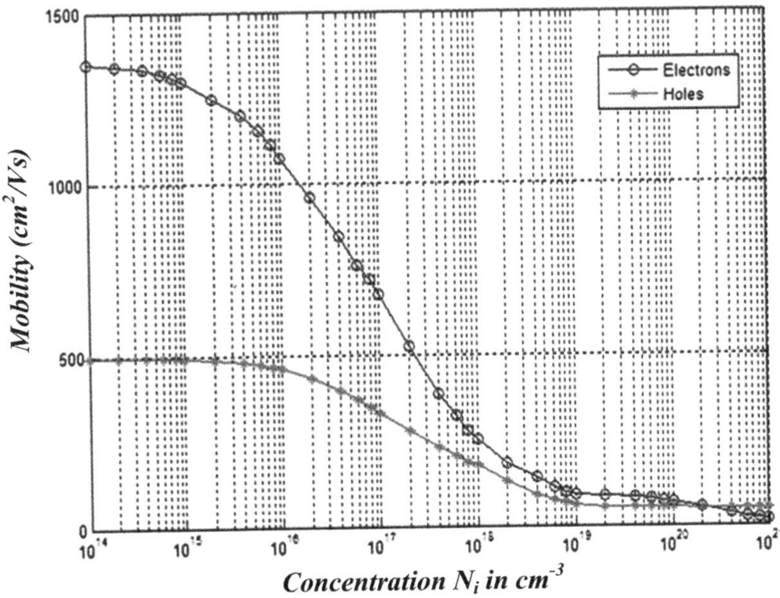

Figure A.7. *Doping dependent mobility at T = 300 K*

A.5.3. *Dopant concentration dependence*

The presence of ionized donor or acceptor atoms in the silicon lattice leads to a reduction in the mobility due to the additional coulombic scattering of the free carriers. Here, we consider the model proposed by Masetti *et al.* (1986) for doping-dependent mobility. The expression of this mobility model could be put in the following form:

$$\mu_{dop} = \mu_{\min 1} \exp\left(-\frac{P_C}{N_i}\right) + \frac{\mu_{const} - \mu_{\min 2}}{1+\left(\frac{N_i}{C_r}\right)^{\alpha}} - \frac{\mu_1}{1+\left(\frac{C_S}{N_i}\right)^{\beta}} \qquad \text{[A.23]}$$

where $N_i = N_D + N_A$ denotes the total concentration of ionized impurities. The model parameters are given in Table A.3. Figure A.8 illustrates the doping-dependent mobility degradation in silicon for electrons and holes. To use it as a standalone mobility model, we use the MASETTI parameter on the MODELS statement for both electron and hole mobility. N.MASETTI on the MOBILITY statement enables it for electrons, and P.MASETTI on the MOBILITY statement enables it for holes.

Symbol	Electrons	Holes	Units
μ_{min1}	52.2	44.9	$cm^2/(Vs)$
μ_{min2}	52.2	0	$cm^2/(Vs)$
μ_1	43.4	29.0	$cm^2/(Vs)$
P_c	0	9.23×10^{16}	cm^{-3}
C_r	9.68×10^{16}	2.23×10^{17}	cm^{-3}
C_s	3.34×10^{20}	6.10×10^{20}	cm^{-3}
α	0.680	0.719	1
β	2.0	2.0	1

Table A.3. *Masetti model: default coefficients*

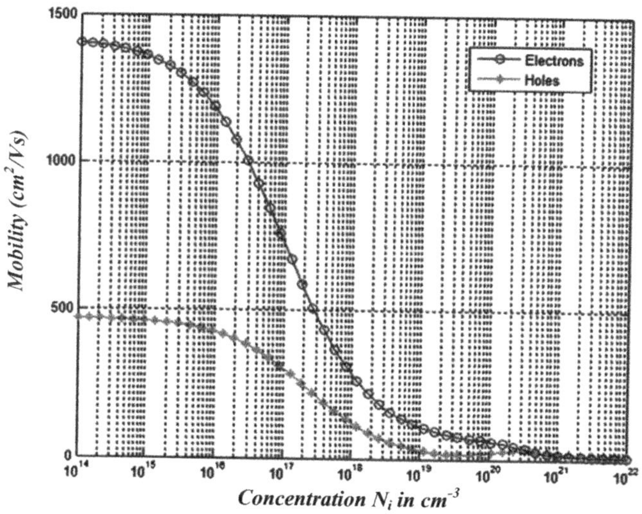

Figure A.8. *Masetti model: doping-dependent mobility at T = 300 K*

A.5.4. *High-field saturation*

In high electric fields, the carrier drift velocity is no longer proportional to the electric field strength, instead the velocity saturates to a finite speed v_{sat}. We will handle the Canali model for the description of this effect. In this model, the mobility is given by

$$\mu(F) = \frac{\mu_{low}}{\left[1+\left(\frac{\mu_{low}F}{v_{sat}}\right)^{\beta}\right]^{1/\beta}} \qquad [A.24]$$

where μ_{low} denotes the low field mobility, F is the driving field and v_{sat} is the saturation velocity. The exponent β is temperature dependent according to

$$\beta = \beta_o\left(\frac{T}{T_o}\right)^{\beta_{exp}} \qquad [A.25]$$

The default parameters of the Canali model are given in Table A.4. Figure A.9 shows the dependence of the electron drift velocity on the field strength for T = 300 K.

Symbol	Electrons	Holes	Units
β_o	1.109	1.213	1
β_{exp}	0.66	0.17	1

Table A.4. *Canali model parameters (default values for silicon)*

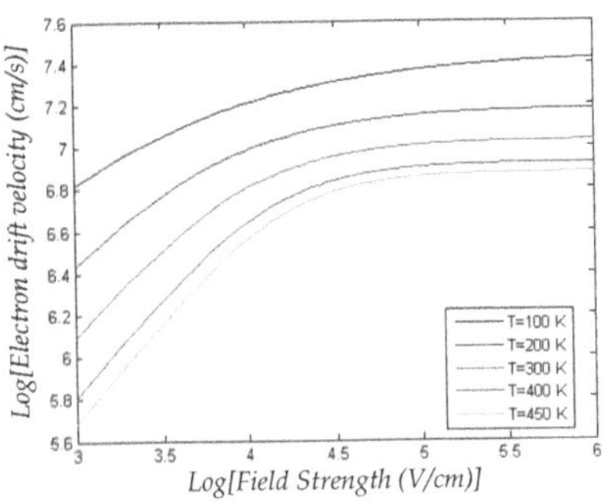

Figure A.9. *Temperature dependence of the electron drift velocity according to the model of Canali* et al. *(1975)*

According to the Canali model, the saturation velocity can be put in the following form:

$$v_{sat} = v_{sat,0}\left(\frac{T_o}{T}\right)^{v_{sat,\exp}} \qquad [A.26]$$

The silicon defaults are given in Table A.5.

Symbol	Electrons	Holes	Units
$v_{sat,0}$	1.07×10^7	8.37×10^6	cm/s
$v_{sat,\exp}$	0.87	0.52	1

Table A.5. *Default velocity saturation parameters (for silicon)*

A.5.5. *Carrier lifetimes and recombination*

Under thermal equilibrium conditions, a continuous balance between the generation and recombination of electron–hole pairs occurs in semiconductors. Any creation of excess carriers by an external stimulus disturbs this equilibrium. On removal of the external excitation, the excess carrier density decays and the carrier concentration returns to the equilibrium value. The lifetime is a measure of the duration of this recovery.

Recombination is the process by which the electrons and holes in a semiconductor are annihilated (see Figure A.10). Recombination takes place, for example, in an optically excited semiconductor or in a forward-biased p-n junction. When energy is supplied to a semiconductor by irradiation, by optical energy, by heating or by subjecting it to an electric field, the densities of the carriers increase above their thermal equilibrium values corresponding to the crystal temperature.

τ_r is the result of Shockley–Read–Hall (SRH) or multi-phonon recombination through intermediate centers (τ_{SRH}), direct band-to-band radiative recombination (τ_{rad}) and Auger or three-particle recombination (τ_{A}). Silicon is an indirect band gap semiconductor and thus τ_{rad}^{-1} is negligibly small. Then, for Si, we can write:

$$\tau_{\mathrm{r}}^{-1} = \tau_{\mathrm{SRH}}^{-1} + \tau_{\mathrm{A}}^{-1} \qquad [A.27]$$

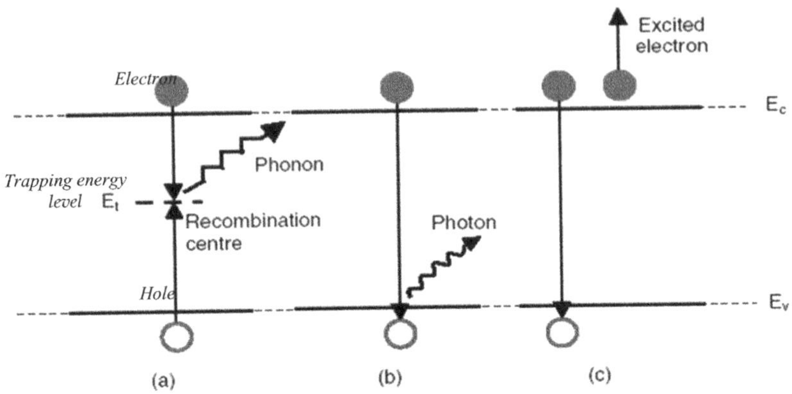

Figure A.10. *The three recombination mechanisms determining recombination lifetime in semiconductors: a) Shockley–Read–Hall recombination, b) radiative recombination and c) Auger recombination*

A.5.6. *SRH recombination*

This lifetime governs various electrical parameters of devices such as the forward voltage drop of power rectifiers at moderate current densities (typically 200–400 A/cm^2), and the current gain of bipolar transistors. This lifetime is a property of the Si material and is inversely proportional to the defect density (N_t). For n-type material, the hole recombination lifetime is given by

$$\tau_{\mathrm{p}} = \tau_{\mathrm{p0}}\left\{1+\frac{1}{1+h}\exp\left(\frac{\mathrm{E_t}-E_F}{kT}\right)\right\} + \tau_{\mathrm{n0}}\left[\frac{h}{1+h}+\frac{1}{1+h}\exp\left\{\left(\frac{E_i-E_t}{kT}\right)+\left(\frac{E_i-E_F}{kT}\right)\right\}\right] \quad \text{[A.28]}$$

where τ_{p0} and τ_{n0} are lifetimes in heavily doped n-type and p-type material, respectively, $h = \Delta n/n_0$ with n_0 the electron density in the n-type semiconductor at thermal equilibrium, E_i is the intrinsic energy level, E_t is the energy level of the recombination center and E_F is the Fermi level. The inverse variation of lifetime with N_t is because τ_{p0} and τ_{n0} are both proportional to $1/N_t$.

For the implementation in device simulation, the net recombination rate of SRH could be written as:

$$R_{net}^{SRH} = \frac{np-n_{i,eff}^2}{\tau_{\mathrm{p}}(n+n_1)+\tau_{\mathrm{n}}(p+p_1)} \quad \text{[A.29]}$$

with $n_1 = n_{i,eff} \exp\left(\frac{E_{trap}}{kT}\right)$, and $p_1 = n_{i,eff} \exp\left(-\frac{E_{trap}}{kT}\right)$

where E_{trap} is the difference between the defect level (E_t) and the intrinsic level (E_i). The silicon default value is $E_{trap} = 0$. The minority lifetimes τ_n and τ_p are modeled as a product of a doping-dependent, a field-dependent and a temperature-dependent factor:

$$\tau_c = \tau_{dop} \frac{f(T)}{1+g_c(F)} \quad [A.30]$$

where $c = n$ for electrons or $c = p$ for holes.

The doping dependence is modeled with the following relation:

$$\tau_{dop}(N_i) = \tau_{min} + \frac{\tau_{max} - \tau_{min}}{1+\left(\frac{N_i}{N_{ref}}\right)^{\gamma}} \quad [A.31]$$

Default values are given in Table A.6. A plot of electron and hole lifetimes as functions of the doping concentration is presented in Figure A.11.

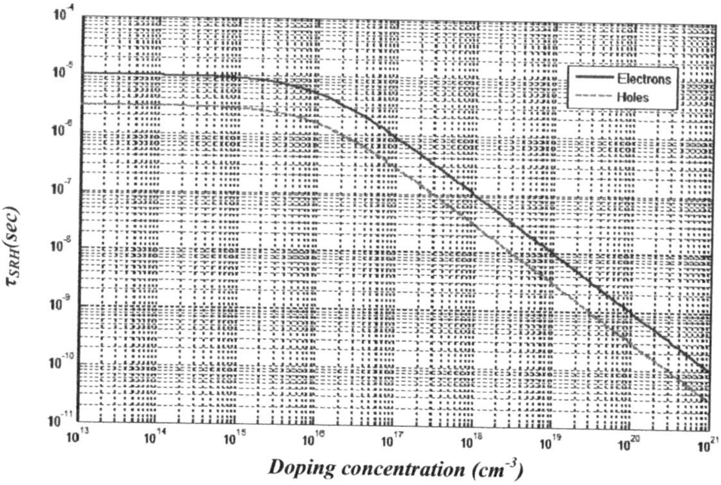

Figure A.11. *Doping dependence of the SRH lifetimes*

Considering temperature dependence, to date there has not been a consensus on the temperature dependence of the SRH lifetimes, which seems to originate from a different understanding of the meaning of "lifetime". From measurements of the recombination lifetime,

$$\tau = \Delta n / R \qquad \text{[A.32]}$$

where Δn is the excess carrier density under neutral conditions, $\Delta n = \Delta p$. It has been concluded that the lifetime increases with rising temperature. Such dependence could be modeled by a power law

$$\tau(T) = \tau_0 \left(\frac{T}{300} \right)^{T_\alpha} \qquad \text{[A.33]}$$

where τ_0 is the lifetime value at 300 K. Table A.6 presents the default parameters for the doping- and temperature-dependent SRH lifetime. The model is activated by using the SRH parameter of the MODELS statement for the concentration-independent case. The electron and hole lifetime parameters, TAUN0 and TAUP0, are user-definable in the MATERIAL statement. For concentration dependence, the model is activated with the CONSRH parameter of the MODELS statement.

Symbol	Electrons	Holes	Units
τ_{min}	0	0	s
τ_{max}	1×10^{-5}	3×10^{-6}	s
N_{ref}	1×10^{16}	1×10^{16}	cm^{-3}
G	1	1	1
T_α	–1.5	–1.5	1
E_{trap}	0	0	eV

Table A.6. *Default parameters for the doping and temperature-dependent SRH lifetime*

A.5.7. *Auger recombination*

The Auger recombination process occurs by the transfer of energy and momentum released by the recombination of an electron–hole pair to a third particle that can be either an electron or a hole. Theoretical investigations of the Auger recombination in indirect-gap semiconductors indicate that the process can occur with or without phonon assistance and may occur directly from band to band or through traps. Furthermore, Auger lifetime is responsible for the emitter injection efficiency of bipolar transistors. It also controls the lifetime values in heavily doped semiconductor regions.

Unlike the SRH lifetime that varies with sample purity, decreasing with the trap density, the Auger lifetime is independent of defects. It decreases very rapidly with injected carrier concentration, varying as:

$$\left(\tau_{\mathrm{p}}\right)_{\mathrm{A}}=\frac{1}{2.8\times10^{-31}n^{2}} \qquad [A.34]$$

for holes injected in n-type Si having majority carrier density n.

For implementation in device simulation, the rate of band-to-band Auger recombination could be written as:

$$R^{A}=\left(C_{n}n+C_{p}p\right)\left(np-n_{i,eff}^{2}\right) \qquad [A.35]$$

with temperature-dependent Auger coefficients:

$$C_{n}\left(T\right)=\left(A_{A,n}+B_{A,n}\left(\frac{T}{T_{o}}\right)+C_{A,n}\left(\frac{T}{T_{o}}\right)^{2}\right)\left(1+H_{n}e^{-\frac{n}{N_{0,n}}}\right) \qquad [A.36]$$

$$C_{p}\left(T\right)=\left(A_{A,p}+B_{A,p}\left(\frac{T}{T_{o}}\right)+C_{A,p}\left(\frac{T}{T_{o}}\right)^{2}\right)\left(1+H_{p}e^{-\frac{p}{N_{0,p}}}\right) \qquad [A.37]$$

where T_0 = 300 K. There is experimental evidence for a decrease in the Auger coefficients at high injection levels. Default values for silicon are listed in Table A.7. To enable the model, we specify AUGER on the MODELS statement.

Symbol	A_A (cm^6/s)	B_A (cm^6/s)	C_A (cm^6/s)	H (1)	N_o (cm^{-3})
Electrons	0.67×10^{-31}	2.45×10^{-31}	-2.2×10^{-32}	3.46667	1×10^{18}
Holes	0.72×10^{-31}	4.59×10^{-33}	2.63×10^{-32}	8.25688	1×10^{18}

Table A.7. *Default coefficients of the Auger recombination model*

Auger recombination becomes important at high current densities when the injected carrier concentration is three to four orders of magnitude higher than background concentration. The limit of the current density at which Auger recombination is dominant varies from 1000 to 3000 A/cm^2 [Ghamdi] or 10^4 A/cm^2 [Baliga]. Actually, these criteria are roughly estimated. The exact procedure to consider Auger recombination is by considering the level of injection of the carrier concentration itself. So, the main parameter for Auger is the carrier concentration directly.

The effective lifetime τ_{EFF} is given by

$$\frac{1}{\tau_{EFF}} = \frac{1}{\tau_{SRH}} + \frac{1}{\tau_A} \qquad \text{[A.38]}$$

Figure A.12 shows the effective lifetime versus carrier concentration. As can be seen, the impact of the Auger recombination process can be substantial under injection levels less than 10^{17} cm^{-3} in the case of SRH lifetimes higher than 10 µs. But, for lower SRH lifetime values, the injection level increases for Auger recombination to be dominated.

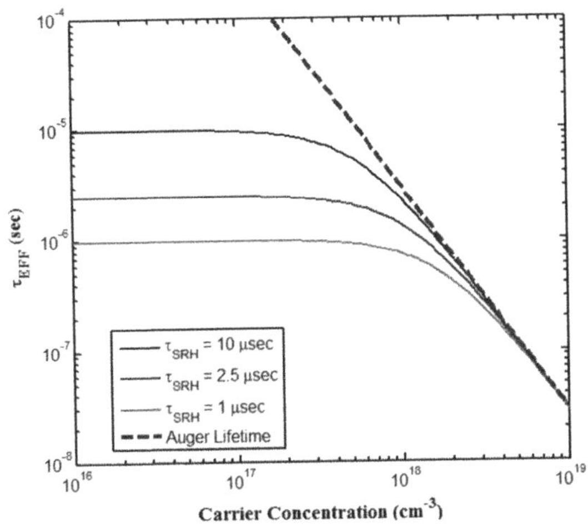

Figure A.12. *Effective lifetime dependence on carrier concentration*

A.5.8. *Band-gap narrowing*

Medium and highly doped semiconductor materials are widely used in device technology. Earlier theoretical approaches to BGN were mostly restricted to the T = 0 K limit. Mahan (1980) performed a Hartree–Fock variational calculation of the ground state energy of the electron-donor system in n-type silicon valid up to concentrations of 10^{20} cm^{-3}, neglecting band tailing and assuming that the donors are distributed on a regular FCC sublattice. He obtained closed form expressions for Si and Ge. Jain and Roulston (1991) treat the problem using the same approach as Mahan and extend the applicability of the expressions to include Si, Ge and GaAs. Schenk (1998) derived an analytical model for Si from a non-self-consistent finite-temperature full RPA formalism. The theoretical results of this model could be applied for finite temperatures rather than the 0 K limit.

There are a lot of experimental results concerning the measurements of the BGN in high doping semiconductors. These results may be recognized as:

i) electrical measurements at room temperature;

ii) photoluminescence (PL) and photoluminescence excitation (PLE) at low temperature and at 300 K;

iii) optical and infrared absorption measurements at low temperatures and room temperature;

iv) theoretical calculations at low temperature and at 300 K.

Statistical analyses of these data show that band gap changes determined by:

i) electrical measurements and optical measurements are not consistent with each other;

ii) low temperature optical absorption measurements and photoluminescence measurements are consistent with each other when the physically correct distorted densities of states are used to interpret the experimental data;

iii) absorption measurements are consistent at both 300 K and at temperatures below 35 K with theoretical calculations.

A.5.8.1. *Optical, rigid and apparent BGN*

The rigid band gap energy is defined as the difference between the minimum of the conduction band and the maximum of the valence band. On the other hand, the optical band gap energy is the same as the rigid band gap if the Fermi level is below the edges of the bands (E_C and E_v). When the Fermi level moves above the conduction band edge (the case of n-type) or below the valance band edge (the case of p-type) or when the quasi Fermi levels move above the two edges (the case of highly excited semiconductors), then the optical band gap energy is greater than the rigid one by the amount of the increase of the Fermi levels. This can be seen from Figure A.13.

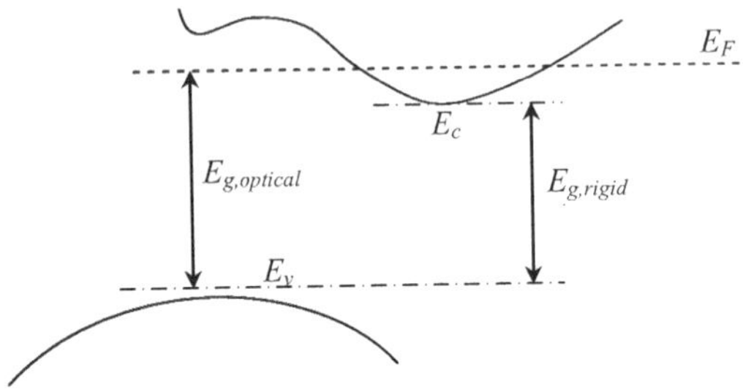

Figure A.13. *The indirect gap in Si, showing the difference between the rigid and optical band gap energy for n-type doping*

According to Figure A.13, we define the optical BGN as

$$\Delta E_g^{op} = E_{gi} - E_{g,optical} = \Delta E_g^{rigid} - \Delta E_F \equiv \Delta E_g - \Delta E_F \qquad [A.39]$$

where E_{gi} is the energy gap of an intrinsic material at thermal equilibrium and ΔE_F could be evaluated (for n-type) as:

$$\Delta E_F = kT\left[F_{1/2}^{-1}\left(\frac{n_o}{N_c}\right)\right] \qquad [A.40]$$

The apparent band-gap narrowing (BGN) is calculated using MB statistics. The difference between the apparent and the rigid BGN is very complicated. It is normally assumed that:

$$\Delta E_g^{app} = \Delta E_g + \Delta E_g^{FD} \qquad [A.41]$$

where ΔE_g^{FD} is a correction factor that could be deduced from the following analysis. When using apparent BGN, the p-n product is defined as:

$$p_o n_o = n_{ie}^2 = n_{io}^2 e^{\Delta E_g^{app}/kT} \qquad [A.42]$$

where n_{io} is the concentration at thermal equilibrium for the intrinsic semiconductor, and n_{ie} is the effective intrinsic concentration that appears as a result of BGN. To calculate the p-n product using the rigid BGN, we should modify equation [A.42] as follows:

$$p_o n_o = n_{ie}^2 = n_{io}^2 \gamma_n \gamma_p e^{\Delta E_g/kT} \qquad [A.43]$$

where

$$\gamma_n = F_{1/2}(\eta_n)e^{-\eta_n}; \eta_n = \frac{E_c - E_F}{kT} \quad [A.44]$$

$$\gamma_p = F_{1/2}(\eta_p)e^{-\eta_p}; \eta_p = \frac{E_F - E_v}{kT} \quad [A.45]$$

From equations [A.41], [A.43], [A.44] and [A.45], we can get an expression for ΔE_g^{FD} for the case of n-type semiconductors (the same expressions hold for p-type replacing n_o by p_o and N_c by N_v) by

$$\Delta E_g^{FD} = kT\left[F_{1/2}^{-1}\left(\frac{n_o}{N_c}\right) - \ln\left(\frac{n_o}{N_c}\right)\right] \quad [A.46]$$

A.5.8.2. *Implementation in device simulators*

In device simulators, BGN is considered to be a function of the total doping concentration N_i (N_D+ N_A) only. Different absorption and luminescence data of heavily doped materials were compared by statistical means, and a simple analytical fit to these data is provided by Bennett (DESSIS-ISE 2000)

$$\Delta E_g = \begin{cases} E_{bgn}\left[\ln\left(\frac{N_i}{N_{ref}}\right)\right]^2 & \text{if } N_i \geq N_{ref} \\ 0 & \text{otherwise} \end{cases} \quad [A.47]$$

Considering the electrical measurements, Slotboom and Klassen's model (Silvaco International n.d.) could give:

$$\Delta E_g = E_{bgn}\left[\ln\left(\frac{N_i}{N_{ref}}\right) + \sqrt{\left(\ln\left(\frac{N_i}{N_{ref}}\right)\right)^2 + 0.5}\right] \quad [A.48]$$

Also, del Alamo's model (DESSIS-ISE 2000) could be written as:

$$\Delta E_g = \begin{cases} E_{bgn}\ln\left(\frac{N_i}{N_{ref}}\right) & \text{if } N_i \geq N_{ref} \\ 0 & \text{otherwise} \end{cases} \quad [A.49]$$

Table A.8 shows the band gap narrowing default parameters for the four different models: Bennett, Slotboom, Klassen and del Alamo (DESSIS-ISE 2000; Silvaco International n.d.). Figure A.14 differentiates between the four electrical models. Bandgap narrowing effects in ATLAS are enabled by specifying the BGN parameter of the MODELS statement. We can specify the parameters in the MATERIAL statement. The Klaassen defaults will be used if we specify the BGN.KLA parameter in the MODELS statement. Otherwise, the Slotboom values will be used (by default).

Symbol	Bennett	Slotboom	Klassen	del Alamo	Units
E_{bgn}	6.84×10^{-3}	9.0×10^{-3}	6.92×10^{-3}	18.7×10^{-3}	eV
N_{ref}	3.162×10^{18}	1.0×10^{17}	1.3×10^{17}	7.0×10^{17}	cm^{-3}

Table A.8. *Band gap narrowing models: default parameters for silicon*

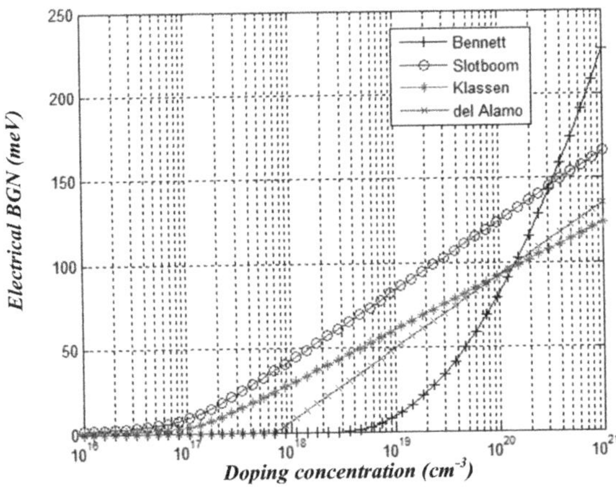

Figure A.14. *Comparison of electrical BGN models*

References

Abouelatta, M. (2010). Designing smart power integrated circuits with LDMOS devices compatible standard BiCMOS technology; substrate coupling. PhD Thesis, INSA Lyon, France.

Altermatt, P.P., Schenk, A., Schmithüsen, B. (2006a). A simulation model for the density of states and for incomplete ionization in crystalline silicon I. Establishing the model in Si:P. *J. Appl. Phys.*, 100, 113714.

Altermatt, P.P., Schenk, A., Schmithüsen, B. (2006b). A simulation model for the density of states and for incomplete ionization in crystalline silicon II. Investigation of Si:As and Si:B and usage in device simulation. *J. Appl. Phys.*, 100, 113715.

Canali, C., Majni, G., Minder, R., Ottaviani, G. (1975). Electron and hole drift velocity measurements in silicon and their empirical relation to electric field and temperature. *IEEE Trans. Electron Dev.*, 22(11), 1045–1047.

COMSOL (2021). COMSOL Multiphysics (Version 6.0). Software [Online]. Available at: http://www.comsol.com.

DESSIS-ISE (2000). ISE TCAD Release 7 User's Manual, Zurich.

Foussier, P. (1998). Contribution à l'intégration des systèmes de commande des machines électriques à courant alternatif. PhD Thesis, INSA Lyon, France.

Gharpurey, R. (1992). Modeling and analysis of substrate coupling in integrated circuits. PhD Thesis, University of California, Berkeley, USA.

Jain, S.C. and Roulston, D.J. (1991). A simple expression for band gap narrowing (BGN) in heavily doped Si, Ge, GaAs and GexSi1-x strained layers. *Solid-State Electronics*, 34(5), 453–465.

Keysight Technologies (2021). PathWave Advanced Design System (ADS). Software [Online]. Available at: https://www.keysight.com/gb/en/products/software/pathwave-design-software/pathwave-advanced-design-system.html.

Li, S.S. and Thurber, W.R. (1977). The dopant density and temperature dependence of electron mobility and resistivity in n-type silicon. *Solid-State Electronics*, 20(7), 609–616.

Ma, Y. and Gontrand, C. (2019). *Power, Thermal, Noise, and Signal Integrity Issues on Substrate/Interconnects Entanglement*. CRC Press, Boca Raton, USA.

Mahan, G.D. (1980). Energy gap in Si and Ge: Impurity dependence. *Journal of Applied Physics*, 51(5), 2634–2646.

Ma, Y., Fakri Bouchet, L., Calmon, F., Gontrand, C. (2016). Electrothermal modeling for three-dimensional nanoscale circuit substrates. *IEEE Transactions on Components, Packaging and Manufacturing*, 6(7), 1040–1050.

Mathworks (2021). MATLAB. Advanced design system. Software [Online]. Available at: http://mathworks.com.

Masetti, G., Severi, M., Solmi, S. (1983). Modeling of carrier mobility against carrier concentration in Arsenic-, Phosphorus- and Boron-doped Silicon. *IEEE Transactions on Electron Devices*, ED 226(30), 764–769.

Schenk, A. (1998). Finite-temperature full random-phase approximation model of band gap narrowing for silicon device simulation. *Journal of Applied Physics*, 84(7), 3684–6395.

Sharma, R. (2014). *Design of 3D Integrated Circuits and Systems*. Taylor & Francis, Oxford, UK.

Silvaco International (n.d.). Silvaco [Online]. Availalbe at: www.silvaco.com.

Todri-Sanial. A. and Tan, C.S. (eds) (2015). *Physical Design for 3D Integrated Circuits*. CRC Press, Boca Raton, USA.

Wiesław, K. (1986). Ionization of impurities in silicon. *Solid-State Electronics*, 29(12), 1223–1227.

Index

G, H

I

L, M

N, O, P

R

S

T

V

Other titles from

iSTE

in

Energy

2021

ROBYNS Benoît, DAVIGNY Arnaud, FRANÇOIS Bruno, HENNETON Antoine, SPROOTEN Jonathan
Electricity Production from Renewable Energies (2nd edition)

2020

BOISGIBAULT Louis, AL KABBANI Fahad
Energy Transition in Metropolises, Rural Areas and Deserts

SOUALHI Abdenour, RAZIK Hubert
Electrical Systems 1: From Diagnosis to Prognosis
Electrical Systems 2: From Diagnosis to Prognosis

2019

BENALLOU Abdelhanine
Energy Transfers by Convection
(Energy Engineering Set – Volume 3)
Energy Transfers by Radiation
(Energy Engineering Set – Volume 4)
Mass Transfers and Physical Data Estimation
(Energy Engineering Set – Volume 5)

LACHAL Bernard
Energy Transition

ROBYNS Benoît, DAVIGNY Arnaud, BARRY Hervé, KAZMIERCZAK Sabine, SAUDEMONT Christophe, ABBES Dhaker, FRANÇOIS Bruno
Electrical Energy Storage for Buildings in Smart Grids

2018

BENALLOU Abdelhanine
Energy and Mass Transfers: Balance Sheet Approach and Basic Concepts
(Energy Engineering Set – Volume 1)
Energy Transfers by Conduction
(Energy Engineering Set – Volume 2)

JEMEÏ Samir
Hybridization, Diagnostic and Prognostic of Proton Exchange Membrane Fuel Cells: Durability and Reliability

RUFFINE Livio, BROSETA Daniel, DESMEDT Arnaud
Gas Hydrates 2: Geoscience Issues and Potential Industrial Applications

VIALLET Virginie, FLEUTOT Benoit
Inorganic Massive Batteries
(Energy Storage – Batteries, Supercapacitors Set – Volume 4)

2017

BROSETA Daniel, RUFFINE Livio, DESMEDT Arnaud
Gas Hydrates 1: Fundamentals, Characterization and Modeling

LA SCALA Massimo
From Smart Grids to Smart Cities: New Challenges in Optimizing Energy Grids

MOLINA Géraldine, MUSY Marjorie, LEFRANC Margot
Building Professionals Facing the Energy Efficiency Challenge

SIMON Patrice, BROUSSE Thierry, FAVIER Frédéric
Supercapacitors Based on Carbon or Pseudocapacitive Materials
(Energy Storage – Batteries, Supercapacitors Set – Volume 3)

2016

ALLARD Bruno
Power Systems-on-Chip: Practical Aspects of Design

ANDRE Michel, SAMARAS Zissis
Energy and Environment

DUFOUR Anthony
Thermochemical Conversion of Biomass for the Production of Energy and Chemicals

2015

CROGUENNEC Laurence, MONCONDUIT Laure, DEDRYVERE Rémi
Electrodes for Li-ion Batteries
(Energy Storage – Batteries, Supercapacitors Set – Volume 2)

LEPRINCE-WANG Yamin
Piezoelectric ZnO Nanostructure for Energy Harvesting
(Nanotechnologies for Energy Recovery Set – Volume 1)

ROBYNS Benoît, FRANÇOIS Bruno, DELILLE Gauthier, SAUDEMONT Christophe
Energy Storage in Electric Power Grids

ROSSI Carole
Al-based Energetic Nanomaterials
(Nanotechnologies for Energy Recovery Set – Volume 2)

TARASCON Jean-Marie, SIMON Patrice
Electrochemical Energy Storage
(Energy Storage – Batteries, Supercapacitors Set – Volume 1)

2013

LALOUI Lyesse, DI DONNA Alice
Energy Geostructures: Innovation in Underground Engineering

2012

BECKERS Benoit
Solar Energy at Urban Scale

ROBYNS Benoît, DAVIGNY Arnaud, FRANÇOIS Bruno, HENNETON Antoine, SPROOTEN Jonathan
Electricity Production from Renewable Energies

2011

GAO Fei, BLUNIER Benjamin, MIRAOUI Abdellatif
Proton Exchange Membrane Fuel Cell Modeling

MULTON Bernard
Marine Renewable Energy Handbook

2010

BRUNET Yves
Energy Storage

2009

SABONNADIÈRE Jean-Claude
Low Emission Power Generation Technologies and Energy Management

SABONNADIÈRE Jean-Claude
Renewable Energy Technologies

Printed and bound by CPI Group (UK) Ltd, Croydon, CR0 4YY
16/09/2026
14963076-0003